P9-BVM-476

The Impact of State and National Standards on K-12 Science Teaching

a volume in
Research in Science Education

Series Editors:
Dennis W. Sunal, *University of Alabama*
Emmett L. Wright, *Kansas State University*

Research in Science Education

Dennis W. Sunal and Emmett L. Wright, Series Editors

Reform in Undergraduate Science Teaching for the 21st Century (2003)
edited by Dennis W. Sunal, Emmett L. Wright, and Jeanelle Bland

The Impact of State and National Standards on K-12 Science Teaching

edited by

Dennis W. Sunal
University of Alabama

and

Emmett L. Wright
Kansas State University

Greenwich, Connecticut • www.infoagepub.com

Library of Congress Cataloging-in-Publication Data

The impact of state and national standards on K-12 science teaching / edited
by Dennis W. Sunal, Emmett L. Wright.
p. cm. — (Research in science education)
Includes bibliographical references.
ISBN 1-59311-364-1 (pbk.) — ISBN 1-59311-365-X (hardcover)
1. Science—Study and teaching—Standards—United States. I. Sunal, Dennis W. II. Wright, Emmett. III. Series.
LB1585.3.I54 2005
507.1'0973—dc22

2005020176

Printed in the United States of America

CONTENTS

PREFACE TO THE SERIES

Science education as a professional field has been changing rapidly over the past 2 decades. Scholars, administrators, practitioners, and students preparing to become teachers of science find it difficult to keep abreast of relevant and applicable knowledge concerning research, leadership, policy, curricula, teaching, and learning that improve science instruction and student science learning. The literature available reports a broad spectrum of diverse science education research, making the search for valid materials on a specific area time-consuming and tedious.

Science education professionals at all levels need to be able to access a comprehensive, timely, and valid source of knowledge about the emerging body of research, theory, and policy in their fields. This body of knowledge would inform researchers about emerging trends in research, research procedures, and technological assistance in key areas of science education. It would inform policy makers in need of information about specific areas in which they make key decisions. It would also help practitioners and students become aware of current research knowledge, policy, and best practice in their fields.

For these reasons, the goal of the book series, *Research in Science Education,* is to provide a comprehensive view of current and emerging knowledge, research strategies, and policy in specific professional fields of science education. This series presents currently unavailable, or difficult to gather, materials from a variety of viewpoints and sources in a usable and organized format.

Each volume in the series presents a juried, scholarly, and accessible review of research, theory, and/or policy in a specific field of science education, K-16. Topics covered in each volume are determined by current issues and trends, as well as generative themes related to up-to-date research findings and accepted theory. Published volumes will include empirical studies, policy analysis, literature reviews and positing of theoretical and conceptual bases.

PREFACE

The Impact of State and National Standards on K-12 Science Teaching examines research, theory, and policy related to reform issues and events surrounding the development, status, influence, and future of science standards in education. The viewpoints of policy makers, teachers, and students are expressed through research involving original documents, interviews, analysis and synthesis of the literature, surveys, case studies, ethnographic and narrative studies, observations of teachers and students, and assessment of student learning outcomes. Volume Two of the series, *Research in Science Education*, also addresses the expectations toward the science standards of various stakeholders including students, parents, teachers, administrators, higher education science and science education faculty members, politicians, governmental and professional agencies, and the business community.

This volume investigates how the science standards have been translated into practice at the K-12 school district level, addressing issues around professional development, curriculum, assessment/evaluation, and accountability. The fundamental questions to be addressed are: (1) What is the response in terms of trends and patterns, of the educational system to the introduction of the national and state science standards since the late 1980s? and (2) What is the impact of the introduction of the science standards on teachers, classrooms, and students?

Volume Two addresses these questions from different viewpoints. Part one of the volume focuses on the historical, professional, political, and economic influences that led to the development and evaluation of the science standards. Part two examines the status and impacts of science standards using professional, political, and social factors that influenced their development, validation, and implementation in classrooms. Part three examines the impact of science standards on teaching. Part four examines science standards across the education continuum, promising practices in teacher preparation, school systems, schools, and classrooms, and the impact on student content learning. These chapters provide an evidence-based view of the current status and future of science standards and the work that still needs to be done to meet the expectations of various constituencies for improving science education at the K-12 level.

ACKNOWLEDGMENTS

The completion of this second volume was made possible by our authors, colleagues, who gave priority in their professional lives to conduct investigations, write, manuscripts, and submit their work to the scrutiny of others, and persisted through many revisions. Our special thanks go out to them because their outstanding professional experiences and expertise makes this volume possible. These author contacts have developed through our membership in professional organizations and activities with numerous individuals in education institutions. Therefore, we wish to acknowledge these organizations and individuals for providing us with a forum to meet, interact, disseminate, and form professional collaboration communities involving individuals with an interest in improving teaching and learning in science. We especially thank the National Association for Research in Science Teaching, Association of Science Teacher Educators, American Association for the Advancement of Science, National Science Teachers Association, and the National Aeronautics and Space Administration NOVA Program and Network of university teams involved in reform. Additional thanks go to Horizon Research, Inc. for their sharing of information used in this volume.

Special recognition is given to the students, teachers, and administrators who were concerned enough to take part in the investigations and contribute their thoughts to our discussions.

We also acknowledge our graduate students at various institutions who in addition to completing their own research and course work undertook tasks that allowed this volume to come to completion. Special thanks go to Jennifer Joiner at the University of Alabama.

Dennis W. Sunal
Emmett L. Wright
February 2005

CHAPTER 1

INTRODUCTION TO THE SCIENCE STANDARDS

Their Impact on K-12 Science Teaching

Dennis W. Sunal and Emmett L. Wright

INTRODUCTION

The term *standards* refer to guidelines that develop an education curriculum defining student learning outcomes and the means to accomplish them. Standards also include an accountability component that involves assessment, rewards, and punishments (Horn, 2004). Beginning with the passage of Title I of the Elementary and Secondary Education Act in 1965 with its focus on equity, there has been a clash of values leading to a pendulum effect, between equity versus excellence driving policy and standards in K-12 education. Between the late 1980s and the present, education policy and standards have moved strongly toward excellence culminating in the passage of the No Child Left Behind (NCLB) Act in 2001. NCLB requires that all states develop standards and ensures compliance through accountability and incentives.

The Impact of State and National Standards on K-12 Science Teaching, 1–4

IMPACT ON THE EDUCATION COMMUNITY

Today, education standards and related assessments drive much of the decision making in education in the United States (Horn, 2004). Every day the lives of students, parents, teachers, and administrators become more intertwined by the actions called for in mandated and voluntary education standards. Assessments that hold these lives accountable for student achievement of the standards effect students, and in turn, their parents, in passing from grade to grade, graduation, the curriculum they receive, acceptance into colleges, and schools they are allowed to attend. Teachers and administrators are affected through the amount of monetary school support and services they receive, curriculum they are allowed to use, job security, advancement, salary, and loss of control of schools. Business is affected through the costly redevelopment of standardized tests and published educational textbooks and materials proposed to meet the education standards and the new assessment of student achievement. College teachers, certification programs and courses, and internships must now relate to new directions, teaching strategies, syllabi, content, and oversight of research that is deemed to be legitimate. The agenda of government agencies and professional organizations all have been changed by the advent of education standards and their assessment.

CONTROVERSY AND POLITICS

The NCLB Act and the push for excellence over equity have not been accepted quietly (Elmore, 2003). Calls for rethinking and revision of current directions have focused on weak research and theory supporting school improvement policies, learning outcomes, and teaching strategies advocated by NCLB. Ill suited incentives and punishments for performance and control of student learning by laws have been widely discussed. Also criticized is the lack of funding for infrastructure and professional development that has resulted since passage of this legislation. As can be seen in the history of education over the past century, the standards movement is a political process whose form conforms to the values and purpose of controlling forces generally outside of the local education community (Horn, 2004).

SCIENCE STANDARDS

Science education in the United States is a complex network of interconnected elements influenced by many forces originating from multiple sources both inside and outside of the educational setting. It is difficult to focus on a single network component without considering the interactive impact of forces and events occurring continuously over time. Beginning in the late 1980s, various governmental agencies and professional organizations developed standards for science curriculum and instruction for the kindergarten through high school setting. Standards were designed to create a vision for what is needed to enable all students to become literate in science (National Research Council, NRC, 1996, p. 2). The standards represented a departure from traditional science in schools. Science standards have grown to include guidelines on content, teaching, professional development, assessment, support, and organization of schools and state school systems. The science standards have been described in documents published by the NRC (1996) and the American Association for the Advancement of Science (1993, 1998, 1999) as well as numerous state and local educational agencies throughout the United States.

IMPACT OF THE SCIENCE STANDARDS

The science standards have impacted the current education setting and have also developed and changed due to the interplay of numerous forces. Reforms created by the science standards have reportedly: (1) been influenced by the other components of the educational system and other standards, (2) made an impact on the educational system, and (3) created changes in the delivery and process of teacher preparation and professional development, science teaching, classroom culture in schools, and student learning outcomes (NRC, 2001). It is useful to understand the real (evidence-based) influence of science standards on policies, programs, practices, and outcomes to determine future courses of action for parents, teachers, schools, school districts and the mix of stakeholders involved in this important process.

RESEARCH ON THE IMPACT OF THE SCIENCE STANDARDS

Conceptual overviews have been proposed and a body of research has emerged, related to the impact of the science standards. One conceptual overview that attempts to guide standards-focused research has been proposed by the NRC (2001). As these reforms enter the classroom, they impact schools and students through multiple and simultaneous paths. One path is through external factors, contextual forces, relating to the response of the system to the standards and includes politicians and policy makers, citizens, business and industry, and professional organizations. A second path is constituted by school system factors including alternative paths of influence involving teacher development, curriculum, and assessment and accountability. An additional path involves classroom factors relating to school culture, teachers, the teaching process, classroom climate, and student learning. Research in each path is necessary to provide an evidence-based view of the current status and future of science standards and the work that still needs to be done to meet the expectations of all stakeholders in America.

REFERENCES

American Association for the Advancement of Science. (1993). *Benchmarks for scientific literacy: Project 2061*. New York: Oxford University Press. Retrieved August 20, 2004, from http://project2061.aaas.org/tools/benchol/bolframe.html

American Association for the Advancement of Science. (1998). *Blueprints for reform: Project 2061*. New York: Oxford University Press. Retrieved August 20, 2004, from http://www.project2061.org/tools/bluepol/blpintro.htm

American Association for the Advancement of Science. (1999). *Designs for science literacy: Project 2061*. New York: Oxford University Press. Retrieved August 20, 2004, from http://www.project2061.org/tools/designs/default.htm

Elmore, R. (2003). A plea for strong practice. *Educational Leadership*, *61*(3), 6-11.

Horn, R. (2004). *Standards*. New York: Peter Lang.

National Research Council. (1996). *National science education standards*. Washington, DC: National Academy Press. Retrieved August 20, 2004, from http://www.nap.edu/readingroom/books/nses/html/

National Research Council. (2001). *Investigating the influence of standards: A framework for research in mathematics, science, and technology education*. Washington, DC: National Academy Press. Retrieved August 20, 2004, from http://books.nap.edu/books/030907276X/html/1.html

PART I

HISTORICAL, PROFESSIONAL, POLITICAL, AND ECONOMIC INFLUENCES

Part I of the volume focuses on the historical, professional, political, and economic influences that led to the development and evaluation of various versions of the science standards. Chapter two examines the historical, professional, political, and social events that influenced the development, validation, and present use of the science standards. George DeBoer explores the differences between two interrelated standards movements in the United States, their impact on science education reform efforts, and what remains to be done to promote further progress toward achieving widespread science literacy. Chapter three provides a much needed research framework for investigating the current science standards. In this chapter Iris Weiss describes three potential channels which national standards might traverse in order to influence teaching and learning: (1) curriculum; (2) teacher development; and (3) assessment and accountability.

CHAPTER 2

HISTORY OF THE SCIENCE STANDARDS MOVEMENT IN THE UNITED STATES

George E. DeBoer

In science, as in other areas of K-12 education, the development of standards has played a significant role in efforts to reform teaching and learning. In this chapter, the author explores the differences between two interrelated *standards movements* in the United States, their impact on science education reform efforts, and what remains to be done to promote further progress toward achieving widespread science literacy. Starting with the 1983 release of *A Nation At Risk,* the author traces the parallel development of standards that require accountability at every level of the education system and standards that describe the content that students need to learn and how best to teach it. While acknowledging the contributions of seminal standards documents such as *Benchmarks for Science Literacy* and *National Science Education Standards*, the author also recognizes the complex and change-resistant contexts—colleges and universities, the textbook publishing industry, classrooms and schools, and state bureaucracies—within which education reform must take place. The author concludes that standards aimed at accountability, such as those embodied in the current No Child Left Behind Act legislation, may be effective in moving people to action, but real and lasting change requires carefully developed content standards that provide

The Impact of State and National Standards on K-12 Science Teaching, 7–49

the vision and the tools that can guide the ongoing reform of science curriculum, instruction, and assessment.

INTRODUCTION

Raymond Williams, in *Keywords*, distinguishes between three uses of the word *standard:* standard past, standard present, and standard future. About standard future, he says: "It is a very interesting use. Instead of referring back to a source of authority, or taking a current measurable state, a standard is set, projected, from ideas about conditions which we have not yet realized but which we think should be realized" (Williams, 1983, p. 299). The distinction nicely summarizes the differences between two closely interconnected standards movements in the United States today. Standards present and past are about accountability measured with respect to a level of achievement determined by authority, whereas standard future is about a vision of what could be. Standard future is squarely in the camp of efforts to improve science education that have taken place continuously throughout the past century and more, culminating in the present focus on science literacy for all. Standards-as-accountability, focusing as it does on a fixed authoritative statement of correctness and competence, is something different. It is something that has the potential to catalyze efforts for good in the ongoing struggle to improve science education, but also something that has the potential to undermine those efforts. The themes of this chapter are the way these two standards movements interact and what still needs to be done to ensure continued progress toward furthering a deeper and more widespread understanding of science through our educational efforts.

REFORM OF SCIENCE EDUCATION IN THE UNITED STATES

It is not possible here to examine in depth the historical roots of the current reform efforts in the United States, but a few historical notes will be helpful in understanding why we are where we are today. First, it is important to recognize that most of what we are tying to do is not new. Many current ideas can be found in an earlier time. The idea that there should be a common curriculum for all, whether the students were bound for college or the world of work, was put forward in the report of the Committee of Ten during the 1890s (National Education Association, NEA, 1893). The idea of *inquiry teaching* can be seen in Herbert Spencer's statement that: "Children should be led to make their own investigations, and to draw their own inferences" (Spencer, 1864, p. 124). The value of teaching for understanding is evident in the warning made by the *Geography Conference of the Committee of Ten* (NEA, 1893) against educational practices that led to memorization without understanding and in Herbart's (1901) theory of instruction that focused on the development of conceptual understanding. The use of standardized testing to measure educational outcomes began in the second decade of the twentieth century (see DeBoer, 1991, pp. 121-124).

A second historical point to note is that curriculum and pedagogy have become "contested terrain" in our society "where warring parties contend for a kind of official sanctification of their deeply held beliefs" (Kliebard, 1995, p. 63). The tension has often been between ideas that are characterized as progressive on the one hand and traditional on the other, focusing respectively on the child or the content. Traditionalists think of themselves as rigorous, concerned about high standards, and valuing the authority of the past while progressives think of themselves as being sympathetic to how children think, how they interact with the world, and with preparing them for an ever-changing world. John Dewey (1938) laid out the parameters of the debate between traditionalists and progressives in the early part of the twentieth century and little has changed since then. Arguments over what standards and rigor mean, what our educational goals should be, or how to implement those goals continue today often in terms of that progressive versus traditionalist debate. If there has been one change that has occurred over the past 100 years it may be the efforts, in true Deweyan spirit, to accommodate the child and the curriculum, to find ways to connect the interests, experiences, and capacities of the child to the subject matter that has been organized by adult minds (Dewey, 1902, 1916). This does not mean that the progressive-traditionalist debate has ended. In fact, that debate is present in education as in all parts of our society. It is that some of the more thoughtful efforts to solve the seemingly impenetrable and intractable problems in science education are now focused on finding a new synthesis out of those opposing ideologies.

A third point to note is that rather than there being a clear trend in thinking throughout the last century, the attractiveness of ideas has tended to ebb and flow, more popular at one point in time and less popular at another (DeBoer, 1991). Typically what has happened is that in reaction to the perceived excesses of one era the emphasis shifts back toward what it had previously been. For example, the 1970s was, as some have described it, an era of "new progressivism" (Ravitch, 1983). Many of the child-centered approaches advocated by progressive educators during the first half of the twentieth century were reshaped in the form of open classrooms, elective courses for students, and other student-centered approaches. That revival of progressive ideas followed closely on the heels of the more content-centered reforms of the 1960s. Then, largely in response to the perceived laxness of the educational system in the 1970s and declining student test scores, there was a renewed call for more rigor in the educational system. T. H. Bell, U.S. Secretary of Education during the Reagan administration, established a National Commission on Excellence in Education (NCEE) to help "define the problems afflicting American education and to provide solutions" (NCEE, 1983, p. iii).

A fourth point relates to the issue of excellence and equity in education, which is at the center of current reform efforts and has a long history as well. There has always been a question of how to deal with the enormous diversity of students who attend public schools. The American high school was established to accommodate that diversity through a variety of academic and vocational curricular tracks and varied opportunities for students to select programs most suited to their interests and projected life work. That system has come under close scrutiny at various times for its apparent unfair allocation of resources toward the most advantaged students (Kozol, 1992). In the early 1980s Mortimer Adler, Chairman of the Board of Editors of the Encyclopedia Britannica and Director of the Institute for Philosophical Research in Chicago, published *The Paideia Proposal: An Education Manifesto* (Adler, 1982). Adler, in the spirit of Robert Maynard Hutchins, activist President of the University of Chicago between 1929 and 1951, called for a required common academic curriculum for all students—with no electives or specialized courses—as a way of ensuring that all citizens would receive the education needed to become capable thinkers and to participate fully in a democratic society. Doing away with a multitrack educational system would prevent students from downgrading their education by their own choice or having others downgrade it for them. According to the *Paideia Proposal*, all people, because of their common humanity, deserve an education that offers a prospect of personal growth and self-improvement, effective citizenship, and the ability to earn a living in our society.

Adler identified three categories of student learning. They are the acquisition of organized knowledge by means of didactic instruction and demonstration, the development of intellectual skills through supervised practice, and an enlarged understanding of ideas and values through Socratic questioning and discussion. Consistent with the ideas of other educators in the early 1980s, Adler emphasized that learning is an active process. "All genuine learning is active, not passive. It involves the use of the mind, not just the memory. It is a process of discovery, in which the student is the main agent, not the teacher" (Adler, 1982, p. 50). The teacher aids discovery and elicits the activity of the student's mind by "inviting and entertaining questions, by encouraging and sustaining inquiry, by supervising helpfully a wide variety of exercises and drills, by leading discussions" (p. 50). Adler was recommending this program for all students, not just the academically inclined or the college bound. The idea was summed up in a phrase attributed to Hutchins: "the best education for the best is the best education for all" (p. 6). Although this idea of a liberal arts type of academic education for all was not new, having been introduced in much the same form by the Committee of Ten in 1893, the idea lost ground in the early twentieth century because of a perceived need to concentrate on vocational education for many of the students. Although sometimes associated with a conservative and elitist approach to education, the idea of a common core of academic learning was resurrected in the 1980s and become one of the cornerstones of the late twentieth century reform movement.

A Nation at Risk

NCEE was established on August 26, 1981, in an environment of low and declining test scores, and in an environment in which the United States was experiencing difficult times economically when compared with other nations that were thriving, especially Germany and Japan. The report of the commission, *A Nation at Risk*, was presented on April 26, 1983. The report was a call to mobilize the efforts of the federal government along with states and local school districts to raise the level of competence of American students in all academic areas but with special emphasis on science and mathematics. This was not the first time that the United States government had intervened so vigorously in American education. There were significant efforts by the federal government—in 1917 with the Smith-Hughes Act on vocational education, and in 1958 with the National Defense Education Act, following the Soviet Union's launch of the earth-orbiting satellite Sputnik—to influence education policy in response to problems that were perceived to be serious enough to require national solutions. The NCEE concluded that we had lost sight of our true educational mission and the need for high expectations for students. They recommended a return to a more academic educational focus and more disciplined effort on the part of students. They said that our international competitors were well-educated and highly motivated and that the United States needed to be as well if it were to compete successfully. The new raw materials of international commerce were knowledge, learning, information, and skilled intelligence. The argument that the federal government had to intervene in order to develop human capital for reasons of international economic competition was the same argument that had been used to support the 1917 Smith-Hughes vocational education act, although the means of accomplishing the goal (academic study vs.

vocational education) was very different. The federal government's intervention in 1958, in contrast, was about national defense and Cold War competition with the Soviet Union.

The NCEE pointed to the importance of a high level of common understanding in a free and diverse democratic society. The concern was not just for competitive success in industry and commerce. It was also for the intellectual, moral, and spiritual strength of the people who form the society. The educational system should contribute to the development of a common culture to help achieve a shared understanding of complex societal issues. The commission pointed to the value of an education where comprehension, analysis, and problem solving were fostered rather than rudimentary knowledge or technical and occupational skills. Schools also should develop a commitment to lifelong learning because education adds value to the quality of life throughout one's life. Schooling provides the foundation, but without lifelong learning the knowledge and skills learned in school will soon become outdated.

The commission argued for genuinely high standards for all. Neither mediocrity nor elitism was acceptable, although the commission did grant the possibility that the curriculum might vary for students of differing capabilities and interests. However, all should be expected to work to their capacity and develop their talents to the fullest. In high school, all students would learn the *new basics*, including English, mathematics, science, social studies, computer science, and, for the college-bound students, 2 years of foreign language.

The commission recommended that schools, colleges, and universities adopt more rigorous and measurable academic standards and that they raise expectations for academic performance and student conduct. Textbooks should be upgraded and updated to assure more rigorous content, and university scientists should be called on to help in this task. In science, students should be introduced to: "(a) the concepts, laws, and processes of the physical sciences; (b) the methods of scientific inquiry and reasoning; (c) the applications of scientific knowledge to everyday life; and (d) the social and environmental implications of scientific and technological development" (NCEE, 1983, p. 25). Textbooks should be chosen by states and school districts on the basis of their ability to present rigorous and challenging material clearly, and textbook publishers should be required to furnish evaluation data on the material's effectiveness. The commission recommended more homework, longer school days and longer school years, better attendance policies, and placement and promotion of students on the basis of academic progress. Higher standards and expectations were needed as well to give students themselves a deep respect for intelligence, achievement, learning, and for self-disciplined effort.

The vision was of an academically educated society, a common culture, rigorous academic standards, and accountability through standardized tests of achievement to be administered at major transition points from one level of schooling to the next, especially from high school to college or the work world. The commission recommended that the tests be administered as part of a national (but not federal) system of state and local standardized tests.

Educating Americans for the 21st Century

Just five months later, on September 12, 1983, the Commission on Precollege Education in Mathematics, Science and Technology of the National Science Board, which acts as an advisory board to the National Science Foundation, issued its report, *Educating Americans for the 21st Century* (National Science Board, 1983). The report echoed many of the ideas in *A Nation at Risk* and provided additional detail on how the vision of improved science education for all could be realized. The problem they identified was the same as that identified by the NCEE. The educational system had undergone a period of neglect, resulting in unacceptably low performance levels in science and mathematics. Unites States national security and economic health depended on its human resource development. And for reasons of national pride and international prestige, a world leader with the stature of the United States should have an educational system that was the finest in the world. Finally, a commitment to academic excellence would put the United States on a firm economic footing in its competition with other countries.

The recommended strategy to accomplish these priorities involved the development of national goals and curricular frameworks, strong national leadership for monitoring the quality of efforts, local responsibility for meeting the goals, and local variation in how they would be implemented. The commission recommended increased student exposure to science; higher standards of participation and achievement (noting comparisons to Japan's system where students spent more time in school), and a system of objective measurement to monitor progress. The belief was that our diverse educational system could be improved by establishing national goals and a system to measure local accomplishment of those goals.

The Commission on Precollege Education in Mathematics, Science, and Engineering provided more detail than the NCEE in its recommendations regarding the content of the science, mathematics, and technology curriculum. They recommended drastically reducing the number of topics that students would study, in part by integrating topics within subject areas and by making connections between subject areas, especially between mathematics, science, and technology. Courses should focus on thinking, communication, and problem solving skills. Students should have early hands-on experiences and formulate questions and seek answers from their observations of natural phenomena. The study of science should provide knowledge that would lead to civic responsibility and the ability to cope in a technological world. The commission recommended that the courses be oriented toward practical problems that "require the collection of data, the communication of results and ideas and the formulation and testing of solutions" (p. 45). Content recommendations were given for each subject area and organized into three grade bands within subject areas. These content recommendations were at the topic level of detail. For example, at the high school level they recommended that biology should emphasize concepts and principles such as "genetics, nutrition, evolution, reproduction of various life forms, structure/function, disease, diversity, integration of life systems, life cycles, and energetics" (p. 98). Specific ideas within these topic areas were not identified.

There was concern within the commission that by emphasizing academic rigor, its recommendations would seem to be intended only for those students who would pursue careers in science, mathematics, and technology and would seem to advocate intellectual elitism. The commission addressed the excellence-equity distinction by saying: "these new basics are needed by *all* students—not only tomorrow's scientists—not only the talented and fortunate" (p. v). "While increasing our concern for the most talented, we must now also attend to the need for early and sustained stimulation and preparation for all students so that we do not unwittingly exclude potential talent" (p. x).

The commission was also careful, as they recommended the development of standards written at the national level, to leave room for variation in the way states and local school districts would implement those standards. "This should not be construed as a suggestion for the establishment of a national curriculum; rather these are guides that state and local officials might use in developing curricula for local use" (p. 41).

> No one course of study is appropriate for all students and all teachers in all schools in all parts of the country. Nor is there just one good curriculum. Various parts of the Nation must develop their total curriculum and revise it repeatedly to keep it suitable for students and teachers. (p. 92)

Reports such as those of the NCEE (*A Nation at Risk*) and the Commission on Precollege Education in Mathematics, Science, and Technology (*Educating Americans for the 21st Century*) set the stage for hundreds of additional reports to be published on education during the 1980s. These two documents, however, aptly summarize a vision and strategy for reform. The vision included an intellectually rigorous common core of science knowledge for all, which would lead to an understanding of science ideas that are personally fulfilling and that can help build a knowledgeable and competent citizenry that is well prepared for life in a free society. The strategy involved national goals, local implementation, and accountability through student testing. Details of the vision and the strategy were to be worked out over time with the help of scientists and professional educators.

Many of the reports that followed, however, simply lamented the poor performance of United States students on national and international tests, especially in mathematics and science, and continued to link the nation's economic problems to the poor quality of the educational system. In response, many of the changes that were initiated by state legislatures and state departments of education during the 1980s were structural in nature. They often did not pay attention to the broader goals of the educational reformers. As Hurd said toward the end of the 1980s: "changes implemented ... include lengthening the school day and year, requiring more science courses, intensifying course rigor, increasing student testing and school assessments, and raising graduation requirements; but, to what ends" (1989, p. 16)?

Science for All Americans

The first detailed and substantive response to the call for a comprehensive statement of what all students should know and be able to do in science came from the American Association for the Advancement of Science (AAAS) through the establishment of Project 2061, a long-term reform effort to define and promote science literacy. Their work began with the publication of *Science for All Americans* (AAAS, 1990) and continued with the development of tools and resources to bring the vision of science for all to full realization.

In 1985, Project 2061 began work on producing a coherent statement of what all adults should know in order to be considered science literate. Science literacy is a term that became popular in the 1970s and 1980s to describe the knowledge people needed to live successfully in a world where science played such a large part. (For a discussion of the multiple meanings of science literacy, see DeBoer, 2000.) The AAAS statement that resulted consolidated reform ideas that had been promoted by science educators during the previous decades, many of which could also be found in *A Nation at Risk* and *Educating Americans for the 21st Century.* The 1989 publication of *Science for All Americans* brought these ideas together in one bold statement. Project 2061 was so named because it was the year that Halley's Comet was to be visible again on earth, 76 years after its appearance in 1985, and the year the project originated. Most of those born in 1985, it was said, would live to see the comet's return in 2061. Enacted reforms could very well touch their entire lives.

Science for All Americans was a vision of adult science literacy, what everyone should know to be able to participate fully in society. That core knowledge included concepts and skills in science, mathematics, technology, and the social sciences. It included knowing about the nature of science, the nature of mathematics, and the nature of the designed world. It also included an understanding of historical perspectives, common themes having to do with systems, models, constancy and change, and issues of scale. It included information on scientific habits of mind and effective teaching and learning. *Science for All Americans* was not, however, a prescription for what students should know at various grade levels. Although it provided much more detail on what students should know and be able to do than had been provided by the National Commission on Precollege Education, this was not a list of content standards.

The language of *Science for All Americans* is inspiring and stresses both personal development and responsible citizenship:

> Education has no higher purpose than preparing people to lead personally fulfilling and responsible lives. For its part, science education—meaning education in science, mathematics, and technology—should help students to develop the understandings and habits of mind they need to become compassionate human beings able to think for themselves and to face life head on. It should equip them also to participate thoughtfully with fellow citizens in building and protecting a society that is open, decent, and vital. America's future—its ability to create a truly just society, to sustain its economic vitality, and to remain secure in a world torn by hostilities—depends more than ever on the character and quality of the education that the nation provides for all of its children. (AAAS, 1990, p. xiii)

In contrast to many of the reports produced during the 1980s, *Science for All Americans* did not propose a get-tough approach or that schools should teach *more* science content. Instead, it suggested that schools should focus on what is essential for science literacy—a common core of ideas and skills that have the greatest scientific, educational, and personal significance—and should teach that science better and for deeper understanding. The recommendations for content and for pedagogy were meant for all students regardless of social circumstances or career ambitions. Criteria for content selection included: (1) the utility of the content for employment, personal decision making, and intelligent participation in society; (2) the intrinsic historical or cultural significance of the knowledge; (3) the potential to inform one's thinking about the enduring questions of human meaning; and (4) the value of the content for the child's life at the present time and not just for the future. Within these parameters, the recommended content in *Science for All Americans* represented a consensus view of what the scientific community thought was important for everyone to know.

There were also recommendations regarding pedagogy. *Science for All Americans* said that in order to teach for understanding "people have to construct their own meaning regardless of how clearly teachers or books tell them things" (AAAS, 1990, p. 198). The student does this by connecting new information to what he or she already knows. Knowledge is remembered best if it is connected with other ideas and encountered in a variety of contexts. When new ideas do not fit within a student's existing knowledge framework, restructuring of existing ideas becomes necessary. This is done by providing students with experiences where they can see how the new information helps them make better sense of the world.

Science for All Americans also suggested that: "Young people learn most readily about things that are tangible and directly accessible to their senses—visual, auditory, tactile, and kinesthetic" (AAAS, 1990, p. 199). Although the ability to think abstractly, to reason logically, and to manipulate symbols develops throughout schooling, most people continue to rely on concrete examples of new ideas throughout their lifetimes. Other pedagogical approaches that would support conceptual understanding include applying ideas in novel situations and giving students practice in doing so themselves, having students express ideas publicly and obtaining feedback from their peers, allowing time to reflect on the feedback they receive, and having the chance to make adjustments and try again. Also noted were the values of self-confidence and the importance of the expectations that others have on one's self-confidence.

In addition to these general pedagogical principles, *Science for All Americans* also made recommendations specifically for the teaching of science, mathematics, and technology. There, too, the recommendations were not new but rather a reaffirmation of many of the ideas that were being suggested by science educators. According to *Science for All Americans*, to appreciate the special modes of thought of science, mathematics, and technology, students should experience the kind of thinking that characterize those fields: "To understand [science, mathematics, and technology] as ways of thinking and doing, as well as bodies of knowledge, requires that students have some experience with the kinds of thought and action that are typical of those fields" (AAAS, 1990, p. 200). *Science for All Americans* also pointed out the value of beginning instruction within the range of concrete experiences that students have already had:

> Sound teaching usually begins with questions and phenomena that are interesting and familiar to students, not with abstractions or phenomena outside their range of perception, understanding, or knowledge. Students need to get acquainted with the things around them—including devices, organisms, materials, shapes, and numbers—and to observe them, collect them, handle them, describe them, become puzzled by them, ask questions about them, argue about them, and then to try to find answers to their questions. (p. 201)

It was also recommended that the content and methods of science be taught together:

> In science, conclusions and the methods that lead to them are tightly coupled. Science teaching that attempts solely to impart to students the accumulated knowledge of a field leads to very little understanding and certainly not to the development of intellectual independence and facility.... Science teachers should help students to acquire both scientific knowledge of the world and scientific habits of mind at the same time. (AAAS, 1990, pp. 201-203)

The suggestion that because content and method are coupled in the doing of science they should also be coupled in the teaching of science presents a significant challenge to educators, one that has historically been easier to state in the abstract than to implement in actual practice.

Science for All Americans also recognized that doing science was a social activity that incorporates a number of human values, and these too should be part of the science curriculum. For example, curiosity, creativity, imagination, skepticism, and the absence of dogmatism should all be acknowledged as important in the conduct of science. Science teaching should encourage students to raise questions about what is being studied, help them frame their questions clearly enough to begin to look for answers to those questions, and support the creative use of imagination. It should promote the idea that one's evidence, logic, and claims will be questioned, and experiments will be subjected to replication. Students should be encouraged to ask: How do we know? What is the evidence? Are there alternative explanations? *Science for All Americans* makes clear that science is a way of extending understanding and not a body of unalterable truth. It also suggests that teachers and textbooks should not be viewed primarily as purveyors of truth. Because science ideas are often modified, an open mind is needed when considering scientific claims.

Originally Project 2061 was designed in three phases. In Phase II, teams of educators and scientists were charged with transforming *Science for All Americans* into blueprints for action. Six school-based teams would produce curriculum models that other school districts and states could use as they undertook the reform of science, mathematics, and technology education. Phase II would also specify the changes needed in research, teacher education, testing, educational technologies as well as the organization of schooling, and state and local policies for the vision to succeed. The idea that local sites could interpret general frameworks and provide the implementation suitable to their own communities was consistent with the general belief in the importance of local control that was reflected in the reports of the early 1980s. That approach called for a centralized vision and decentralized implementation.

In Phase III, there was to be collaboration with scientific societies, educational organizations and institutions, and other groups involved in the reform of science, mathematics, and technology education to turn the Phase II models into educational practice (AAAS, 1990, p. 221). Project 2061 was hopeful that *Science for All Americans* would be part of the policy discussions that would ultimately lead to a specification of the science content that needed to be taught to achieve science literacy for all. Project 2061 recommended that the President, the U.S. Secretary of Education, Congress, state governors, business and labor leaders, and the news media use the *Science for All Americans* report along with other reports to stimulate discussion and debate on the goal of scientific literacy for all. As noted earlier, *Science for All Americans* did not, however, lay out grade-by-grade learning expectations for students.

The first organization that did produce such grade-by-grade learning expectations for students in the form of content standards was the National Council of Teachers of Mathematics (NCTM). In 1986, the NCTM board of directors created a Commission on Standards for School Mathematics to "create a coherent vision of what it means to be mathematically literate" and to "create a set of standards to guide the revision of the school mathematics curriculum" (NCTM, 1989, p. 7). In 1989, NCTM produced a document specifying content standards for students in school mathematics at four grade bands (K-2, 3-5, 6-8, and 9-12). The standards were offered as "statements of what is valued." The standards were to serve: "(1) to ensure quality, (2) to indicate goals, and (3) to promote change" (NCTM, 1989, p. 2). The NCTM report was well received by the education community as a model for standards-based reform. Kirst and Bird (cited in Collins, 1998, p. 713) listed four reasons for the positive reception that the NCTM standards received: (1) the preparation time taken to develop the intellectual groundwork for mathematics reform, (2) broad involvement in the development process including significant roles for educators and subject-matter specialists, (3) a far-reaching review and feedback process, and (4) continued robust efforts to establish consensus and build capacity.

The National Governors' Conference

In September 1989, President George H. W. Bush met with the state governors in Charlottesville, Virginia, to discuss a national agenda for education. At this meeting the president and the governors agreed to establish clear national performance goals and strategies to ensure United States international competitiveness. They also agreed that there should be annual reporting on progress toward meeting those goals and that in return the states would be given greater flexibility in the use of federal resources to meet the goals. The goals themselves were to be announced in early 1990. Structural changes that would be needed to implement the goals would include

> a system of accountability that focuses on results … and decentralization of authority and decision-making responsibility to the school site, so that educators are empowered to determine the means for achieving the goals and to be held accountable for accomplishing them. (U.S. Department of Education, 1991, p. 79)

The strategy of centralized goal setting and decentralized implementation that had been introduced in the early 1980s was finding favor across the country.

On April 18, 1991, President Bush released *AMERICA 2000: An education strategy* (U.S. Department of Education, 1991) which described a plan for moving the nation toward the national goals adopted the previous year. In his opening remarks, the President said: "If we want to keep America competitive in the coming century, we must stop convening panels to report on ourselves.… We must accept responsibility for educating everyone among us, regardless of background or disability" (p. 2). "It's time to turn things around—to focus on students, to set standards for our schools—and let teachers and principals figure out how best to meet them" (p. 4). President Bush spoke also of the importance of character education, school choice, lifelong learning, and the cultivation of communities where learning was valued and available throughout one's lifetime. Forged in cooperation with the nation's governors, six goals were identified that would be accomplished by the year 2000: (1) ensure that every child starts school ready to learn, (2) raise the high school graduation rate to 90%; (3) ensure that each American student leaving the fourth, eighth, and 12th grades can demonstrate competence in core subjects; (4) make United States students first in the world in math and science achievement; (5) ensure that every American adult is literate and has the skills necessary to compete in a global economy and exercise the rights and responsibilities of citizenship; and (6) liberate every American school from drugs and violence so that schools can encourage learning (p. 4).

This was to be a national program but it would not be dictated by the federal government. According to President Bush,

> America 2000 is a national strategy, not a federal program. It honors local control, relies on local initiative, affirms states and localities as the senior partners in paying for education... It recognizes that real education reform happens community by community, school by school, and only when people come to understand what they must do for themselves and their children and set about to do it. The federal government's role in this strategy is limited.... But that role will be played vigorously. Washington can help by setting standards, highlighting examples, contributing some funds, providing flexibility in exchange for accountability and pushing and prodding. (pp. 11-12)

The strategy for accomplishing the national goals was to include a 15-point accountability package that would encourage schools and communities to measure and compare results, and insist on change when the results were not good enough. The package included national standards, national tests, reporting mechanisms, and various incentives. Content standards in each of five core subject areas would be developed in conjunction with a national education goals panel. "These standards will incorporate both knowledge and skills, to ensure that, when they leave school, young Americans are prepared for further study and the work force" (p. 21). Tests to measure achievement of that content would also be developed in conjunction with the national education goals panel. These tests would be national but voluntary and tied to the national standards. Colleges would be encouraged to use the tests in admissions, and employers would be encouraged to use them in hiring. To ensure accountability to the public, school districts would issue report cards on results to provide clear (and comparable) information on how they were doing. The President's proposals called for Congress to authorize the National Assessment of Educational Progress

> regularly to collect state-level data in grades four, eight and twelve in all five core subjects, beginning in 1994. Congress will also be asked to permit the use of National Assessment tests at district and school levels by states that wish to do so. (p. 22)

Finally, the President argued for a decentralization of authority and decision making in which the local school was identified as the site of reform.

> Because real education improvement happens school by school, the teachers, principals and parents in each school must be given the authority—and the responsibility—to make important decisions about how the school will

operate. Federal and state red tape that gets in the way needs to be cut. (p. 23).

Recognizing that in 1991 national content goals in the subject areas were not yet available, the President said:

> First, what students need to know must be defined. In some cases, there is a solid basis on which to build. For example, the National Council of Teachers of Mathematics and the Mathematical Sciences Education Board have done important work in defining what all students must know and be able to do in order to be mathematically competent. A major effort for science has been initiated by the American Association for the Advancement of Science. These efforts must be expanded and extended to other subject areas. (p. 70)

Benchmarks for Science Literacy

In 1991, Project 2061 was engaged in Phase II of its reform efforts, the development of curriculum models in six school districts around the country. Over the course of four summer workshops between 1989 and 1992 and 40 days of release time during each school year, teams from each location worked on developing these curriculum models. During the first summer, a number of cross-team working groups were formed as well. One was a *strand* group, whose purpose was to begin the process of determining what knowledge and experiences a student had to have in grades K-12 in order to achieve the goals in *Science for All Americans*. It had become clear to everyone that it would not be possible to directly translate the content described in *Science for All Americans* into a K-12 curriculum without intermediate learning goals. The intent was that the product of the efforts of this strand group would be an internal document that would aid the curriculum development teams in their work. The process was called *backmapping* and the products were called *maps* or *strand maps*. These maps showed the progression of ideas through four grade bands. They were based on (1) the logical development of the science ideas and (2) the age at which students would be expected to understand each idea. During the second summer workshop in 1990, all of the teams were assigned to work on these maps because the maps were proving to be so useful to the schools.

With President Bush's declaration in his *America 2000* report in the spring of 1991 that standards "will be developed ... for each of the five core subjects," and his explicit mention of the work of AAAS, the focus at Project 2061 shifted from the further development of school-based curriculum models to the transformation of back-mapped ideas into "benchmarks." By 1993 that work was complete and in print, and *Benchmarks for Science Literacy* (AAAS, 1993) was hailed by many as the long-awaited national content standards in science (Atkin, 1997, p. 168). Although the original intent of the back-mapping was not to produce content standards, but rather to produce an ordering of ideas that could be used in the development of curriculum models, Project 2061 decided to use the work they had already done in creating the benchmarks to more directly impact the national debate on science literacy and ultimately to influence what became the *National Science Education Standards*.

The decision to produce and publish *Benchmarks* thrust Project 2061 into the center of the standards movement. But according to an external evaluation of Project 2061 prepared by SRI International in 1996, even before *Benchmarks* was published in 1993 "both the National Science Foundation and the U.S. Department of Education urged states and local districts to incorporate or demonstrate consistency with Project 2061's vision of science literacy in their proposals for important federal initiatives" (SRI International, 1996a, pp. 10-11). For example, the statewide systemic initiatives (SSI) program, a National Science Foundation program begun in 1990, was based on the idea of state-level, systemic, standards-based reform. According to the SRI report, *Science for All Americans* was explicitly named as a useful model for supporting a systemic initiative in elementary and secondary science education. The U.S. Department of Education also referred to Project 2061 as a model to support standards-based or systemic reform. In the national portion of the Department of Education's Eisenhower National Mathematics and Science Program that began in 1992 to promote the development and implementation of state curriculum frameworks in mathematics and science, "*Science for All Americans* [was] noted as one of a very small number of documents that could serve as a foundation on which the states applying for federal funds could build their frameworks" (SRI International, 1996a, p. 11).

Especially after *Benchmarks* was published, Project 2061, which was meant to be part of a long-term reform effort bringing together many of the values and ideals of enlightened science educators, soon became identified with the standards movement. *Benchmarks* became a *standards* document in the minds of most people, in part because the nation was poised to receive national standards. By the early 1990s, the idea of national goals (along with local implementation) that had been discussed throughout the 1980s had become a widely accepted approach to reform throughout the country.

At the same time that work was progressing at Project 2061 to create benchmarks for science literacy, the national education goals panel was established in 1991 to measure progress toward the broad national goals that had been established by the Bush administration. Also that year, consistent with the charge from the goals panel to establish national content standards in five content areas, the U.S. Department of Education and the National Science Foundation made a decision to fund the National

Research Council (NRC) to create national standards in science. According to Collins (1998), in receiving the award "the NRC was encouraged to draw on expertise and experience from both the AAAS and NSTA [National Science Teachers Association]" (Collins, 1998, p. 715), and the advisory board for the project included representatives of both NSTA and AAAS. In the fall 1991 issue of its newsletter, *Project 2061 Today*, the response of Project 2061 to this decision was clear:

> [*Science for All Americans* and *Benchmarks*] will contribute to the formulation of national standards for science education, an enterprise to be orchestrated by the National Research Council and funded by the U.S. Department of Education. This welcome opportunity for working together will produce a much-needed resource for the nation. ("Developing Standards," 1991, p. 3)

National Science Education Standards

National Science Education Standards was written in direct response to a call from the federal government to establish content standards in each of five disciplinary areas that could be used to measure progress toward the national goals. (In addition to writing content standards, the NRC also wrote standards for teaching, professional development, assessment, science education programs, and science education systems.) The national standards in each disciplinary area were to offer a means to judge the quality of student learning and educational programs. In the words of the NRC: "Science education standards provide criteria to judge progress toward a national vision of learning and teaching science" (p. 12). As that statement makes clear, however, the *National Science Education Standards* represents a vision as well as a set of criteria for making judgments. *Standards* are forward looking even as they provide a means for measuring progress toward attaining that vision. In fact, from the beginning to the end of the document, it is the forward-looking vision of science literacy that comes through most strongly in the *Standards*.

The tension between a statement that represents a fixed standard that can be used for accountability purposes and one that represents a flexible and ever-evolving vision of reform is difficult to resolve. That tension is evident in Collins' (1998) discussion of the political nature of the process of developing the *Standards*. One place this is particularly apparent is in how *standard* is defined. In the NRC *Standards*, there are seven content areas at each grade band, each divided into several bulleted topics, which in turn are elaborated into a small number of more detailed fundamental abilities and concepts. For example, for grades 5-8, Content Standard B (Physical Science) says: "As a result of their activities in grades 5-8, all students should develop an understanding of (1) properties and changes of properties of matter, (2) motions and forces, and (3) transfer of energy"

(NRC, 1996, p. 149). There are then six "fundamental concepts and principles" that underlie the "transfer of energy" part of this standard, one of which that states: "Heat moves in predictable ways, flowing from warmer objects to cooler ones, until both reach the same temperature" (NRC, p. 155). Clearly, when considered at the topic level (transfer of energy) there is a lot of discretion granted for making decisions about content, instructional activities, and assessment, and not much guidance regarding exactly what to teach or what to hold students accountable for. At the level of the detailed knowledge statements (heat moves in predictable ways, flowing from warmer objects to cooler ones, until both reach the same temperature), there is more specificity and less flexibility. The position taken in the *Standards* is captured in the statement: "the discussion of each standard concludes with a guide to the fundamental ideas that underlie that standard, but these ideas are meant to be illustrative of the standard, not part of the standard itself" (pp. 6-7). This interpretation of what a standard is suggests that in the development of the *Standards*, accountability with respect to a fixed standard was not the first priority.

Additional evidence that the *Standards* was not attempting to be a present day accountability document is the strong position that the NRC took with respect to inquiry as pedagogy, even though such approaches had met with mixed success throughout the twentieth century, most notably in the reform efforts of the 1960s (Welch, Klopfer, Aikenhead, & Robinson, 1981, p. 40). In its "Call to Action," the NRC said that the *Standards* "emphasize a new way of teaching and learning about science that reflects how science itself is done, emphasizing inquiry as a way of achieving knowledge and understanding *about* the world" (p. ix). Later, when recommending changing directions for science content, *inquiry* is the dominant theme, including the statement that there should be greater emphasis on "doing more investigations in order to develop understanding, ability, values of inquiry and knowledge of science content" (p. 113). In its "Overview," the NRC said: "Implementing the *Standards* will require major changes in much of this country's science education.... Inquiry is central to science learning" (p. 2). Although these statements make it clear that the "importance of inquiry does not imply that all teachers should pursue a single approach to teaching science" (p. 23), nevertheless, inquiry pedagogy is so deeply imbedded in the NRC *Standards*, that many have come to equate standards-based science education with "inquiry-based" science education. The point is that given the historic difficulty of implementing inquiry-based pedagogies, inquiry could not reasonably be used as an accountability criterion at this time. It must be viewed as a vision for the future.

Goals 2000

The specific proposals in President Bush's *America 2000* report were never enacted into law during his presidency because in November 1992 President Bush was defeated in his reelection attempt by Bill Clinton, then governor of Arkansas. But because many of the ideas in *America 2000* had come from the governors at the education summit that Bush had convened shortly after he took office in 1989, including Clinton as governor of Arkansas, the general strategy for reforming education remained intact and the legislation that President Bush had proposed was resurrected by the Democrats as Goals 2000.

On March 31, 1994, President Clinton signed the Goals 2000: Educate America Act. The act retained the six goals from *America 2000* and added two new goals, one on teacher professional development and another on promoting the involvement of parents in their children's education. The rationale for the eight goals centered on educating workers for productive employment, with special reference to competition in international trade. Again, as throughout the 1980s, the 1960s, and earlier in the century, the government's primary interest in education was the development of human capital so that the U.S. could remain competitive internationally (Spring, 2001, p. 434). In addition to stating national goals, the Goals 2000 legislation also created the National Education Standards Council, which had the authority to approve or reject the states' content standards. This body subsequently dissolved following the 1994 midterm elections when the Republicans took control of Congress and voiced objections to the increasing intrusion of the federal government in education (National Conference of State Legislatures Report, 2004). Also in 1994, President Clinton signed the Improving America's Schools Act (IASA), which was a reauthorization of the original Elementary and Secondary Education Act of 1965, first enacted as part of President Johnson's War on Poverty and intended to improve education for disadvantaged children in poor areas. Under the IASA, states had to: (1) develop challenging content standards for what students should know in mathematics and language arts; (2) develop performance standards representing three levels of proficiency for each of those content standards—partially proficient, proficient, and advanced; (3) develop and implement assessments aligned with the content and performance standards in at least mathematics and language arts at the third through fifth, sixth through ninth and 10th through 12th grade spans; (4) use the same standards and assessment system to measure Title I students as the state uses to measure the performance of all other students; and (5) use performance standards to establish a benchmark for improvement referred to as *adequate yearly progress*. All schools were to show continuous progress or face possible consequences, such as having to offer supplemental services and school choice options to students or replacing the existing staff (National Conference of State Legislatures, n.d.).

The trend toward holding schools accountable for their students' performance through standards setting and assessments that began in the early 1980s was continued and strengthened with this legislation. It played a significant role in the development of the standards movement because it moved the emphasis away from national standards and voluntary national testing to a state-by-state system of standards setting and accountability. And as we will soon see, it provided the basis for the No Child Left Behind Act of 2001. But before moving forward to 2001, it is important to note that not everyone in the education community in the early 1990s was in favor of the standards-based accountability movement, especially when it spoke of national accountability and hinted at the possibility of a national curriculum. Although the movement was generating bipartisan support in Congress and in statehouses across the country, concerns persisted about the *nationalization* of education and were, in part, why attention shifted, particularly legislatively, from a focus on national goals and accountability to state-by-state accountability programs in the mid to late 1990s.

Reaction to the Idea of National Standards

In June 1993 the American Educational Research Association (AERA) sponsored an invitational conference to explore the implications of the new standards-based accountability movement. Papers were commissioned as part of the conference and published as the first in a series of AERA public service monographs in 1995. In general, the papers were critical of what was seen as a move toward a national curriculum and toward greater federal control over education. The fear was that aligning national tests (albeit voluntary) to national goals was a recipe for a mandated curriculum, and the country would inevitably move in the direction of a national curriculum. The proposal that there would be a national test was the most troubling aspect of the plan, for it was felt that a national test would create a de facto mandated curriculum (Zumwalt, 1995).

Although some felt there were likely to be advantages to having a national curriculum, there were many concerns that such centralization would take control away from local communities which, it was thought, had the best insights into what was important for their students to know. There were also concerns that common goals and accountability through testing would lead to a narrowing of the curriculum, a de-skilling of teachers, a focus on external academic knowledge, and a move toward direct instruction (Kellaghan & Madaus, 1995; McNeil, 1995). It was also feared that these changes would threaten the "soul of the curriculum" (aesthetics, for example) and draw attention away from caring about children as individuals (Kellaghan & Madaus, 1995; McNeil, 1995; Zumwalt, 1995). Some questioned whether it was desirable to specify what all students should know and whether it was possible to validate such knowledge given the vast array of social and cultural contexts from which people have come and in which they now live. In his concluding statement, Asa Hilliard, said: "The problem is at its base one of curriculum validity.... Multiethnic, multinational, multidisciplinary scholars must be consulted to review every area of the curriculum to cleanse it of error and to enrich it with diverse perspectives" (1995, pp. 153-154). Others questioned the overall effectiveness of this approach to reform, especially if it was even possible to create unity out of such a diverse system.

No Child Left Behind

Consistent with a move toward state-level standard setting and accountability through testing, on January 8, 2002, President George W. Bush signed into law the No Child Left Behind Act of 2001 (NCLB), a bill to extend and revise the Elementary and Secondary Education Act of 1965. The changes in this legislation over the 1994 reauthorization are significant because the new law emphasizes even greater public accountability, with funding tied directly to meeting expectations. In the words of one observer: "This landmark event certainly punctuated the power of assessment in the lives of students, teachers, parents, and others with deep investments in the American educational system" (Jorgensen & Hoffmann, 2003).

NCLB requires states to build assessment systems to track the achievement of students in their state against a common set of state-derived standards. By the 2005-2006 school year, states are required to test students annually in reading and mathematics between grades 3 and 8 using statewide tests, and to test students at least once during grades 10 through 12. The tests must provide individual student scores. By the 2007-2008 school year, students must be tested in science at three grade bands. Every 2 years, states must also administer the mathematics and reading tests of the National Assessment of Educational Progress to a sample of students in grades 4 and 8. This allows the states to check the rigor of their own tests and to make national comparisons. The law also requires schools, school districts, and states to disaggregate the average test score results for major racial and ethnic groups, income groups, students with disabilities, and students with limited English proficiency. Starting in the 2001-2002 school year, states have 12 years to achieve the goal of having all groups of students meet their own state's benchmark for proficiency in reading and mathematics. As with the 1994 Improving America's Schools

Act, the new legislation calls for states to define three levels of proficiency and to measure students with respect to those proficiency levels. The goal of NCLB is to raise reading and math proficiency to 100% for all students in the country by 2014. NCLB measures the performance of subgroups of students in reading and math and requires all groups—defined by race, ethnicity, income, and other characteristics—to keep improving until all groups reach the 100% goal. Failure to make *adequate yearly progress* toward meeting these goals results in various actions intended to help a school improve. In addition to technical assistance, staff changes, and the possibility of private or state takeover of the failing school, students in schools that do not meet their target goals are able to transfer to another school or use their Title I funds to pay for tutoring or other supplemental services.

NCLB continued a standards-based accountability movement that began in the early 1980s and that was strengthened with passage of the Improving America's Schools Act (1994) and the Goals 2000: Educate America Act (1994). The NCLB legislation of 2001, however, took the focus on standards-based accountability significantly farther through the specificity of its requirements and its sanctions. Standards present and past were now firmly entrenched in federal law.

IMPACT OF THE STANDARDS MOVEMENT

It is clear that the standards movement that began in the early 1980s represents one of the most significant changes in education policy in the United States to date. It is reasonable, then, to ask what the impact of that change in policy has been. Immediately, though, we are faced with the question: Which standards movement are we talking about? There are at least four ways to think about the *impact of standards*, depending on whether the standards movement is defined as (1) the general goal of raising the rigor of the educational experience for all, to be accomplished by requiring all students to take more science courses, offering students more academically challenging experiences, and increasing the length of the school day and the school year; (2) an accountability strategy involving testing and public reporting; (3) a detailed specification of what all students should know and be able to do; or (4) an elaboration of a vision of science literacy for all.

It is not my intent to provide a detailed discussion of each of these meanings or their implications here. A brief response to each of the questions below provides an overview of the widely acknowledged impact of standards before moving to a discussion of what next steps should be taken to move the standards-based reform agenda forward.

Question 1: What has been the effect of calls for more rigor? There is no question that the ideas presented in *A Nation at Risk* have resonated with many people in this country over the past two decades. States were quick to raise science course requirements, and some increased either the length of the school year or the school day. Between 1982 and 1994 the percentage of high school students taking science courses increased substantially from 77% to 93% in biology, from 31% to 56% in chemistry, and from 14% to 25% in physics (see Demarest, 2002, p. 332). During the 1980s and 1990s states and local school districts raised their requirements for the number of science courses students had to take. Some states also increased the length of the school day and the school year. There is no question that schools today are valued for the number of rigorous and challenging courses that they offer, especially honors and advanced placement courses. Holding students back when they do not perform at grade level is frequently seen as a way to maintain standards, and some states have officially ended the practice of "social promotion" (Neal & Poole, 2004). "Standards as rigor" is very much a part of the present educational environment.

Question 2: What has been the effect of the standards movement as an accountability strategy that uses high-stakes testing and external reporting? The conventional wisdom is that high-stakes tests concentrate energy and resources toward the things being tested. This is becoming particularly evident under NCLB legislation that uses test results in mathematics and reading as part of annual yearly progress reports, but does not include test results from any other subject areas (Perlstein, 2004). The power of the test to focus attention on what is being tested applies both at the school wide curriculum level and it applies at the level of the content of individual courses. Teachers are beginning to limit their instruction to those ideas that are going to be on the state's test. The issue of *teaching to the test* raises questions about whether such a practice is helpful or detrimental to good education.

Question 3: What has been the impact of defining—much more precisely than had been done in the past—what all students should know and be able to do? This is an extremely difficult question to answer by itself. Especially at the state level, specification is accompanied by accountability testing, so it is difficult to tell whether the effect, if any, is due to the testing or to the specification. Specification of content was intended to be a positive feature of reform. The specified content would represent the most important things for students to know, including ideas that were functional in students' lives and that taught them important aspects of the structure of science itself—a *less is more* approach. One place to look for the impact of content specification is textbooks. In general, there has been little if any effect on textbooks of the efforts to identify the most important content for students to study. Textbooks continue to cover as much content as possible. Another place to look for the effect of greater specification of content is student learning. Whether a sharper focus on important science ideas has affected either the depth or breadth of student understanding, we do not yet know. In fact, the assessment instruments that would provide an answer to that question do not exist.

Question 4: What has been the contribution of standards-based reform as presented in the *National Science Education Standards* and *Benchmarks for Science Literacy* to the on-going efforts to achieve science literacy for all? Where will these documents stand in the record of important efforts to improve science education in the United States? At one level their contribution has already been impressive. These national standards documents were used as models to write state curriculum frameworks to guide the development of the science curriculum in each state. This was the first time that the science content most worth knowing had been specified with such great precision. In the past, guidance had been given at a more general level of specification, and no one had provided a system of interconnected ideas at this level of detail. In addition, both national documents consolidated the best thinking about *how* science should be taught. There is no question that through their influence on the states, that they have had a major contribution to national attempts to achieve science literacy for all.

NEXT STEPS IN THE REFORM OF SCIENCE EDUCATION

It is this fourth category of impact concerning the vision of science literacy for all that I will address in the remainder of this paper. It is this vision that is the focus of standards future, and that is ultimately what reform today in science education should be about. Accountability through assessment is simply a strategy for motivating people to act. Reform, on the other hand, is a continuing process of change. The standards documents in science education embodied a vision for improvement; they were not authoritative pronouncements of a standard that all must meet. They were meant to inform and enliven reform efforts. Although established by means of broad-based authoritative consensus, it was anticipated that within certain boundaries consistent with that vision, the standards documents were expected to be used by people who would apply that vision of reform to their own practice, resulting in products that would be shaped by their own particular situations.

The goals of the reform movement that appear in the standards documents are: (1) decreased content coverage to increase student understanding of the ideas being taught, (2) science for all students, (3) the integration of ideas within subject areas and across subject areas to increase meaning and aid in retention, (4) knowledge that is useful for personal growth and development, and (5) content that takes into account both the products of science and the way that knowledge was developed. These are noble goals worthy of pursuit.

In the early and mid1990s, the assumption was that it was possible for the educational system to take content standards and turn them into curriculum as long as important elements of the educational system were considered and brought into the process. Working through issues of school organization, curricular connections, equity, and finance seemed doable (AAAS, 1998). That idea began to change as it became clear that the content standards alone were not enough to guide reform. An entire coordinated and integrated system of clarified and elaborated content standards, more detailed guidance regarding effective classroom practices, curriculum materials aligned to the content standards that incorporate pedagogical support for teachers, and student assessment aligned to those content standards needed to be developed before significant numbers of people in state education departments, higher education, the business world, and family and community would be convinced of the soundness of the reform message. Without further elaboration of the intent of the reform movement and models of how to get there, it was unlikely that the desired changes would occur.

At a minimum, three additional steps of considerable magnitude still need to be taken if we are to build a standards-based educational system that works and that the educational community will embrace. The standards documents that we now have lay out a comprehensive vision for science literacy, but they do not adequately operationalize that vision in terms of teacher practice, curriculum materials, or assessment. A number of principles of effective instruction are included in the standards documents as well as some guidelines regarding assessment, but what we do not yet have is: (1) models of what textbooks actually look like that are at one with the vision of science literacy and that can be successfully used on a wide-scale basis; (2) a thorough and convincing statement, along with examples, regarding the range of research-based teaching practices or classroom activities that are consistent with the reform agenda; and (3) an assessment system that is aligned with the content standards that is laid out in the *Standards* and in *Benchmarks for Science Literacy.*

Fostering Implementation

In theory it is possible to construct an educational system made up of a set of interconnected parts each of which is aligned with the content standards. But even with clearly elaborated content standards in place, accompanied by curriculum materials aligned with those content standards, teaching that is consistent with the goals of reform, and student assessments aligned to the content standards, we still need to consider the contexts in which these parts will have to be implemented and the various players who can help or hinder progress. There is a real world of existing organizational structures, values, and practices that needs to be considered both during development and during implementation. In 2002, a NRC panel produced a report on the influence of standards on the educational system in the United States. The panel did not attempt to answer the question of what the influence of standards had been to date; rather, it created a framework for conducting future research on the impact of standards. (See chapter 2, A Framework for Investigating the Influence of Standards, for a detailed discussion of the panel report.) As part of that framework, the panel identified three "channels of influence" —assessment, professional development, and curriculum materials—that potentially can leverage change in the system.

These channels of influence can also be viewed as contexts for reform, that is, elements in the system whose preexisting interests and capacities need to be accommodated if change is to occur. Those who control these channels of influence, who have interests and capacities of their own, can provide support for reform, remain uninvolved and disinterested bystanders, or even counteract the efforts of reform. The point is that these are not necessarily freely flowing and readily available channels waiting to serve a new interest. They all represent well-established systems of their own, and may not be easily changed. Tradition and resistance to change has proved to be very powerful throughout our educational history.

In European countries that have a more centralized means of enacting and monitoring education, the channels of influence are very similar to those in the United States, but the way education is controlled is very different. According to Cohen and Spillane (1992), channels of influence in centralized systems can be used to direct what happens in classrooms by controlling instructional materials, preservice and in-service education of teachers, classroom instruction, and assessment of student achievement (cited in Kellaghan & Madaus, 1995, p. 88). In centralized systems

> there are definite procedures to reinforce the messages about content and pedagogy conveyed in curriculum documents unlike the practice in America, where the dispersed organization of education might have 'rendered the connections between policy and instruction inconsequential for most of our history.' (p. 87)

The decentralized nature of the U.S. educational system, both in its structural organization and its beliefs about local control, raises significant challenges to any national efforts of reform.

ELEMENTS IN AN EDUCATIONAL SYSTEM IMPORTANT IN IMPLEMENTING REFORM

In the following sections, I will discuss some parts of the educational system that need to be kept in mind when thinking about implementing a reform strategy, and I will offer suggestions as to what is needed to move the reform agenda forward.

State Education Departments and State Legislatures

At the present time, states are struggling with the requirements of NCLB. Under this legislation, each state sets its own standards for proficiency and is required to assess its students with respect to those standards. This federal requirement has the potential to have enormous impact on the educational system. At present, the federal mandates require annual testing in mathematics and reading in grades four through eight and once between grades 10 through 12. Science will be tested at three grade levels in 2007-2008 school year, but under present legislation, science will not be part of the annual yearly progress determination, which affects the designation of the school as meeting or not meeting expectations. In addition, states are not required to provide content standards consistent with any nationally recognized vision of what quality science education is. As noted earlier, the primary goal of NCLB is to raise reading and math proficiency to 100% for all populations of students in the country by 2014. In addition to the requirements of the federal legislation, some states have made their own decision to test science throughout the grades and to use those results for making high-stakes decisions, including graduation from high school.

There is no question that states have made good use of and will most likely continue to consult the national standards as they develop and revise their own state content standards in science. But most states have chosen to create something unique to their state rather than to appropriate the national standards without modification. There is often an antipathy in state legislatures to anything *national* when it comes to education, especially when it suggests the possibility of a national curriculum. When the Colorado standards were drafted between 1994-1995, the drafting committee used *Benchmarks for Science Literacy* as a guide but they also used current drafts of the NRC *Standards* as well as the California and New Jersey frameworks. According to an SRI report: "Key drafters pointed to the usefulness of *Benchmarks* as an intellectual guide, but the committee did not feel obliged to deal with all of *Benchmarks* and certainly did not believe that they had to justify excluding something included in *Benchmarks*.... Furthermore, the day-to-day criteria the committee used in drafting the standards were independent of national standards" (SRI International, 1996b, p. 4). Specifically, Colorado placed restrictions on the number of standards taught, on the degree of specificity of content to be covered, and on recommendations concerning pedagogy. According to SRI, "For political reasons, it was important that these should be seen as model content standards, that they not tell teachers how to teach, and there not be teaching standards" (p. 4).

Although state standards vary greatly in their degree of consistency with the national standards, an examination of current state content standards and the accompanying assessment instruments reveals that the general vision of science literacy is present in most of them, and many of the specific content areas that appear in the national standards are in the state standards as well. One observer concluded that the state frameworks capture "the spirit and essence of the national standards and the *Benchmarks* remarkably well" (SRI International, 1996a, pp. 36-37). From the perspective of national reform efforts, this is an encouraging observation. However that does not mean that state framework documents treat the full range of concepts found in the national standards or approach all concepts in the same manner as the national standards do. One problem found by the SRI evaluators was that of overgeneralization of the content. In some cases "the standards had become so diluted by generalization that they doubted a teacher or other practitioner could make use of them and certainly could not develop assessment tasks to measure the stated standard" (p. 39). An example of a typical overgeneralized standard said that students should be able to identify the characteristics of and understand the relationships among heat, light, sound, magnetism, and electricity (p. 38). Specifically *which characteristics* and *which relationships* were not noted in the state standards document.

Given the decentralized nature of schooling in this country and the lack of a national infrastructure for implementation, influencing the quality of education nationally must depend on state and local efforts. There is, for example, no tradition of school inspectorates in the United States to monitor what happens in classrooms as there is in many European countries. In fact, states themselves often do not exert much centralized control over education. In some states basic curricular decisions are made at the district or even the individual school level and pedagogy is almost always an individual matter for each teacher to decide. It is not clear that a centralized approach would ever work in this country. According to Atkin (1997):

> With the country's traditions of local autonomy, the increased assertiveness of teachers, and the changes taking place in science itself, it would be surprising in American science education reform if any central agency ... will be able to maintain strong and direct influence over actual classroom practices for very long. (p. 237)

According to another observer:

> Standards-based reform that begins with developing a set of ambitious goals for what students should know and be able to do is a very rational theory of change. However, the political and educational systems operate in ways that are not always describable as rational. (SRI International, 1996a, p. 20)

Thus, national efforts to reform science education are inevitably going to encounter obstacles "due to the intrinsic difficulty of changing such a large, complex enterprise as American education" (SRI International, 1996b, p. 89).

But, in general, states do want coherent, integrated educational systems. Therefore, the potential still exists for national groups to have a significant impact on the continuing development of state standards and resources for implementing those standards through the creation of educational products in which standards, instructional practices, curriculum materials, and assessment are aligned.

Universities, Colleges, and the Professional Development Community

Unlike countries that have a national education system, where a ministry of education can directly influence preservice and in-service education programs and, therefore, what teachers know and can do, there is no centralized control over what happens in preservice teacher education or in-service teacher development programs in the United States. There are thousands of college- and university-based preservice teacher education programs, all operating independently. The only centralized control of these programs is the general oversight that each state has over its colleges and universities, and direct control over the curriculum in colleges and universities is rare. In the case of preservice teacher education programs, where state certification and licensure is an issue, the states establish standards that colleges and universities must meet with respect to these programs, but rarely at the level of what gets taught in any particular course.

Similarly, in-service professional development is a decentralized activity in the United States, controlled to some extent by states, but largely an activity organized by local school districts or voluntarily selected by teachers as part of their own personal growth as teachers. There is little hope of broadly impacting such a system; any influence will most likely be at the level of individual school districts. Even in New York, which recently enacted a statewide requirement of 175 hours of professional development every 5 years to retain licensure for teachers newly certified in 2004, the options for meeting that requirement are many (New York State Education Department, n.d.). Often agreements of this kind between teachers and the states are negotiated as part of collective bargaining agreements.

The degree of flexibility in the many ways to meet the professional development requirement in New York raises the question of the role of professional development for teachers. Is its main purpose the general enrichment of teachers with decisions about those experiences left largely to their personal preferences, or should professional development be used as a tool for directly affecting the educational system? According to Elmore (2002), there is disagreement in the field on this issue. He says that whereas some educators believe that schools should take a more instrumental view of professional development many "argue that teachers, as professionals, should be given much more discretion and control as individuals and in collegial groups in deciding the purpose and content of professional development" (p. 32).

In either case, to achieve their approval and full participation, professional development should be planned with the real-world necessities of teachers in mind, and as much as possible, teachers should be engaged in the planning of the professional development experiences. If the experience is related to the work they are doing, and it has a high probability of impacting student learning, teachers are more likely to participate. In Elmore's words: "The work itself, then, is the primary motivator for learning and improvement. If the work is not engaging and if it is not demonstrably beneficial to student learning, then any incentives are likely to produce weak and unreliable effects" (2002, p. 21).

For reform of science education to become a reality, teachers and the college and university students who will become teachers must understand the reform vision, but they also need to have available to them specific tools to help them implement that vision. This means that the people who provide preservice and in-service educational experiences to teachers must be well-informed of the vision themselves and have ways to communicate it clearly and evidence that it can be accomplished successfully. With no centralized system for delivering the knowledge needed to advance the reform agenda, progress depends on the quality of the reform ideas and their fit with existing professional development delivery systems.

Curriculum Materials Developers

For the most part, commercial textbook publishers have not yet accepted the ideals of the current reform movement. The textbook publishing industry is a profit-driven and demand-driven enterprise. Publishers are unlikely to make changes unless they are certain that those changes are what the purchasers of textbooks want. According to SRI: "To protect their competitive position, publishers are not willing to align textbooks with reforms that a majority of teachers have not fully accepted" (SRI International, 1996a, p. 18) and publishers have little incentive to "take on what they perceive as greater business risk in order to be aligned with reform documents that, according to them, have not taken hold within the teaching community—at least at this point in time" (p. 19).

Most textbooks are still very large and cover many more topics than the national standards call for. To quote SRI:

> Voluminous biology texts reflect the organization of high schools around science disciplines. Technical vocabulary abounds (in each case, the glossaries contain about 1,000 technical terms that are used in the body of the book), and the authors have not eliminated material extraneous to what is required to meaningfully convey key concepts.... Not only are the textbooks 'designed to cover the standards, [they] cover everything.' (SRI International, 1996a, p. 17)

By providing coverage of many topics, publishers can appeal to a wider range of states that may have differing content lists.

> Publishers admit that texts continue to be laden with superfluous details because 'Even though people ask for 'less is more,' when they go to make their decision, they want everything.... From a business point of view, we can't make the decision to cut content. Every state looks at content differently ... to cut content would be financial suicide.' (p. 18)

To date, the changes in curriculum materials have been small at best. Some of the ideas present in the national standards have been incorporated, such as suggestions to include historical episodes and strategies for linking science, mathematics, and technology (SRI International, 1996a, p. 15). The publishers, however, have made little attempt to incorporate the pedagogical supports that are consistent with the vision of teaching that the reform documents encourage (Kesidou & Roseman, 2002). Most experimentation with curriculum materials has come with the support of federal funding, primarily through the projects sponsored by the instructional materials development programs of the National Science Foundation. Some of these innovative materials and ideas are finding their way into the marketplace as commercial publishing companies often offer one or more textbooks that appeal to the reform market, but the impact so far has been small. For the innovative federally funded materials development projects to have an impact, they will have to appeal to teachers and help teachers meet the new accountability demands being placed on them.

As with other elements of the educational system, there is no centralized infrastructure for affecting the textbook publishing industry. The closest thing is the statewide adoption committees of some states and, to a lesser extent, district and local adoption committees. By controlling the list of approved materials, these committees have the potential to significantly affect what gets taught and how it gets taught. Educational reformers can influence these decision makers by providing clear and compelling guidelines for choosing materials that are consistent with the reform vision, along with accompanying documentation of the effectiveness of those materials to advance student learning in the areas recommended in those reform documents.

Teachers

As Atkin (1997) noted: "One of the clearest lessons of successful reforms is the importance of according considerable weight to the insights and initiatives of those closest to the point of provision of educational services" (p. 219). Without the support of teachers and a vision that is consistent with their values and capabilities, change is unlikely. Teachers control the classroom. Regardless of the curriculum they are given, they enact the curriculum the way they see fit. We also know that, historically, classroom instruction is a story of diversity. Teachers are unlikely to march lockstep with respect to any innovation. Teachers develop and follow their own guidelines (Demarest, 2002, p. 319).

The extent to which teachers have accepted the vision of reform in the national content standards is not clear. A 2000 survey of teachers conducted by Horizon Research, Inc. showed that about 25% of respondents felt that they were prepared to explain the NRC *Standards* to their colleagues; 20% felt that the standards had been thoroughly discussed in their schools; approximately 39% believed that teachers had implemented the *Standards* in their teaching, and less than 30% believed that their principal was well-informed about the *Standards* (Weiss, Banilower, McMahon, & Smith, 2001). Other responses suggest that teachers have not yet embraced, or perhaps do not yet know about, national content standards. But whether they are familiar with the standards themselves or not, what is more important is whether they are sympathetic to the goals found in them. There is some evidence that they are, but there are reasons for concern as well.

In a survey of participants in Project 2061 professional development workshops over a 3-year period, 85% of teachers strongly agreed with the statement that "science teaching should be consistent with the nature of scientific inquiry" and 69% strongly agreed that "students should learn scientific 'habits of mind' that explicitly include ways to assess the validity of claims and arguments." However, only 33% agreed strongly that "most students are not currently learning science well, even in the best schools," 53.6% that "changes in the education system should be driven by important, specific goals to be achieved by all students," and 56.8% that "identification of appropriate materials for curriculum and instruction requires detailed study of how they relate to specific learning goals." In addition, 60.7% of teachers agreed strongly that "improving student understanding of science requires reducing in number and detail the science topics in the current curriculum, leaving time for more effective teaching of important ideas" (SRI International, 1996a, p. 69). The SRI report concluded:

> When one notes that only 61% of teachers strongly agree that the level of detail taught to students must be reduced, this is cause for real concern. Considering that a much smaller percentage of the teachers who did not attend Project 2061 workshops are likely to agree, the data suggest that there is much work to be done if teachers' classroom practice is to become consistent with the philosophy that 'less is more.' Furthermore, only 54% of teachers strongly agree that education system reform should be driven by specific goals for all students. This would be consistent with the finding … that teachers as a group, particularly those at the high school level, are unwilling to discard much of the content that is now taught and do not, as yet, fully embrace standards-based reform efforts. (SRI International, 1996a, p. 87)

There is some evidence that teachers view most of the statements in national reform documents as being too general and vague for their purposes. Thus they are influenced in what they teach primarily by the more specific requirements of their district administrators and building principals, the curriculum materials that they use, and the culture of the school. State testing may also have an impact on what teachers teach especially in school districts where many students are likely to do poorly.

When it comes to how teachers teach, high-stakes testing is a major factor in their willingness to try alternative instructional approaches. In a high-stakes testing environment where student performance on the tests is a concern, teachers are very reluctant to modify their teaching unless it can be demonstrated that the new approach is clearly superior in achieving improved test scores. In schools where students generally do well on tests, testing pressures are minimal, but there, as well, teachers are not likely to change how they teach without good reason. As Elmore notes,

> Few people willfully engage in practices that they know to be ineffective; most educators have good reasons to think that they are doing the best work they can under the circumstances. Asking them to engage in work that is significantly different from what they are already doing requires a strong rationale and incentive. (2002, p. 20)

According to Demarest (2002), it is not surprising that an attitude of caution or even cynicism is sometimes present among teachers, bombarded as they have been by "waves of reformist exhortations" and what some teachers perceive to be "faddish pendulum swings." For this reason, it is essential that proposals to change teaching should be driven by "foundational beliefs and theories" (p. 94).

Besides a clear foundational rationale for change, teachers also need to see clear models of what is expected. The Project 2061 vision of reform did not specify a particular approach to pedagogy. Teachers were expected to interpret the reform vision and implement it in their own way. This was consistent with a widely held view in the U.S that teachers should develop and follow their own approaches to teaching. *Science for All Americans* identified general principles of good teaching, such as the importance of teaching for understanding and making science teaching consistent with the nature of science, but it did not tell teachers in any detail how to do that. The absence of detailed guidelines allows the teacher flexibility in meeting the goals of the reformers, but it also means that teachers must know what "teach for understanding" means and have a solid grasp of what they can do to achieve such understanding in their students. Without detailed prescriptions of what to do, teachers need examples of teaching practices and curriculum materials that are effective and consistent with the reform vision.

In contrast to the lack of specification regarding pedagogy taken by Project 2061, the NRC took a very strong position on inquiry teaching as a favored pedagogy and recommended that student investigations become a central part of science classrooms. This raises additional challenges for teachers because, historically, teachers have found project-based work difficult to deal with in the classroom. Teachers often see such lessons as too "difficult and troublesome" (St. John, 1987; and Weiss, 1987, as cited in Demarest, 2002, p. 328).

Ultimately, then, the impact of any reform effort comes down to the individual classroom, and with so many teachers and so many classrooms, it is impossible to directly reach each one of them individually. The impact of national reform efforts on teachers' behavior is more likely to be indirect rather than direct. It will come through changes in professional development programs, curriculum materials, and assessment systems. But without a consideration of the values and capabilities of teachers who have so much independent control over what happens in the classroom, efforts at reform will not be productive.

Students

Attempts to change the goals and methods of education must also be consistent with the deep educational purposes that are imbedded in our culture with respect to our children and youth. Is education primarily about acquiring skills and information or is it about the development of social competence, identity, and belonging? (See Wenger, 1998, as cited in Brown, Demarest, Freeman, & Dalton, 2002, p. 193). Brown et al. (2002) argue that

> educators are unlikely to be successful with students—especially with the full diversity of students who attend today's schools—unless they broaden their concept of education so it attends not only to academic competence, but to social competence, interpersonal connection, and identity as well. (p. 194)

According to sociocultural theories, people are motivated by a desire to pursue their own interests, to develop competence, and to build on their own life experiences. Competent performance in most real-life situations requires individuals to draw on knowledge not only in the cognitive domain, but also from the social and affective domains. "Schools are expected to prepare young people to become adults who are capable of participating fully in and contributing to the economic, civic, and social life of their communities and the nation" (Demarest, 2002, p. 243). Many teachers devote a considerable amount of energy to the human interactive aspects of life in the classroom. This means that if national content standards in science are going to take hold, consideration must be given to students in the classroom, where interpersonal competence and personal identity are important to the participants, including the students, their teachers, and their parents. It needs to be asked if the new demands being placed on students are consistent with the identity and social competence needs of students.

It is clear that the standards-based reform movement and accompanying pedagogies do place new demands on students. Students are expected to know more than facts and information. They must also know processes and causal connections, and they must be able to use those ideas to identify and solve nontraditional problems. According to Wilson and Peterson (2002): "They need to learn about the ideas, theories, facts, and procedures of a discipline. They need to become fluent with the linguistic systems of a field, with developing the skill and knowledge associated with inquiry in that field" (p. 108). In reform-based classrooms, students are kept mentally engaged through questions and probes of what they know and how they think. They are asked to provide explanations and rationales, give reasons, make their ideas public, accept the critique of their peers, and offer comment on the thinking of others. Although these expectations and practices are not new and have characterized some classrooms for many years, it is not the predominant mode of teaching in most classrooms. To make it so is an enormous challenge, and to make it fit the learning styles of all students is ambitious to say the least.

When we consider the combined efforts of the science education reform agenda and the accountability movement brought on by federal legislation, the potential impact on students' lives in classrooms is enormous. Student performance is the measure of school success under the present federal legislation. Accountability and reporting is at the school level, but the data comes from student performance on tests. It is they who ultimately will feel the pressure to reach higher levels of performance. As with teachers, the support of students is crucial to the success of any reform efforts. Methods that are not consistent with the nature of children and adolescents, their needs as young people, and their sense of who they are and the importance of what they are learning will be ineffective. It is essential that proposals to reform educational goals and methods, as well as plans to develop curriculum materials and to design activities for classroom use, take the lives of children and youth into account.

Schools

Reform of science education takes place within a context of classrooms, state bureaucracies, professional development systems, and university pre-service teacher education programs. It also takes place within a context of school-based accountability. Under current federal legislation, accountability is at the school level and is based on individual student performance. Science education reformers must realize that for reform to be successful, implementation takes place in an environment where each school is being held accountable for the performance of each student. Just as teachers need new knowledge and an understanding and acceptance of the goals and methods of the reform movement, so too must schools *buy in* to the reform agenda. What kind of institution is a school? How equipped is it to effect the kinds of changes being proposed? According to Elmore,

> American schools and the people who work in them are being asked to do something new—to engage in systematic, continuous improvement in the quality of the educational experience of students and to subject themselves to the discipline of measuring their success by the metric of students' academic performance. Most people who currently work in public schools weren't hired to do this work, nor have they been adequately prepared to do it either by their professional education or by their prior experience in schools. (2002, p. 3)

Elmore goes on to say that

> there are few portals through which new knowledge about teaching and learning can enter schools; few structures or processes in which teachers and administrators can assimilate, adapt and polish new ideas and practices; and few sources of assistance for those who are struggling to understand the connection between the academic performance of their students and the practices in which they engage. (p. 5)

When considering the knowledge needed to reform science education, most school personnel currently have very little detailed knowledge of what the reform agenda in science entails. Weiss et al. (2001), for example, found in a 2000 survey of teachers that, depending on whether it was elementary, middle, or high school teachers who were responding, only 19% to 29% believed their principal was well-informed about the NRC *Standards*. A similar number of teachers felt their district superintendent was well-informed.

In discussing why it is so difficult for fundamental change to take place in schools, Cuban (1993) provides a number of possible reasons. One of these is the "inattention of policymakers to the details of implementing reforms" (p. 251). In describing past efforts at reform, Cuban says: "Absent, more often than not, were administrative mechanisms to dispense information, organizational linkages between school practices and district-wide goals, and teacher participation in the process" (1993, p. 252).

So, in addition to taking the other parts of the educational system into account, reformers also need to look at the schools themselves and consider their ability to allow information to enter, to apply that knowledge effectively, and to connect new ideas to changes in students' performance. These are enormous challenges for schools, but essential if reform is going to become a reality.

SUMMARY

What I have tried to do in this chapter is to provide historical background for the current science education reform movement and to show how closely interconnected, but perhaps diverging in their fundamental mission, are a standards future vision of reform and the federally driven standards present accountability movement. I have also looked at the contexts in which reform must take place, where existing interests, needs, and capabilities of participants play a major role in any efforts at reform.

It is clear from this analysis that there are a number of things that still need to be done to move the reform agenda forward. To begin, we need to provide models of excellence that will be perceived as such by the educational community. Because we are operating in a decentralized, free market system where ideas compete with each other for acceptance, implementation cannot be effected through tightly controlled bureaucratic channels. In a sense we are left with a hopeful *build it and they will come* approach. If the product is good enough, if it is convincing enough, if it is integrated and coherent, if it can be implemented with reasonable effort and resources, and if it can deliver anticipated results, then the hope is that it will be used.

But, even then, without a clear and compelling reason to change, most people will continue to behave as they have behaved in the past. Most people act as they do because they believe it is the right thing to do. Our efforts should continue to go into building a system of well-connected and coordinated parts. But that system must be built with a full awareness of the values and capabilities of the many participants in the system, and, as much as possible, with a level of involvement that achieves a sense of ownership for those participants. It should also be built with an awareness of how the many parts of the system work together. But obviously, we cannot wait for everything to be completed before the new model goes on display. As innovative curriculum materials, instructional strategies, and student assessments that are aligned with content standards and sound pedagogical principles become available, we can expect there to be steady movement toward the reform vision. An internally consistent set of reform-based instructional tools and resources built on the foundation of carefully developed content standards should have great appeal to practitioners in the field. It is also important that state education departments, local school districts, teachers, and professional organizations that are concerned with education issues be kept up to date on our reform efforts.

We now know that the changes being proposed will take time, much more time than originally imagined. And we know that there is no single and final solution to our problems. Our challenge is long-term and far-reaching: We must promote a clear and uncomplicated message of reform and engage the education system at all levels in ongoing and practical efforts to offer the best science education possible to the greatest number of people. We should not be trapped into a narrow and short-sighted standards past or standards present accountability mode, but rather keep our eyes looking toward the future.

REFERENCES

Adler, M. (1982). *The Paideia proposal: An educational manifesto.* New York: Collier Books.

American Association for the Advancement of Science. (1990). *Science for all Americans.* New York: Oxford University Press.

American Association for the Advancement of Science. (1993). *Benchmarks for science literacy.* New York: Oxford University Press.

American Association for the Advancement of Science. (1998). *Blueprints for reform.* New York: Oxford University Press.

Atkin, J., Bianchini, J., & Holthuis, N. (1997). The different worlds of Project 2061. In S. Raizen & E. Britton (Eds.), *Bold ventures: Volume 2. Case studies of innovation in science education* (pp. 131-145). Dordrecht, Netherlands: Kluwer.

Brown, B., Demarest, E. J., Freeman, H. S., & Dalton, S. S. (2002). The challenge to educate all students: Fostering competence, identity, and connection. In E. J. Demarest (Ed.), *Benchmarks for excellence: Learning-centered classrooms* (pp. 191-261). Unpublished manuscript.

Cohen, D., & Spillane, J. (1992). Policy and practice: The relation between governance and instruction. *Review of Research in Education, 18*, 3-49.

Collins, A. (1998). National science education standards: A political document. *Journal of Research in Science Teaching, 35* (7), 711-727.

Cuban, L. (1993). *How teachers taught: Constancy and change in American classrooms 1880-1990* (2nd ed.). New York: Teachers College Press.

DeBoer, G. (1991). *A history of ideas in science education: Implications for practice.* New York: Columbia University Teachers College Press.

DeBoer, G. (2000). Scientific literacy: Another look at its historical and contemporary meanings and its relationship to science education reform. *Journal of Research in Science Teaching, 37*(6), 582-601.

Demarest, E. J. (2002). *Benchmarks for excellence: Learning-centered classrooms.* Unpublished manuscript.

Developing standards. (1991, Fall). *Project 2061 Today, 1*(3).

Dewey, J. (1902). *The child and the curriculum*. Chicago: University of Chicago.

Dewey, J. (1916). *Democracy and education*. New York: Macmillan.

Dewey, J. (1938). *Experience and education*. New York: Macmillan.

Elementary and Secondary Education Act of 1965, 20 U.S.C. § 6301 *et. seq.* (1965).

Elmore, R. (2002). *Bridging the gap between standards and achievement: The imperative for professional development in education*. Washington, DC: The Albert Shanker Institute.

Goals 2000: Educate America Act, 20 U.S.C. § 5801 *et. seq.* (1994).

Herbart, J. (1901) *Outlines of educational doctrine* (C. DeGarmo, Ed., & A. Lange, Trans.). New York: Macmillan. (Original work published 1835)

Hilliard, A. (1995). Modifying national goals is not enough. In L. McNeil (Ed.), *The hidden consequences of a national curriculum* (pp. 145-156). Washington, DC: American Educational Research Association.

Hurd, P. (1989). Science education and the nation's economy. In A. Champagne, B. Lovitts, & B. Calinger (Eds.), *Scientific literacy* (pp. 15-40). Washington, DC: American Association for the Advancement of Science.

Improving America's Schools Act of 1994, 20 U.S.C. § 8001 *et. seq.* (1994).

Jorgensen, M. A., & Hoffmann, J. (2003). *History of the No Child Left Behind Act of 2001*. Orlando, FL: Harcourt Assessment.

Kellaghan, T., & Madaus, G. (1995). National curricula in European countries. In L. McNeil (Ed.), *The hidden consequences of a national curriculum* (pp. 79-118). Washington, DC: American Educational Research Association.

Kesidou, S., & Roseman, J. (2002). How well do middle school science programs measure up? Findings from Project 2061's curriculum review study. *Journal of Research in Science Teaching, 39*(6), 522-549.

Kliebard, H. (1995). The national interest and a national curriculum: Two historical precedents and their implications. In L. McNeil (Ed.), *The hidden consequences of a national curriculum* (pp. 63-78). Washington, DC: American Educational Research Association.

Kozol, J. (1992). *Savage inequalities: Children in America's schools*. New York: Harper-Perennial.

McNeil, L. (Ed.). (1995). Local reform initiatives and a national curriculum: Where are the children? In *The hidden consequences of a national curriculum* (pp. 13-46). Washington, DC: American Educational Research Association.

National Commission on Excellence in Education. (1983). *A nation at risk: The imperative for educational reform*. Washington, DC: U.S. Department of Education.

National Conference of State Legislatures. (n.d.). *No Child Left Behind: History.* Retrieved March 29, 2004, from http://www.ncsl.org/programs/educ/NCLB-History.htm

National Council for Teachers of Mathematics. (1989). *Curriculum and evaluation standards for school mathematics.* Reston, VA: Author.

National Defense Education Act of 1958, 1 U.S.C. § 101, 72 Stat. 1581 (1958).

National Education Association. (1893). *Report of the committee on secondary school studies*. Washington, DC: U.S. Government Printing Office.

National Research Council. (1996). *National science education standards*. Washington, DC: National Academy Press.

National Research Council. (2002). *Investigating the influence of standards: A framework for research in mathematics, science, and technology education*. Washington, DC: National Academy Press.

National Science Board Commission on Precollege Education in Mathematics, Science and Technology. (1983). *Educating Americans for the 21st century: A report to the American people and the national science board*. Washington, DC: National Science Foundation.

National Vocational Education (Smith-Hughes) Act of 1917, 20 U.S.C. § 11 (1917).

Neal, T., & Poole, J. (2004, June 15). *A test in Florida*. Retrieved July 4, 2004, from http://www.washingtonpost.com

New York State Education Department. (n.d.). *Investing in our workforce: Professional development, teacher education, student achievement*. Retrieved July 7, 2004, from http://www.emsc.nysed.gov/development/requirements_summary-PDP.htm

No Child Left Behind Act of 2001, 20 U.S.C. § 6301 *et seq.* (2002).

Perlstein, L. (2004, May 31). *School pushes reading, writing, reform*. Retrieved July 4, 2004, from http://www.washingtonpost.com

Ravitch, D. (1983). *The troubled crusade*. New York: Basic Books.

Spencer, H. (1864). *Education: Intellectual, moral, and physical*. New York: Appleton.

Spring, J. (2001). *The American school: 1642-2000*. New York: McGraw-Hill.

SRI International. (1996a). *Evaluation of the American association for the advancement of science's Project 2061, Volume I: Technical report*. Menlo Park, CA: Author.

SRI International. (1996b). Evaluation *of the American association for the advancement of science's Project 2061, Volume II: Appendices*. Menlo Park, CA: Author.

St. John, M. (1987). *An assessment of the school in the exploratorium program*. Inverness, CA: Inverness Research Associates.

Weiss, I. (1987). *Report of the 1985-86 national survey of science and mathematics education*. Chapel Hill, NC: Horizon Research.

Weiss, I., Banilower, E., McMahon, K., & Smith, P. S. (2001). *Report of the 2000 national survey of science and mathematics education*. Chapel Hill, NC: Horizon Research.

Welch, W., Klopfer, L., Aikenhead, G., & Robinson, J. (1981). The role of inquiry in science education: Analysis and recommendations. *Science Education, 65*, 33-50.

Wenger, E. (1998). *Communities of practice: Learning, meaning, and identity*. New York: Cambridge University Press.

Williams, R. (1983). *Keywords: A vocabulary of culture and society*. New York: Oxford University Press.

Wilson, S. M., & Peterson, P. L. (2002). Theories of learning and teaching: What do they mean for educators? In E. J. Demarest (Ed.), *Benchmarks for excellence: Learning-centered classrooms* (pp. 94-146). Unpublished manuscript.

U.S. Department of Education. (1991). *America 2000: An education strategy*. Washington, DC: Author.

Zumwalt, K. (1995). What's a national curriculum anyway? In L. McNeil (Ed.), *The hidden consequences of a national curriculum* (pp. 1-12). Washington, DC: American Educational Research Association.

CHAPTER 3

A FRAMEWORK FOR INVESTIGATING THE INFLUENCE OF THE NATIONAL SCIENCE STANDARDS

Iris R. Weiss

This chapter introduces a framework for investigating the influence of national standards. It describes three potential channels which national standards might traverse in order to influence teaching and learning: (1) curriculum, which would include state and district policy decisions; (2) teacher development, which would include both initial preparation and professional development; and (3) assessment and accountability, which would include both the tests that are used at the state, district and classroom levels, and the stakes that are attached to the results. Questions that would be asked in order to assess the impact of standards within the education system and in its context include: How are nationally-developed standards being received and interpreted; what actions have been taken in response; and what components of the system have been affected, and how? To assess the impact of standards on teachers and teaching practice, the focus would be both on the extent to which teachers have been exposed to the standards, and the nature and extent of the changes they have made in response. Finally, to assess the impact of the standards on students, the focus would be on the extent of student learning, including who has been affected, and

The Impact of State and National Standards on K-12 Science Teaching, 51–79

how. The framework describes a complex set of interacting forces and conditions that affect teaching and learning, any number of which could be influenced by nationally-developed standards. Various types of studies, each guided by its own appropriate methodology, are needed to establish the nature and extent of the influences of nationally-developed standards.

INTRODUCTION

The American Association for the Advancement of Science's (AAAS) Project 2061 began efforts to identify desired learning goals in science in the mid1980s. As its title implies, *Science for All Americans* (AAAS, 1989) reflected the consensus of much of the scientific community regarding a common core of learning for everyone in science, mathematics, and technology. Then, based on cognitive research and the expertise of teachers and teacher leaders, *Benchmarks for Science Literacy* (American Association for the Advancement of Science, 1993) described how those core concepts can be introduced and developed within the grade level spans of K–12 schooling. In 1989, the National Science Teachers Association (NSTA) started its scope, sequence, and coordination project, which sought to delineate a multigrade sequencing of concepts across scientific disciplines within the secondary-school curriculum (National Science Teachers Association, 1992).

Also in 1989, the National Council of Teachers of Mathematics (NCTM) published the *Curriculum and Evaluation Standards for School Mathematics* (National Council of Teachers of Mathematics, 1989). In 1991, the National Research Council (NRC) agreed to coordinate development of national science education standards, supported by funding from the National Science Foundation, U.S. Department of Education, National Aeronautics and Space Administration, and National Institutes of Health. The *National Science Education Standards* (NRC, 1996), informed by the earlier work of NCTM, AAAS, and NSTA, emerged as the central product of that collaborative effort.

NRC standards offered a vision of science education for all students, including what they should know, understand, and be able to do within particular K–12 grade intervals. In addition to physical, life, earth, and space science concepts, the content standards addressed science as inquiry, unifying concepts and processes (such as systems and the nature of models), science and technology, science in personal and social perspectives, and the history and nature of science. The document takes a systemic perspective, including standards that address science teaching, professional development, and assessment at classroom, district, state, and national levels. It also includes standards that address the necessary components of a comprehensive school science program, and policies and resources deemed necessary from all components of the education system to attain science literacy for all students.

National science education standards call for changes not only in what students learn, but also in how that content is taught. According to the national standards documents, teachers should have deep understanding of the science content they teach, recognize and address common student preconceptions, design classroom experiences that actively engage students in building their understanding, emphasize the use and application of what is learned, and use assessment as an integral part of instruction. Teachers should also listen carefully to students' ideas, recognize and respond to student diversity, facilitate and encourage student discussions, model the skills and strategies of scientific inquiry, and help students cultivate those skills and behaviors. In so doing, teachers should establish a classroom climate that supports learning, encourages respect for the ideas of others, and values curiosity, skepticism, and diverse viewpoints. In addition, teachers should participate in ongoing planning and development of science programs in their schools and should seek and promote professional growth opportunities for themselves and their colleagues.

In 2000, the NRC convened the Committee on Understanding the Influence of Standards in Science, Mathematics, and Technology Education. The committee was charged to develop a framework that could be used to understand the influence of science, mathematics, and technology education standards on programs, policies, and practices. From the beginning, the committee recognized that in a system as complex as the United States education system, it is difficult to focus on any particular component without considering how it is influenced by, and in turn influences other parts of the educational system. For example, what students learn is clearly related to what they are taught, which in turn depends on the intended curriculum, how teachers elect to use that curriculum, the kinds of resources teachers have for their instructional work, what the community values regarding student learning, and how local, state, and national assessments influence instructional practice. The committee also acknowledged early in its work that a body of research related to education standards is emerging, work that addresses questions of student learning and other aspects of the education system. However, no comprehensive map or conceptual overview was available to guide the efforts of producers, interpreters, and consumers of that standards-focused research.

The framework the committee developed was intended to address that need—by providing guidance for the design, conduct, and interpretation of research focused on influences of nationally-developed standards on student learning in mathematics, science, and technology. The framework describes key leverage points, identifies questions that need answers, and considers how evidence can be assembled to address those questions. It neither advocates nor criticizes the standards, and does not attempt to synthesize or interpret existing research concerning influences of standards. Rather, the framework offers guidance and perspective both to the research community and to the policymakers, scholars, and practitioners who use the results of such research to help formulate, conduct, and interpret research about influences on student learning—either positive or negative—of nationally-developed standards.

The framework for investigating the influence of education standards focuses on two questions: *How has the system responded to the introduction of nationally-developed standards?* and *What are the consequences for student learning?*

THE EDUCATION SYSTEM IN THE UNITED STATES

In attempting to investigate the influence of national science education standards, it is important to first consider the system that national standards are trying to influence. The United States education system is large, diverse, and complex. Approximately 2.7 million teachers are responsible for the education of more than 47 million pupils in nearly 90,000 public schools. Another 6 million students attend private schools (National Center for Education Statistics, NCES, 2000b). Such aggregated nationwide data, however, fail to reveal the variation and increasing diversity of student bodies and impoverished communities. Student populations in urban schools are particularly diverse. A large majority of urban students have nonwhite ethnic backgrounds, and increasing numbers are recent immigrants not yet proficient in English (NCES, 1997, 1999).

The United States teacher population also brings an array of different knowledge bases, expectations, cultural backgrounds, and beliefs to classrooms. Since nearly 90% of United States K–12 teachers are white (NCES, 2000b), teachers in some schools are demographically quite different than their students.

The individual classrooms in which teachers and students interact constitute the core of the educational system. At the same time, what happens in a classroom is significantly affected by decisions made in other layers of this loosely coupled system. First, there is the school as an educational unit. Setting expectations in certain content areas, the principal, department chairs, or team leaders can affect beliefs about teaching and learning priorities. They can also establish a climate that encourages or discourages particular pedagogical approaches, collegial interactions, or in-service programs (Little, 1993; McLaughlin, 1993;

Talbert & McLaughlin, 1993). A school's level of commitment to equity and to providing opportunities for all students to learn the same core content can influence how students are scheduled into classes, which teachers are assigned to teach particular classes, and how instructional resources are identified and allocated.

In the next layer of the system, school districts are responsible for ensuring the implementation of state and federal education policies, and often create additional, local education policy. District leaders set instructional priorities, provide instructional guidance, create incentive structures, and may influence the willingness and capacity of schools and teachers to explore and implement different instructional techniques.

The state level is a particularly important one. In the United States, states are constitutionally responsible for elementary and secondary education, and they play major roles in funding and regulating education, providing nearly half of all public school revenues (NCES, 2000a). Each state is responsible for developing and administering its own policies for standards, curriculum, materials selection and adoption, teacher licensure, student assessment, and educational accountability. Across states, the authority of schools and districts to enact policy varies considerably. In states with *local control*, more power resides at the district level than is found in states with centralized control.

Although the federal government contributes less than 10% of all funds invested by states and local districts in education (U.S. Department of Education, 2000), it influences education at all levels through a combination of regulations, public advocacy, and monetary incentives. For example, the U.S. Department of Education creates mandates for serving special-needs students, provides aid for districts serving disadvantaged students, and distributes funds to support professional development (through Title I and Title II of the Elementary and Secondary Education Act). In addition, the National Science Foundation and other federal agencies award competitive grants that address targeted educational priorities in science.

Based on research, interactions with practitioners in the field, and members' own experiences, the committee chose to represent the United States education system as shown in Figure 3.1. The figure highlights the layers of governance described above and identifies three main routes or *channels* through which national reform ideas might flow to various layers of the system and eventually influence teaching and learning. It also includes the social and political contexts within which the Unites States educational system operates. Other factors such as organizational development, could have been selected as system components, but the committee agreed that the elements identified in Figure 3.1 are most relevant to

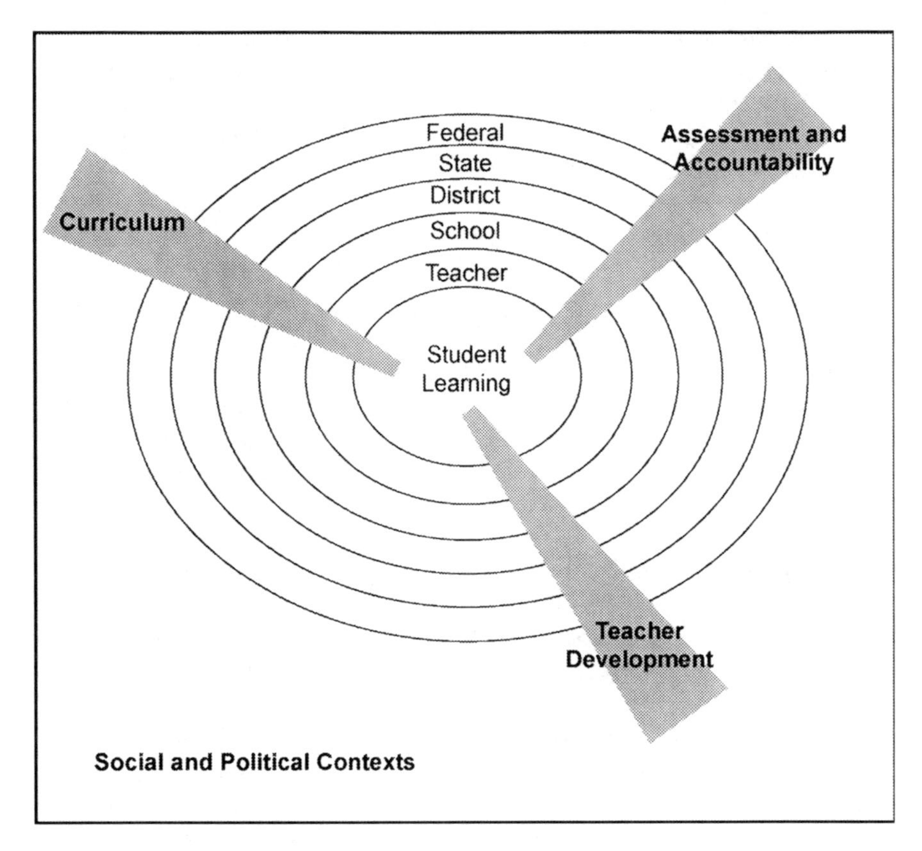

Figure 3.1. The layers of education governance and channels through which reform might flow.

tracing potential effects of nationally-developed standards on the education system and in particular, on student learning.

As reforms (such as standards) enter the education system and traverse one or more channels, they may affect policies, programs, and practices within various jurisdictional layers. The channels are:

- *Curriculum.* Mandates and resources from legislative bodies, and decisions and developmental work by teachers, school and district curriculum coordinators, state agencies, curriculum development organizations, and textbook publishers all collectively define what teachers should teach and students should learn. Nationally-developed standards, as well as state and local standards, typically play roles in this process, and thus may help to define the content of instruction.

- *Teacher Development.* School districts, institutions of higher education, state agencies, and other entities recruit, prepare, license, and evaluate teachers, as well as provide an array of opportunities for continued professional learning. Nationally-developed standards can inform these processes in many ways, influencing the content and expectations for teacher preparation and for their career-long professional growth.
- *Assessment and Accountability.* Student assessment practices—created by teachers, district or state agencies, assessment developers, postsecondary institutions, and others—establish ways that student learning is monitored, and, in so doing, may operationally define the classroom content that matters most. Based on assessment results, accountability mechanisms often establish consequences for students, teachers, and schools. Nationally-developed standards may define the content domain that assessments address, as well as prompt development of new forms of assessment.

Standards may also have an impact on education's social and political contexts, spurring those outside the education system to influence, both directly and indirectly, what happens in classrooms. For example, what parents and other members of the public, their political representatives, the media, and relevant professional organizations say and do can influence the practice of public education. How stakeholders outside the education system understand and interpret standards may therefore influence how, and whether, standards ultimately cause changes in classroom teaching and learning.

THE ELEMENTS OF THE FRAMEWORK

Based on the committee's view of the education system, described above, the committee developed a framework that consists first, of a conceptual map that shows the contextual forces and channels through which nationally-developed standards may influence teachers and student learning (see Figure 3.2). Second, the framework includes a set of guiding questions that can be applied to various policies, programs, and practices within the system and to outside influences that may affect the system (see Table 3.1).[1]

As configured, the framework provides conceptual guideposts for those attempting to trace the influence of nationally-developed science standards and to gauge the magnitude or direction of that influence on the education system and on student learning. In other words, the framework

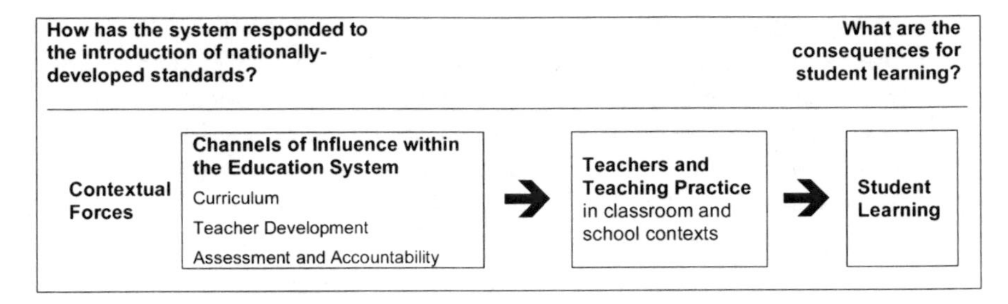

Figure 3.2. A conceptual map for investigating the influence of nationally-developed standards.

Table 3.1. A Set of Guiding Questions for Investigating the Influences of Nationally-Developed Standards

Within the education system and in its context …	*Among teachers who have been exposed to nationally-developed standards …*	*Among students who have been exposed to standards-based practice …*
• How are nationally-developed standards being received and interpreted? • What actions have been taken in response? • What has changed as a result? • What components of the system have been affected and how?	• How have they received and interpreted those standards? • What actions have they taken in response? • What, if anything, about their classroom practice has changed? • Who has been affected and how?	• How have student learning and achievement change? • Who has been affected and how?

is intended to guide inquiries within the education territory encompassed by the map.

As is true for all models, the system represented in the framework is greatly simplified. Those simplifications, however, should not obscure these important realities:

- The channels of influence are complex and interactive, both with other components of the education system and among different levels of jurisdiction. For example, changes in the curriculum framework of a state may affect a district's teacher development program.
- The time needed for the influences of any set of standards to traverse the system may be long. One of the principal players in the

development of the science education standards wrote that the estimate of "a decade or longer" to implement the standards was "modest" (Collins, 1997).

- Reform ideas may be altered or ignored for various reasons (including prior beliefs and ongoing debate) as they work their way through the education system. Thus, nationally-developed standards may stimulate the intended changes, create a backlash, or result in no changes at all.
- Local, state, and regional variability within the U.S. education system all imply that teachers and students are likely to be influenced differently within different locales, depending on available resources, participant backgrounds, and other factors.

The framework is designed to guide inquiry into the influence of standards on various parts and levels of the education system. Those investigations may be centered on one or more of these key questions:

1. *How are nationally-developed standards being received and interpreted?* Because the vision expressed in the standards for student learning, teaching practice, and system behavior is conveyed through broadly framed statements, it is subject to interpretation. Accordingly, individuals throughout the system will necessarily engage in various forms of sense making, drawing on prior beliefs, knowledge, and priorities, as they give educational and operational meaning to the standards (Spillane & Callahan, 2000). Thus, to understand anything about the influence of standards, answers to this first central question are needed. The answers will reveal much about how expectations embedded in nationally-developed standards are understood, and whether they are accepted, rejected, or altered in that interpretive process.
2. *What actions have been taken?* What have curriculum developers, teacher educators, and assessment designers done in response to standards? Actions taken by individuals or entities with respect to the standards will depend on their interpretations and on their capacities and determination. Variations in resources, professional expertise, structural features, working cultures, and values will affect their motivation and ability to implement nationally-developed standards in some form or other. Enactment of standards represents an unfolding story of reform intentions interacting with the multiple contexts within which teachers work and learners learn (Talbert & McLaughlin, 1993). That story will unfold differently in particular states and localities depending on what educa-

tors support, seek, and are able to accomplish. As decades of research on policy and program implementation attest (Anderson, 1996; Anderson & Helms, 2001; McLaughlin, 1987, 1991), it is likely that enactments of nationally-developed standards will take on very different forms as implementation proceeds.

3. *What has changed as a result?* What new policies, programs, or practices can be attributed to the influence of standards? Attempts to implement national standards, whether faithful to their original intentions or to alternative interpretations, do not guarantee educational improvement. Furthermore, as the framework implies, incorporation of standards into one part of the system may or may not lead to programs and practices in other parts of the system that mirror the intent of the nationally-developed standards. Ultimately, what matters is how student learning is affected—or to be more precise, whether standards-based changes in the education system and in teaching practice have led to improvements in student learning.
4. *Who has been affected and how?* In specific terms, how has the learning of students who have been exposed to standards-based practice been affected, and do these effects vary across groups or types of students? For the student population, or subsets of it, do effects on learning represent an improvement? Substantial inequities continue to be documented within U.S. education in general (e.g., Darling-Hammond, 2000) and within science education (Martin, Mullis, Gonzalez, O'Connor, Chrostowski, Gregory, Smith, & Garden, 2001; National Commission on Mathematics and Science Teaching for the 21st Century, 2000). Thus, it is entirely possible that nationally-developed standards or other educational interventions may engender practices that differentially benefit (or harm) some segments of the student population, or that benefit some schools or communities more than others. National science education standards explicitly call for reform in policies and practice leading to literacy for *all* students. It is imperative that investigations of the influence of nationally-developed standards address this critical question when examining particular elements of the system and when gathering evidence regarding student learning.

If these four central questions within the framework are used in the context of particular investigations, both producers and consumers of research can acquire important insights into possible benefits and limitations of nationally-developed standards. The following sections provide more detail regarding the channels and outside forces through which standards may influence the education system.

CHANNELS OF INFLUENCE

The channels set forth in the framework, through which reform ideas may flow, have different properties and points of interface with classroom practice.

Curriculum

The influence of nationally-developed standards on what students are to learn is filtered through the forces and conditions that define the science curriculum and instructional materials. What is actually taught in classrooms in the United States is shaped by decisions made at multiple levels such as the federal government, states, districts, schools, and individual teachers. Exploring what is taught to whom and why involves addressing the implications of a myriad of policy decisions that affect curriculum and resources to support the curriculum, the development of instructional materials and programs, and the processes and criteria for selecting instructional materials that help determine what students will learn in a particular classroom.

Nationally-developed standards describe the organization, balance, and presentation of important science content. The standards intentionally do not prescribe a specific curriculum, but provide criteria for designing a curriculum framework or selecting instructional materials. If standards are influencing what is taught to which students, then curriculum policy, the design and development of instructional materials, and the processes and criteria by which such materials are selected and implemented in classrooms would reflect the content described in the standards. Enacted policies and funded programs defining curriculum would align with those relating to standards-based instruction and assessment. State content standards would be consistent with content specified by the nationally-developed standards, providing comprehensive guidance on what should be taught at each grade level, stimulating creation or adoption of curricular materials and textbooks at the local level that embody the standards' vision, and providing direction to needed curricular guidance and support. Graduation requirements would reinforce the curricular recommendations of the standards, and postsecondary institutions would recognize and accommodate students who successfully complete standards-based school programs.

If standards are influencing the curriculum, both the intended and enacted curriculum would increasingly focus on science learning goals specified in the standards. K–12 programs would be coordinated system-wide both within and across grades and more aligned with the content as outlined in the standards documents. Schools, districts, and states would have an infrastructure supporting delivery of standards-based curricula in science, including programs to support teachers' instructional needs in relation to those curricula. Instructional materials and textbooks would be developed by people who understand the standards, and that understanding would be reflected both in the content they include and the nature of the tasks they use to develop student knowledge of that content. Textbook adoption processes would be carried out by selection committees knowledgeable about standards-based materials. Textbook adoption criteria would be based on features congruent with the standards, such as inquiry-based learning, an emphasis on problem solving, and an emphasis on conceptual understanding as well as skill development. Teachers would have appropriate resources for teaching standards-based curricula, including laboratory equipment and supplies, and support for learning to use them effectively.

Enrollment patterns in schools would reveal whether the vision expressed by the standards applies to all students. If standards are permeating the system, opportunities for taking challenging science courses would be open to every student, and resources needed to implement a robust standards-based curriculum would be allocated in equitable ways. Dual-language materials would be available, as well as other resources designed to accommodate diverse learners to support the standards' focus on all students having access to opportunities to learn important science concepts and skills.

Teacher Development

The teacher preparation and development components within the education system provide channels through which nationally-developed standards might influence how teachers learn to teach both initially and throughout their careers. The policies, practices, and programs at local, state, and federal levels determine investments made in teaching prospective teachers and in molding the ways they continue to develop their skills as classroom teachers. Teachers' science content and pedagogical knowledge are shaped by their initial exposure to science and the ways it is taught prior to and during their formal teacher preparation program and by the requirements for certification and licensure. Teachers' continuing professional learning may be enhanced or constrained by the setting within which they work and by the opportunities available to them.

If nationally-developed standards are influencing the preparation of new teachers, there would be increased alignment of policies and practice with the standards. States, districts, and postsecondary institutions would create systems that enable prospective teachers to gain the knowledge and skills needed to help students meet standards-based learning goals. In particular, analysis of teacher preparation programs and course artifacts would verify that the professional development standards are being interpreted and implemented as intended. Evidence would also confirm that college and university educators are aligning the content and pedagogy of undergraduate courses, conventional teacher preparation programs, and alternate certification programs with expectations of the national standards. State licensure systems would set criteria for initial certification that require graduates to demonstrate their understanding of the standards, knowledge of the content and pedagogy described therein, and ability to implement standards-based instructional programs.

Policies and fiscal investments at local, state, and federal levels would focus on recertification criteria, professional development opportunities, and system-wide support strategies aligned with nationally-developed standards. States and localities would provide a rich *infrastructure* to support standards-based science teaching. Administrators at school and district levels would possess the skills, commitment, and capabilities to promote collegial planning and dialogue about content, teaching, and assessment as called for in the national standards. Experienced teachers well-versed in the teaching, assessment, and professional development standards would be offered leadership roles to assist schools in implementing needed reforms. Teachers would be motivated to enhance their understanding of standards-based content, ways to arrange appropriate learning experiences, and techniques for assessing what students understand. Recertification criteria and teacher evaluations would focus on evidence verifying the knowledge, skills, and practices advocated by the standards.

If standard-based visions of equity are being implemented, teacher preparation programs would prepare prospective teachers to teach in diverse classrooms, and teachers skilled in implementing standards-based education would be distributed so that all learners have access to high-quality learning opportunities.

Assessment and Accountability

As the standards movement has gained strength across the United States, assessment and accountability, which are two distinct but related concepts, have become linked as a way to realize the standards, and as such constitute a channel through which reform might flow. Assessments of various kinds provide systematic means of informing students, teachers, parents, the public, and policy makers about student performance. Accountability mechanisms linked to some or all of these assessments provide incentives to change behavior, by using information from assessments to make consequential decisions about students, teachers, schools, or districts. Thus, consideration of assessment involves a careful study of how assessment interacts with accountability, how teachers conduct and use classroom assessment, how states and districts use assessment for accountability, and how assessment influences choices in postsecondary education.

If nationally-developed standards are influencing assessment policies and practices, assessments would be aligned with learning outcomes embodied in the standards. In particular, if state assessments and standards are aligned with the nationally-developed standards, assessment at all levels would include problem solving and inquiry in addition to other skills and knowledge. Teachers would use classroom assessment results to inform instructional decisions and to provide feedback to students about their learning. Teachers, administrators, and policy makers would employ multiple sources of evidence regarding what a student knows and is able to do, as is called for in the standards, rather than relying on a single source.

Developers of student assessments would be familiar with nationally-developed assessment and content standards and create assessment materials that reflect the standards by having appropriate items, clear examples of the kinds of performance that students are expected to demonstrate, criteria by which these performances are evaluated, and reports that inform instruction as well as measure achievement. Assessment results would be reported in language accessible to parents and other stakeholders, helping them to understand what the tests measure and how results labeled as *proficient* or *basic* should be interpreted.

States and districts would have comprehensive plans for administering the array of assessments they use with students, and the plans would enable teachers to pursue the vision of the standards as well as prepare students to take those assessments that are high stakes. Incentives linked to accountability would encourage standards-based reforms, with policies in place to ensure that schools and teachers have standards-based professional development opportunities, instructional materials, and appropriate resources to enhance their efforts to raise performance levels of their students. Finally, college entrance and placement tests would measure content that is valued by standards created at the national level and contain tasks aligned with those standards.

Contextual Forces

Decision-making within the education system is, in large part, a political process, involving key players such as legislators, government officials, and citizen groups, in addition to educators. Educational concerns may motivate professional organizations, parents, and others to lobby for certain decisions or work toward particular goals. Educational policy decisions may also be influenced by media that convey information and shape public perceptions. In addition to exerting influence through the political system, some businesses, education and professional organizations, and others may influence the education system directly, for example, by supporting ongoing teacher professional development efforts.

Standards are more likely to have an influence on the education system if they are supported by the *outside* forces, rather than being ignored or even opposed. If the standards are influencing individuals and groups external to the education system as intended, decisions enacted by elected officials and policy makers would show support for standards-based reforms. Professional associations in the forefront of the development of national standards for science would lead national and local efforts to implement the standards, as well as work with elected officials and leaders to build a consensus in support of institutionalizing standards-based reforms.

The traditional school priorities of reading, writing, and arithmetic would be joined by science. State and local school boards, reflecting and responding to constituents' views, would ensure that schools have adequate funding to provide students with learning experiences that will enable them to meet the nationally-developed standards.

Professional associations would join together and collaborate with decision makers in establishing assessment and accountability programs that draw on multiple measures and address the full range of standards-based content and skills. The public would be informed of standards-based progress and supportive of continuing efforts. Attempts to weaken or dismantle standards-based education—whether to de-emphasize the place of science in the curriculum, to limit assessment primarily to vocabulary, or to reduce funding for professional development focused on standards-based instruction would be met with vocal public criticism and opposed by policy makers.

On the other hand, nationally-developed standards may generate resistance and opposition by individuals and groups outside the system. In that case, scientists who disagree with the standards' vision of science education would argue, for example, that standards exclude important content or lack rigor. Such groups would work to influence views of policy makers or the public at large, affecting decisions and actions within the education system.

Opponents would encourage funding or programmatic decisions regarding curriculum, professional development, and accountability practices that inhibit implementation of the nationally-developed standards, working to convince legislators, governors, and school boards that the fiscal, resource, or political costs associated with changes urged by the standards are inappropriate.

SUMMARY OF THE FRAMEWORK

The logic implicit in the framework can be summarized through a group of interrelated propositions:

1. *Nationally-developed science standards represent a set of fundamental changes in the way science has traditionally been taught, placing new demands on teachers and students.* The changes needed to move from established modes of teaching and learning to those advocated by the standards imply considerable new learning for both teachers and students. Standards-based practice presumes that teachers understand and have internalized much of what is asserted by national standards documents (e.g., that a core set of important ideas and skills can be identified and that all students can master those fundamental expectations).
2. *The expected influence of nationally-developed standards on teaching practice and student learning is likely to be (a) indirect, taking place through proximate effects on other parts of the education system; (b) entangled (and sometimes confused) with other influential forces and conditions, such as broader state standards-based reforms; and (c) slowly realized and long term.* In other words, within the nation's decentralized system of education, notions of teaching and learning embedded in nationally-developed standards do not have immediate pathways into classrooms. Rather, as these ideas move through channels that cross multiple levels of governance, various forces can alter how the standards are understood and acted on. State-level standards-based reform movements, for example, have introduced numerous interpretations of *standards* and *assessments*, some of which may not be in accord with ideas conveyed in the national standards documents. Given the amount of new learning implied and the complexity of the education system, it would take a long time, if ever, before the visions conveyed by national standards documents would be fully realized.

3. *Three core channels exist within the education system through which nationally-developed standards can influence teaching and learning*. These channels of influence are (a) curriculum, (b) teacher development, and (c) assessment and accountability.
4. *The channels of influence are complex and interactive*. Jointly or separately, the channels may alter the way standards are understood and realized. Public, political, and professional reactions can also affect these channels and shape the way standards reach and influence teaching and learning.
5. *Variability within the education system implies that students and teachers are likely to experience different influences, depending on locality, resources, participant background, and other factors*. Consequently, educational effects of national standards are unlikely to be monolithic. Instead, there may be effects that are constructive and others that are counterproductive, some weak, and others strong.
6. *The task for research—and hence for the framework—is to help identify and document significant standards-based effects, as well as overall trends and patterns among those effects*. That is, the task is to provide evidence-based descriptions of the channels and mechanisms through which those effects take place and determine what conditions may be associated with particular effects.
7. *The ultimate focus is on the changes in students' knowledge and abilities that have occurred since standards have entered the system and that can be reasonably attributed to the influence of the standards. As part of this, it is essential to consider how standards have affected the achievement of all students, including those who were previously underrepresented in science.*
8. *Eventually, nationally-developed standards will be judged effective if resources, requirements, and practices throughout the system align with the standards and if students in standards-based classrooms demonstrate high achievement in knowledge and skills deemed important*. Although there may be other grounds on which individuals or groups elect to accept or reject the standards, the only empirical approach for making that judgment presumes that standards have had opportunities to permeate the education system, and, having done so, are associated with student-learning outcomes that can be judged as desirable or undesirable.

HOW THE FRAMEWORK CAN BE USED

The framework developed by the NRC committee is intended to help guide the sponsorship, design, and interpretation of research on nationally-developed standards. The challenge is far from simple. The framework lays out a complex domain of interacting forces and conditions that affect teaching and learning, any number of which can be touched by the influence of standards. Thus, no single study can investigate all the ways that national standards are, or could be, part of the education reform story. Rather, various types of studies, each guided by its own appropriate methodologies, will be needed to establish the scale and scope of influences, identify routes by which standards actually exert influence, and ascertain the direction and educational consequences of those influences.

Table 3.2 contains several hypothetical examples that illustrate how different macro and micro studies can *cover* the terrain of the framework, and respond to one or more questions posed earlier in this document. Each of those hypothetical studies addresses only part of the broad territory embraced by the framework. Multiple studies could collectively paint a more satisfactory picture of the effects of nationally-developed standards if they were designed to generate complementary databases and were carefully synthesized.

In carrying out such research, the framework offers assistance in several important ways: (1) situating existing studies within the educational terrain relevant to the standards, (2) providing a conceptual tool for analyzing claims and inferences made by these studies, and (3) generating questions and hypotheses to be explored by future studies.

Situating Current Studies

The framework can assist researchers in locating their work within a particular frame of reference and may highlight possible connections or lack of connections to other parts of the education system. Sponsors, investigators, and consumers of research findings should keep in mind aspects of the education territory that may and may not be addressed by particular studies or programs of investigation. Consider this study of implementation of standards-based instruction in several districts:

> *A Close Look at Effects on Classroom Practice and Student Performance* (Consortium for Policy Research in Education, 1999). The Merck Institute for Science Education (MISE) was created in 1993. With support from Merck, and additional funding from the National Science Foundation, MISE formed partnerships with four public school districts to improve science curriculum and instruction in grades K–6. The vision of quality science instruction was based on national and state standards, and all of the partnership activities—selection of instructional materials, professional development for teachers and administrators, review of district and school policies, developing community support for science reform—were aligned with that vision. An evalu-

Table 3.2. Hypothetical Studies that Address One or More of the Framework Questions

	How are the nationally-developed standards being received and interpreted?	*What actions have been taken in response to the nationally-developed standards?*	*What has changed as a result of nationally-developed standards?*	*Who has been affected and how?*
• Analysis of media coverage of science education can determine how often national standards are mentioned, which components of the standards are highlighted, how they are interpreted, and what value is attached to each.	✓			
• Expert reviews of documents can identify ways that state standards, assessments, and accountability systems may and may not reflect the content advocated in the national science standards	✓			
• National teacher surveys can reveal how aware teachers are of national standards, whether—and in what ways—they believe they are orienting their professional practices to these standards, and in what ways they are supported in their efforts to realize the standards.	✓	✓	✓	
• Comparative studies of reform-based science curricula at a particular grade level can build understanding about how curriculum developers interpret the standards and how those interpretations may affect what students have opportunities to learn.	✓	✓		
• Observational studies can reveal whether—and how—science standards are realized in classroom practice within particular kinds of school settings.		✓	✓	
• Case studies of district reform can explore the alignment of science curriculum and assessment policies with national standards and the nature and extent of district support teachers receive for teaching in standards-aligned ways.		✓		
• Quasi-experimental design studies can compare teachers' responses to standards and their students' performance in settings with differing degrees of exposure to, and support for, standards-based practices.			✓	✓
• Case studies of standards-based classrooms can explore whether teachers adjust science instruction appropriately, with respect for students' cultural backgrounds as well as ways of learning.			✓	✓

ation team from the Consortium for Policy Research in Education (CPRE) studied the relationship between extent of participation in professional development and classroom practice, based both on teacher questionnaire responses and on classroom observations.

This work can be mapped onto the framework as shown in Figure 3.3. The shaded areas identify aspects of the framework and related questions that are addressed in this analysis.

Studies with different purposes, designs, and evidence bases would cover different parts of the framework. Consider this analysis:

Mathematics and Science Content Standards and Curriculum Frameworks (Council of Chief State School Officers, 1997). An expert panel reviewed state frameworks, standards documents, and related materials developed or revised during the period 1994 to 1997. The analysis sought to determine the extent that state curriculum frameworks, standards, and other materials were consistent with NCTM's *Curriculum and Evaluation Standards for School Mathematics* (NCTM, 1989), NRC's *National Science Education Standards* (NRC, 1996), and AAAS's *Benchmarks for Science Literacy* (AAAS, 1993). The analysis also considered differences in content found in state mathematics and science frameworks or standards documents and main ideas and categories found in corresponding national standards, noting omissions and additions. The analysis pointed out how state documents acted as a "bridge" between nationally-developed standards and local efforts to improve teaching and learning in these subject areas.

In contrast to the more broad-based CPRE study, the council's investigation focused in depth on one framework component: state policy decisions within the curriculum channel. Thus, shading would highlight only that particular feature of the framework.

Examining Claims and Inferences Reported in Current Studies

In addition to helping locate relevant areas of research, the framework offers a conceptual tool for assessing claims made by researchers. By highlighting multiple influences on teaching and learning, the framework can suggest plausible alternative explanations for research findings. Referring to the framework, scholars and other consumers of research can decide whether investigators accounted for all the plausible channels of influence on teaching and learning within the settings under study.

Without reference to a conceptual map such as the framework, weak inferences may arise about the influence of standards. Ultimately, strong claims about positive or negative effects of nationally-developed stan-

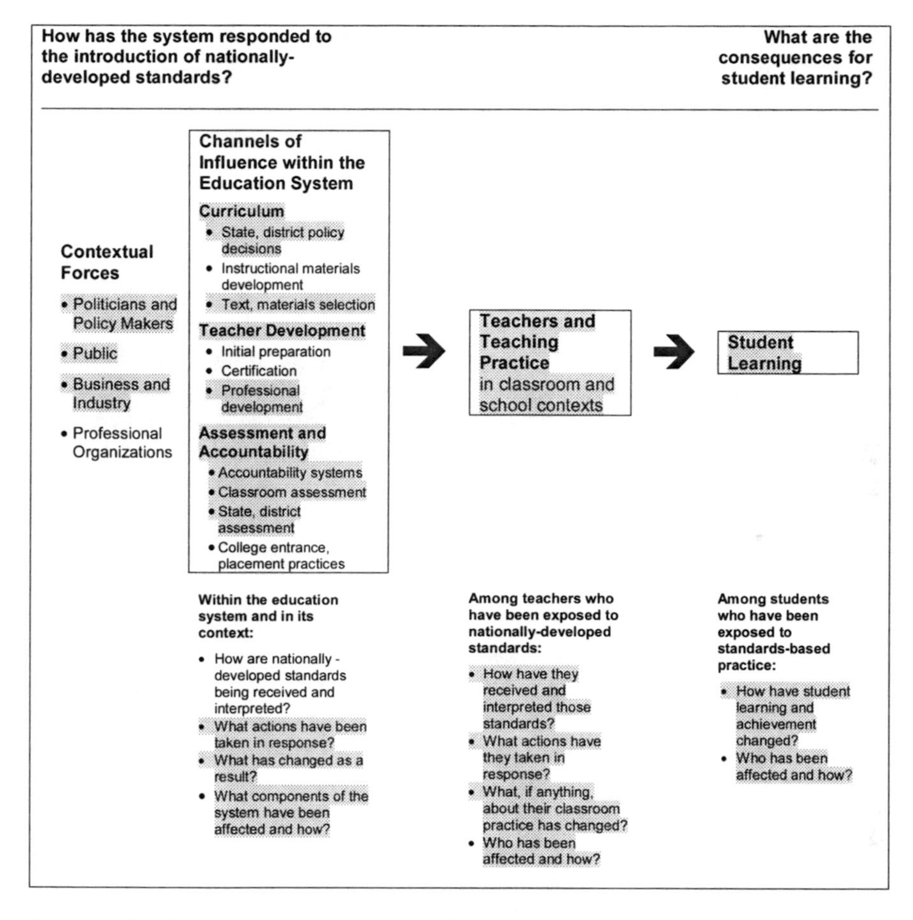

Figure 3.3. Parts of the framework addressed in the CPRE Study.

dards depend on a chain of evidence and inference linking promulgation of standards (at the national level) to particular sites (in schools and classrooms) within which standards-based ideas may be found to exert influence. As noted earlier, developing such a chain of evidence and inference will usually require multiple, coordinated studies.

There is an important caution, however. The framework provides only a *conceptual* scheme for considering research claims, not *all* the information needed to assess research-based claims fully. A full analysis must include a host of *technical* considerations, such as standards of evidence, quality of measurement, and appropriateness of the research design. All of these concerns must be addressed in deciding whether particular research conclusions are trustworthy and rigorous.

Assuming comparable technical quality, several hypothetical examples illustrate how the framework can help determine the soundness of research-based inferences.

- *A study of standards-based classroom practice*. Imagine an investigation of standards-based science teaching practice in a high socioeconomic environment that supports this kind of instruction, using a curriculum that embodies the principles of the NRC standards. Assume that teachers have been well trained in this form of teaching and are committed to it. If after sufficient time passes for the curriculum to have affected student learning across grades, a well-designed study documents indifferent or poor student results on assessments keyed to NRC standards, it would be reasonable to infer that national science standards contributed little to student learning—or might even have detracted from it. That inference could be further substantiated if other school settings less committed to NRC standards produced more favorable results with comparable students.
- *A study of district investment in teacher professional development aligned with nationally-developed science content standards.* In a group of districts heavily emphasizing the principles and themes of NRC content standards in their professional development and support programs, assessments of teacher knowledge and pedagogical approaches might show that they have acquired the intended knowledge and skills, and that the teachers are attempting to realize these ideas in their teaching. Assuming a well-designed investigation into teachers' participation in standards-oriented professional development and the outcomes of that participation (including direct observations in their classrooms), it would be reasonable to infer that NRC content standards had contributed to changes in those teachers' thinking and practice. (Establishing this particular claim does not necessarily imply that students learned more, that inference would require a different study, or an additional component to this investigation.) Once again, comparisons with other school sites less invested in standards-related content would help to establish the claim.

Note that the framework helps to establish the conceptual soundness of research inferences by highlighting elements of the domain that, through a reasonable chain of evidence and inference, link national standards to classroom outcomes.

By contrast, the following hypothetical examples involve unwarranted conceptual leaps in their reported conclusions:

- *Analysis of student achievement gains in states that align their science standards with NRC standards.* Impressive student achievement gains in states that apparently embrace national science standards invite the possible conclusion that the standards contributed to the improvements in student performance. But even assuming a technically sound analysis of test score trends that took into account known correlates of student achievement scores (e.g., student socioeconomic status), the inference is weak at best, or even fallacious, if the analysis did not consider other components highlighted by the framework. Those conditions include alignment of the science achievement measures with the standards, local interpretation of state and national curricular guidance, and the extent of standards-based classroom practice. In the absence of those considerations, there are too many other plausible explanations for the achievement gains to place any confidence in the inference that national standards had anything to do with them.
- *An investigation of declining science scores in a district committed to NRC standards.* Declines in student performance on district science assessments within a setting that has tried to encourage standards-related instruction may suggest to observers that the national standards are detrimental to student learning. Even if the investigation were carefully designed and executed, it would not support that conclusion unless relevant components highlighted by the framework were taken into account. These include alignment between the district's science assessments and the curriculum, teachers' interpretations of the standards and attempts to realize them in classroom instruction, and the extent of professional development for teachers unfamiliar with standards-based classroom practice.

In short, consumers of research, with the framework in hand, can examine the results and conclusions of studies—or sets of studies—guided by questions such as, does the study:

- *Establish a plausible, evidence-based chain of influence that connects nationally-developed standards to particular elements of the system under investigation?* The framework highlights components that might be part of that chain of influence.
- *Address plausible alternative explanations that could be advanced to account for observed effects or outcomes?* The framework highlights alternative forces and conditions that may influence effects or outcomes.

- *Consider interactions among different channels of influence that can convey either mutually reinforcing or contradictory messages to teachers and schools about standards-based practices?* The framework lays out the three primary channels of influence by which national standards could affect teaching and learning and notes ways these can interact with one another.
- *Allow sufficient time for the education components under investigation to have been affected by nationally-developed standards?* The framework demonstrates the complexity of the system through which messages about standards-based practice must move, thus the effects may become visible only after an extended period of time.

Again, these are not all of the important questions to be asked about the findings and conclusions of research related to nationally-developed standards in science education. Other important questions include congruence of the research design with the research questions, execution of the design, adequacy of the database, and quality of data analysis approaches. Still, the framework establishes a conceptual map that provides relevance and meaning for answers to such additional questions.

Generating Questions and Hypotheses for Future Investigations

The framework can help pinpoint areas of potential influences operating within the education system that may or may not have been considered by particular studies. This third main application of the framework has two parts:

- *Assembling knowledge.* The framework offers a basis for assembling knowledge gained from existing studies. Using framework components as organizers, research syntheses and reviews can summarize what has been learned about the extent to which the education system has changed in response to nationally-developed standards, particularly in terms of classroom practice and student learning.
- *Identifying gaps.* Gaps in current research can be identified by considering questions that could be (but have not been) asked about elements and relationships within the framework. By highlighting where research attention has been most and least focused, the framework can help researchers—and sponsors of research—target issues and areas of concern that merit more study. For example, relatively few studies have investigated the relationships among professional development, teacher knowledge, instructional practice,

and student achievement, either generally or with regard to national standards (Kennedy, 1998; Wilson & Berne, 1999). This paucity of studies regarding potentially important avenues for standards to reach classroom practice and student learning may signal a need to fill the gap.

Even in areas where substantial numbers of studies have been completed, the framework can highlight additional questions not yet extensively posed or answered. For example, a nagging concern within the broader standards-based reform movement regarding the equitable distribution of standards-based practice and equitable accountability systems (McKeon, Dianda, & McLaren, 2001) suggests an aspect of the story about the influence of national standards that may deserve greater attention. The relevant question in the framework—*Who is affected and how?*—encourages researchers to explore possible differential effects of standards within diverse student populations and settings, while taking into account the varied capacities of teachers, schools, and districts to engage in standards-based practice.

Other examples can be readily envisioned. One important advance in cumulative understanding of nationally-developed standards would be to assemble and map current knowledge using the framework so that gaps and opportunities for further study emerge.

ASPIRATIONS FOR FRAMEWORK-DRIVEN RESEARCH ON NATIONALLY-DEVELOPED STANDARDS

The framework was created in the hope that it would be useful to producers, consumers, and sponsors of research regarding central questions about the influence of nationally-developed standards. Applications of the framework described earlier will help to inform opinions and debate about those standards.

Three major aspirations of the committee regarding use of the framework are highlighted below.

1. *The framework should be regarded as an evolving conceptual picture, rather than a definitive final statement.* In that spirit, the framework should continue to evolve, informed by accumulating knowledge about standards-based reforms. It is essential that researchers build their understanding of the influence of nationally-developed standards in terms of *some* overarching model of the education system (or subsystem) within which standards play out. The schematic of the framework presented in this document can be regarded as one

sketch of such a system, including the dynamics of influence contained within it.

2. *The framework should stimulate different forms of inquiry into influences of nationally-developed standards.* Given the complex and interactive nature of the territory within which standards have been enacted, a mosaic of evidence from many different types of studies is more likely to build overall understanding of the influence of standards than the results of a few purportedly comprehensive studies.
3. *The framework should help guard against the superficiality that often permeates debate about high-visibility national policies by stimulating a critical view of claims regarding either the success or the failure of the standards.* Strong conclusions about effects or implications of nationally-developed standards presume an understanding of the entire education system (encompassed by the framework) and presentation of a chain of evidence that connects the emergence of particular education practices, policies, or learning outcomes to the influence of standards.

All United States youth deserve access to the best possible education. In pursuit of that goal, the education community should complete a comprehensive, critical appraisal of the power and limits of nationally-developed standards. That appraisal is still far from being realized. Public conversations about the worth and impact of standards—or about standards-based reforms in general—will continue. The framework developed by the NRC committee is intended to help the education research community contribute to that debate with reasoned voices based on evidence and sound inference.

APPENDIX:
A FRAMEWORK FOR INVESTIGATING THE INFLUENCE OF NATIONALLY-DEVELOPED STANDARDS FOR SCIENCE EDUCATION

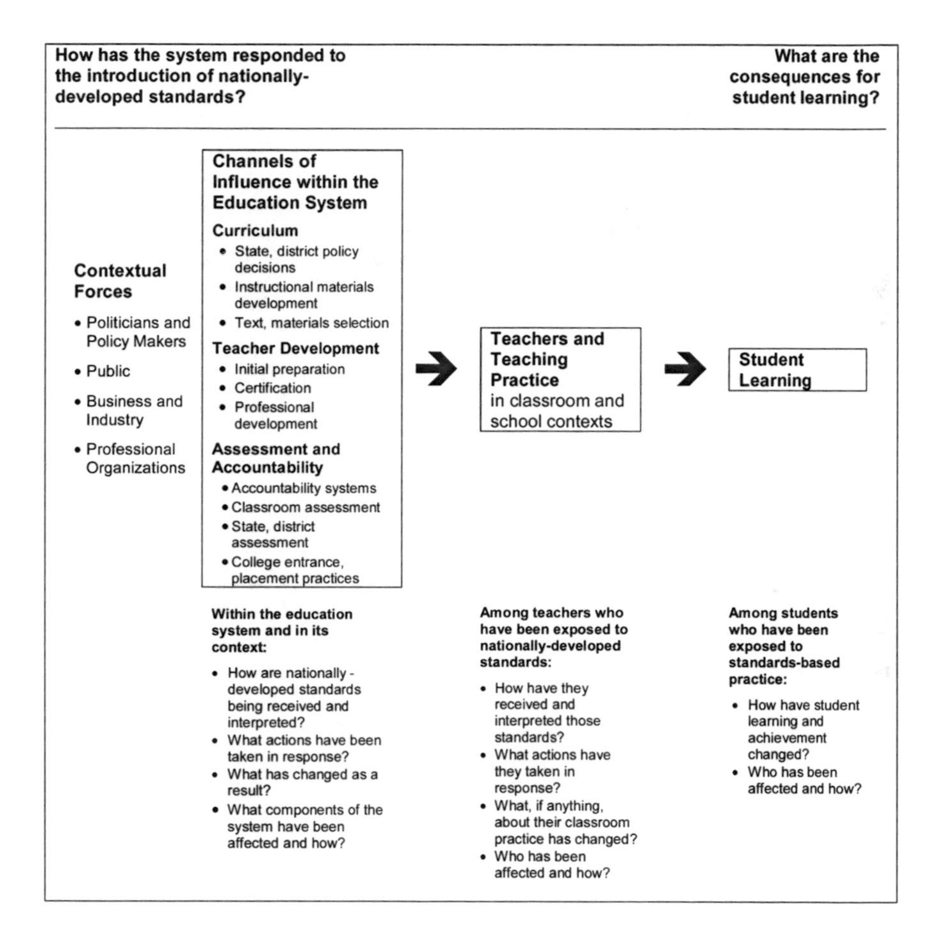

AUTHOR'S NOTE

This chapter was adapted from a report of the NRC Committee on Understanding the Influence of Standards in Mathematics, Science, and Technology Education, Chaired by Iris R. Weiss. Used with permission from *Investigating the Influence of Standards* by the National Academy of Sciences, courtesy of the National Academies Press, Washington, DC.

NOTE

1. A copy of the framework that includes both the conceptual map and the guiding questions is included in the Appendix.

REFERENCES

American Association for the Advancement of Science. (1989). *Science for all Americans: A Project 2061 report on literacy goals in science, mathematics, and technology.* Washington, DC: Author.

American Association for the Advancement of Science. (1993). *Benchmarks for science literacy.* New York: Oxford University Press.

Anderson, R. D. (1996). *Final technical research report: Study of curriculum reform: Findings and conclusions, Volume 1.* Washington, DC: U.S. Government Printing Office.

Anderson, R. D., & Helms, J. V. (2001). The ideal of standards and the reality of schools: Needed research. *Journal of Research in Science Teaching, 38*(1), 3–16.

Collins, A. (1997). National science education standards: Looking backward and forward. *The Elementary School Journal, 97*(4), 299–313.

Consortium for Policy Research in Education. (1999). *1997–1998 Annual report: A close look at effects on classroom practice and student performance: A report on the fifth year of the Merck Institute for Science Education.* Philadelphia: Consortium for Policy Research in Education.

Council of Chief State School Officers. (1997). *Mathematics and science content standards and curriculum frameworks: States progress on development and implementation.* Washington, DC: Council of Chief State School Officers.

Darling-Hammond, L. (2000, January). *Teaching for America's future: A progress report for the millennium, congressional program.* Paper presented at the Congressional Seminar for the Aspen Institute, Aspen, CO.

Kennedy, M. (1998). *Form and substance in inservice teacher education.* Research Monograph No. 13. Madison, WI: University of Wisconsin, National Institute for Science Education.

Little, J. W. (1993). Teachers' professional development in a climate of educational reform. *Educational Evaluation and Policy Analysis, 15*(2), 129–151.

Martin, M. O., Mullis, I. V. S., Gonzalez, E. J., O'Connor, K. M., Chrostowski, S. J., Gregory, K. D., Smith, T. A., & Garden, R. A. (2001). *Science benchmarking report: TIMSS 1999—eighth grade.* Boston, MA: International Association of the Evaluation of Educational Achievement.

McKeon, D., Dianda, M., & McLaren, A. (2001). *Advancing standards: A national call for midcourse corrections and next steps.* Washington, DC: National Education Association.

McLaughlin, M. W. (1987). Learning from experience: Lessons from policy implementation. *Educational Evaluation and Policy Analysis, 9*(2), 171–178.

McLaughlin, M. W. (1991). The Rand change agent study: Ten years later. In A. R. Odden (Ed.), *Education policy implementation* (pp. 143-155). Albany: State University of New York Press.

McLaughlin, M. W. (1993). What matters most in teachers' workplace context? In J. W. Little & M. W. McLaughlin (Eds.), *Teachers work: Individuals, colleagues, and contexts* (pp. 79-103). New York: Teachers College Press.

National Center for Education Statistics. (1997). *The condition of education 1997.* Washington, DC: U.S. Department of Education.

National Center for Education Statistics. (1999). *Characteristics of the 100 largest public elementary and secondary school districts in the United States: 1997–98*. Washington, DC: U.S. Department of Education.

National Center for Education Statistics. (2000a). *The condition of education 2000.* Washington, DC: U.S. Department of Education.

National Center for Education Statistics. (2000b). *Digest of education statistics, 1999.* Washington, DC: U.S. Department of Education.

National Commission on Mathematics and Science Teaching for the 21st Century. (2000). *Before it's too late: A report to the nation from the National Commission on Mathematics and Science Teaching for the 21st Century.* Washington, DC: U.S. Department of Education.

National Council of Teachers of Mathematics. (1989). *Curriculum and evaluation standards for school mathematics.* Reston, VA: Author

National Research Council. (1996). *National science education standards.* Washington, DC: National Academy Press.

I. R. Weiss, M. S. Knapp, K. S. Hollweg, & G. Burrill. (2002). *Investigating the influence of standards: A framework for research in mathematics, science, and technology education.* Washington, DC: National Academy Press.

National Science Teachers Association. (1992). *Scope, sequence, and coordination of secondary school science. Volume 1: The content core: A guide for curriculum developers.* Washington, DC: Author.

Spillane, J. P., & Callahan, K. A. (2000). Implementing state standards for science education: What district policymakers make of the hoopla. *Journal of Research in Science Teaching, 37*(5), 401–425.

Talbert, J. E., & McLaughlin, M. W. (1993). Understanding teaching in context. In D. K. Cohen, M. W. McLaughlin, & J. E. Talbert, (Eds.), *Teaching for understanding: Challenges for policy and practice* (pp. 167-206). San Francisco, CA: Jossey-Bass.

U.S. Department of Education. (2000). *The federal role in education*. Washington, DC: Author.

Wilson, S. M., & Berne, J. (1999). Teacher learning and the acquisition of professional knowledge: An examination of research on contemporary professional development. *Review of Research in Education, 24*, 173–209.

PART II

THE IMPACT OF SCIENCE STANDARDS ON CLASSROOMS AND TEACHERS

Part two examines the status and impacts of science standards by investigating the professional, political, and social factors that influenced their development, validation, and present use in K-12 classrooms. Chapter four reports on a national study of classroom practice under the influence of the science standards. Eric Banilower, P. Sean Smith, Iris Weiss, and Joan Pasley discuss the findings in terms of the strengths and weakness of science lessons taught by teachers and the likely impact of the lessons on students' understanding. Results are also discussed in terms of alignment of lessons with the content and pedagogical practices described in the *National Science Education Standards.* Chapters five and six explore the impact of the science standards on the understanding and decision making of teachers and administrators in two case studies. In chapter five, Dennis Sunal and Emmett Wright surveyed K-12 teachers in the state of Alabama to determine perceptions and expectations of the national and state science standards. In chapter six, Stephan Marlette and M. Janice Goldston surveyed elementary principals and teachers in the state of Kansas to gain insight into the implementation of science standards. They examined four characteristics of the change process that affect implementation—need, clarity, complexity, and the quality. Chapter seven describes how national science standards have influenced other policies, programs, and practices. Using Kansas as an example, the John Staver documents the principal influence of national standards as increasing the alignment of curriculum, instruction, and assessment with state standards.

CHAPTER 4

THE STATUS OF K–12 SCIENCE TEACHING IN THE UNITED STATES

Results from a National Observation Survey

Eric R. Banilower, P. Sean Smith, Iris R. Weiss, and Joan D. Pasley

The *National Science Education Standards* (*NSES*) (National Research Council, NRC, 1996) describe a vision of science instruction in terms of both content and pedagogical strategies. Almost all of what is known about how closely actual instruction mirrors the *NSES* comes from surveys of teachers, few of which are nationally representative. Such surveys provide valuable information about the frequency of instructional activities and topics addressed, but little insight into the quality of instruction. This chapter reports findings from classroom observations of a nationally representative sample of 180 K-12 science teachers in the United States. Within a theoretical framework of teaching for understanding, the study addressed such questions as: (1) How does science instruction *look* in the nation's classrooms? (2) To what extent are science lessons likely to engage students intellectually with important science disciplinary content? (3) Are students actively engaged in pursuing questions of interest to them? (4) Do teachers display an understanding of

The Impact of State and National Standards on K-12 Science Teaching, 83–49

science concepts? (5) Are adequate time and structure provided for student reflection and sense-making? (6) To what extent is there a climate of respect for students' ideas, questions, and contributions? (7) Are students encouraged to generate ideas, questions, and conjectures? Findings are reported in terms of the most prevalent strengths and weakness of lessons and the likely impact of lessons on students' understanding. Results are also discussed in terms of alignment of lessons with the content and pedagogical practices described in the *NSES*.

INTRODUCTION

For the last 50 years, the nation's attention has focused with varying intensity on the quality of science education. The NRC's *NSES* (NRC, 1996) represent a major and potentially influential effort to describe a vision of excellent science instruction. The *NSES* calls for teaching for understanding by engaging students with meaningful ideas in science at a conceptual level, rather than focusing on terminology and the ability to recall facts.

"Understanding" has been defined as "a matter of being able to do a variety of thought-demanding things with a topic such as explaining, finding evidence and examples, generalizing, applying, analogizing, and representing the topic in a new way" (Perkins & Blythe, 1994, p. 5) and as "the capacity to use current knowledge, concepts, and skills to illuminate new problems" (Gardner & Boix-Mansilla., 1994, p. 199). Piaget and other like-minded theorists suggested that learning "involves the acquisition of organized knowledge structures ... [and] the gradual acquisition of strategies for remembering, understanding, and solving problems" (Bransford, Brown, & Cocking, 2003, p. 80). Ausubel suggested that regardless of whether one experiences "reception learning" (the acquisition of information through lecture, print, image, etc.) or "discovery learning" (through which the principal content must be discerned by the learner), the learner must be able to relate new information to existing cognitive structures in order for learning to be meaningful (Ausubel, 1967b).

Students clearly enter the classroom with knowledge and ideas about the world (Bransford et al., 2003). Under the right conditions, students integrate new concepts and information with these ideas and arrive at a deeper level of understanding. This is what Ausubel termed "meaningful learning" (Ausubel, 1967b). If students' "initial understanding is not engaged, they may fail to grasp the new concepts and information that are taught or they may learn them for purposes of a test but revert to their preconceptions outside the classroom" (Bransford et al., 2003, p. 14).

To teach for understanding, teachers must situate factual knowledge within a conceptual framework and help students organize knowledge for retrieval and application (Bransford et al., 2003; Gallagher, 2000). Factual knowledge is important for understanding. However, it is as a means, rather than as an end in itself, for students to construct deep understanding. Facts are essential, but without a broader framework, they lose their power. Similarly, when teachers facilitate students' inquiries or investigations into a topic, it is important that those experiences be meaningful, relevant, and again, situated within a broader conceptual framework (Ausubel, 1967a, 1967b; Bransford et al., 2003; NRC, 1996; Wong, Pugh, & the Dewey Ideas Group, 2001). Finally, teachers should link the information to the real world and to other disciplines by encouraging questioning and discussion because "[s]tudents need to be explicitly helped in extending new ideas to different situations" (Tytler, 2002, p. 30).

Teaching for understanding places demands on teachers and learners that exceed those associated with either direct instruction or open inquiry. Cohen, McLaughlin, and Talbert (1993) identified three critical components teachers must possess. These are the profound knowledge of subject matter, the ability to represent the subject matter in multiple ways during instruction, and classroom management skills that facilitate active participation by students in learning. A fundamental question of the study reported in this chapter is: To what extent is the kind of teaching envisioned in the *NSES*—teaching for understanding—evident in science classrooms in the United States?

WHAT IS KNOWN FROM OTHER STUDIES ABOUT THE STATUS OF SCIENCE TEACHING IN THE U.S.?

Much of what is known about science teaching in the U.S. comes from large national surveys of teachers, such as the National Assessment of Education Progress (O'Sullivan, Lauko, Grigg, Qian, & Zhang, 2003) and the 2000 National Survey of Science and Mathematics Education (Weiss, Banilower, McMahon, & Smith, 2001). The value in these surveys is that they gather information from thousands of teachers, allowing national estimates of important indicators. Previous research has demonstrated that teachers' self-reports on the frequency of reform-oriented instructional practices meet reasonable standards of validity and reliability. However, teachers are clearly not in a position to judge the quality of their own instruction (Burstein et al., 1995; Mayer, 1999; Porter, Kirst, Osthoff, Smithson, & Schneider, 1993; Spillane & Zeuli, 1999). Some areas of instructional practice, such as discourse practices, and coherence of presentations, can only be measured by actual observations in classrooms (Burstein et al, 1995). Numerous observational studies have been conducted, but these have typically been small, or conducted in the context of the evaluation of a reform initiative, in both cases limiting the generalizability of the results.

There has not been a national observation study since the Case Studies in Science Education conducted in the late 1970s (Stake & Easley, 1978a). Sponsored by the National Science Foundation (NSF), this study involved in-depth case studies of a cross-section of 11 United States school districts, describing the condition and needs of science, mathematics, and social studies education. The authors noted that the quality of science instruction students experienced was quite varied; while some of the observed science classes stressed important science ideas and were described as interesting for students, most "overemphasized facts and memorization" and were not seen as relevant by the students.

As noted above, teaching for understanding requires that teachers be aware of and take into account students' prior knowledge and understanding (including misconceptions) when planning and implementing instruction. Nationally, roughly three out of four grade K–12 science teachers indicate that they feel well prepared to consider students' prior conceptions when planning curriculum and instruction (Weiss et al., 2001). A similar proportion of the teachers participating in the NSF Local Systemic Change (LSC) through teacher enhancement initiative noted feeling well prepared in this regard (Horizon Research, Inc., HRI, 2003). External observers often do not see this preparedness playing out in classroom instruction. For example, evaluators who observed lessons taught by LSC teachers found that far fewer teachers (57%) adequately read the students' level of understanding and adjusted instruction accordingly (HRI, 2003).

Given the defining role of inquiry in science, students need to understand that scientific knowledge is generated through investigation. There is considerable information from national surveys about class activities in science, for example, the frequency of hands-on activities, but until recently, very little was known about the nature of those activities and the extent to which students are intellectually engaged in investigating natural phenomena of interest to them as opposed to following the predetermined steps in an investigation of no interest to them. Based on national survey data, teachers are far more likely to have students follow specific instructions in an activity or investigation than to have them design their own investigations (Weiss, et al., 2001). In evaluating some reform projects, observers encountered some teachers who equated any kind of hands-on instruction with inquiry-based learning, thereby missing key opportunities to make conceptual connections. Others were too procedure-focused and reduced the investigative nature of the materials (Boyd, Banilower, Pasley, & Weiss, 2003).

PROBLEM AND RESEARCH QUESTION

The research and policy communities are interested in learning about classroom practice not only from the perspective of the classroom teacher, but also through the eyes of external observers. The need for up-to-date information on the nature and quality of K–12 lessons is particularly acute given the current emphasis on science education reform, yet until recently there have been no national efforts along these lines since the Stake and Easley case studies of 1978. The *Inside the Classroom* study (Weiss et al., 2003), completed by HRI in 2002, included observations of a nationally representative sample of science lessons taught by teachers in a variety of contexts in the United States. Science instruction was assessed by trained researchers using an adapted version of the classroom observation protocol developed and validated as part of the core evaluation of NSF's LSC initiative.

The *Inside the Classroom* study was driven by the following research questions:

- How does science instruction *look* in the nation's classrooms? To what extent is science portrayed as an inert collection of facts and algorithms, as opposed to a dynamic body of knowledge continually enriched by conjecture, investigation, analysis, and proof/justification?
- Are students actively engaged in pursuing questions of interest to them, or simply *going through the motions*?
- To what extent do science lessons engage students intellectually with important science disciplinary content?
- Is teacher-presented information accurate? Do teachers display an understanding of science concepts in their dialogue with students?
- When teachers ask questions, are they posed in a way that is likely to enhance the development of student conceptual understanding?
- Are adequate time and structure provided for student reflection and sense-making?
- To what extent is there a climate of respect for students' ideas, questions, and contributions? Are students encouraged to generate ideas, questions, and conjectures?

METHODOLOGY

For the *Inside the Classroom* study, HRI staff and consultants conducted observations and interviews with 180 grade K–12 science teachers between November 2000 and April 2002. The study design involved selecting a sample of schools to be representative of all schools in the United States by gaining school cooperation, sampling classes to be observed, collecting observation and teacher interview data, and weighting and analyzing the data appropriately to provide estimates for science lessons in the nation as a whole.

A subset of 40 middle schools was selected from the sample of schools participating in the 2000 National Survey of Science and Mathematics Education, also conducted by HRI. Systematic sampling with implicit stratification was used to ensure that the 40 sites would be as representative of the nation as possible. When a middle school agreed to participate, the study coordinators identified the elementary schools and high schools in the same feeder pattern and randomly sampled one of each. For classroom observations, a simple random sample was drawn from among the science teachers in the sampled school.

HRI encountered some resistance in securing cooperation of the sampled sites. Therefore, when roughly half of the observations had been completed, the study coordinators inspected the demographic characteristics of the observed sites to confirm that they were representative of schools in the nation. Noting some gaps, HRI drew a new random subsample of middle schools from the national survey schools and handpicked a subgroup of 14 sites that would round out the sample in terms of demographic characteristics.

Due to time and resource constraints, HRI ended the observation phase of the study having visited 31 sites. These sites and the sampled schools were representative of districts and schools in the nation. Observed teachers completed a shortened version of the teacher questionnaire used in the 2000 national survey. Based on their responses, the observed teachers and classrooms were also representative of those in the nation in terms of teacher backgrounds, instructional objectives, and instructional activities.

Researchers observed one lesson taught by each teacher and conducted an in-depth interview with the teacher following the observation. Observers did not record audio or video of the lesson, but rather took detailed field notes. Based on their observation and the teacher interview (which was recorded), observers rated each lesson on 29 indicators in four categories—lesson design, lesson implementation, lesson content, and the culture of the classroom. Researchers also provided a single overall rating of lesson quality on a seven-point scale. Detailed descriptions of each scale point are provided in the Appendix.

Data from the classroom observations were weighted in order to yield unbiased estimates of all science lessons in the nation. Each sampled teacher was assigned to a cell determined by school urbanicity (rural vs. urban vs. suburban) and sample grade range (K–5 vs. 6–8 vs. 9–12). Urbanicity was determined from the National Center for Education Statistics' Common Core of Data. Grade range was determined by the grade of the lesson observed. The 180 teachers observed represent just over one million science teachers in the nation, including those in self-contained settings. All sampled teachers in a cell were then given the same weight such that the sum of weights of the sampled teachers equaled the number of teachers in the nation in that cell. These weights were multiplied by the average number of science classes taught by teachers in the nation. To avoid underestimating the standard errors used in tests of statistical significance, the weights were normalized, effectively returning the weighted *N* to the actual sample size. In this chapter, all reported percentages are calculated using these design weights, while counts are based upon the actual number of observations.

The remainder of this chapter reports results of the study in three broad categories. First, data are presented about the quality of science instruction by grade range (elementary, middle, and high school) in the nation as a whole. Second, lesson quality is discussed in relation to the science content of the lessons; that is, to what extent is science instruction in the United States aligned with the content in the *NSES*, and does lesson quality vary depending on whether or not the lesson content can be found in the *NSES*? Third, lesson quality is discussed in relation to the instructional strategies that were used; that is, to what extent is science instruction in the United States aligned with the *NSES* in terms of instructional strategies, and does lesson quality vary depending on which instructional strategies are employed?

RESULTS

The vision of teaching for understanding that guided this study considers the primary goals of science education to be helping students understand important science concepts and deepening students' abilities to successfully engage in the processes of science. To achieve these goals, not only do lessons need to provide students opportunities to learn, but teachers also need to be very clear about the purposes of each lesson in relation to the specific concepts being addressed in order to help guide students in their learning. The observation protocol developed for the *Inside the Classroom* study was designed to assess the quality of lessons in relation to this vision of effective science instruction. In addition to rating specific components of the lessons, such as the accuracy of the science content and the quality of the teachers' questioning, observers rated the likely impact of each lesson on students' conceptual understanding.

Table 4.1. Lessons Utilizing Various Instructional Strategies, by Grade Level

	Percent of Lessons			
	K–12 (N = 180)	*K–5 (N = 55)*	*6–8 (N = 64)*	*9–12 (N = 61)*
Lecture/discussion	74	69	78	84
Working in small groups	39	43	40	28
Hands-on/laboratory activities	38	51	21	21
Worksheet/textbook problems	29	25	36	31
Reading about science	19	19	25	14

In grade K–12 science instruction, lecture/discussion is by far the most commonly used instructional strategy, found in nearly three out of four lessons (see Table 4.1). Small group work and hands-on/laboratory activities are each utilized in about 40% of all science lessons. About one in three lessons include students completing worksheets/answering textbook problems and roughly one in five include students reading about science, either from a textbook or from another source.

Not surprisingly, the use of lecture/discussion rises steadily from the elementary through the secondary grades. In addition, there is also a dramatic decrease in the use of hands-on/laboratory activities between grades K–5, where about one-half of the lessons include this strategy, and grades 6–12, where less than one-quarter of lessons include a hands-on/laboratory activity.

As can be seen in Table 4.2, based on observers' judgments, only about a third of the science lessons nationally are likely to have a positive impact on student understanding of science concepts, and between 10 and 20% are likely to have a negative effect on their understanding the remaining lessons would likely have no effect, or both positive and negative effects. Lessons at the elementary level are almost twice as likely as those in middle and high school to be judged likely to have a positive effect on students' interest in and/or appreciation for the discipline. The same is true of impacts on students' (1) understanding of science as a dynamic body of knowledge generated and enriched by investigations and (2) capacity to carry out their own investigation. These differences are likely due to the more frequent use of hands-on/laboratory activities in the elementary grades (see Table 4.1).

The scale observers used to provide an overall assessment of the quality and likely impact of the lesson is divided into the following categories:

Table 4.2. Likely Impacts of Science Lesson

	Percent of Lessons		
	Negative Effect	*Mixed or Neutral Effect*	*Positive Effect*
Grades K–12			
Students' ability to apply or generalize skills and concepts to other areas of science, other disciplines, and/or real-life situations	11	61	28
Students' capacity to carry out their own inquiries	15	59	26
Students' interest in and/or appreciation for the discipline	23	44	33
Students' self-confidence in doing science	17	59	24
Students' understanding of important science concepts	15	52	33
Students' understanding of science as a dynamic body of knowledge generated and enriched by investigation	23	52	25
Grades K–5			
Students' ability to apply or generalize skills and concepts to other areas of science, other disciplines, and/or real-life situations	9	58	32
Students' capacity to carry out their own inquiries	13	53	34
Students' interest in and/or appreciation for the discipline	14	42	43
Students' self-confidence in doing science	13	57	31
Students' understanding of important science concepts	15	51	34
Students' understanding of science as a dynamic body of knowledge generated and enriched by investigation	18	49	33
Grades 6–8			
Students' ability to apply or generalize skills and concepts to other areas of science, other disciplines, and/or real-life situations	11	65	23
Students' capacity to carry out their own inquiries	17	67	16
Students' interest in and/or appreciation for the discipline	33	45	22
Students' self-confidence in doing science	22	65	13
Students' understanding of important science concepts	19	57	24
Students' understanding of science as a dynamic body of knowledge generated and enriched by investigation	28	58	14
Grades 9–12			
Students' ability to apply or generalize skills and concepts to other areas of science, other disciplines, and/or real-life situations	14	63	23
Students' capacity to carry out their own inquiries	18	69	14
Students' interest in and/or appreciation for the discipline	37	47	16
Students' self-confidence in doing science	24	60	16
Students' understanding of important science concepts	14	48	38
Students' understanding of science as a dynamic body of knowledge generated and enriched by investigation	33	55	12

Level 1: Ineffective instruction

a. "passive learning"
b. "activity for activity's sake"

Level 2: Elements of effective instruction
Level 3: Beginning stages of effective instruction (low, solid, high)
Level 4: Accomplished, effective instruction
Level 5: Exemplary instruction

Lessons are also broadly categorized in this chapter as low in quality (1a, 1b, 2); medium in quality (low 3, solid 3), and high in quality (high 3, 4, 5). Lessons judged to be low in quality are unlikely to enhance students' understanding of important science content or provide them with abilities to engage successfully in the process of science. While low quality lessons fall down in numerous areas, their overarching downfall tends to be the students' lack of engagement with important science. At the other end of the scale, high quality lessons are structured and implemented in a manner which engages students with important science concepts and are very likely to enhance their understanding of these concepts and to develop their capacity to do science successfully. Regardless of the pedagogy (e.g., investigations, teacher presentations, discussions with each other or the teacher, reading), high quality lessons provide opportunities for students to interact purposefully with science content and are focused on the overall learning goals of the concept. In the middle are lessons that are purposeful and include some elements of effective practice, but also include substantial weaknesses that limit the potential impact for students. The specific areas where *middle quality* lessons fall down varies widely and could be related to the content that is the focus of the lesson, how the lesson is designed and implemented, and/or the classroom culture.

The majority (62%) of K–12 science lessons are low in quality (i.e., the likelihood that the lesson would lead to student understanding), while few (14%) are considered high quality. Figure 4.1 shows the complete distribution of observers' ratings, both overall and by grade level. Although there appears to be some variation in the ratings across the grade levels, the differences among the distributions of low, medium, and high ratings are not statistically significant ($\chi^2 = 7.48$, $df = 4$, $p = 0.112$). However, this lack of significance may be partly due to the size of the observation sample.

FINDINGS: STRENGTHS AND WEAKNESSES OF LESSONS

Clearly the quality of the lessons teachers design and enact to help students learn science content varies considerably. At one end of the spectrum, researchers saw classrooms where the students were fully and

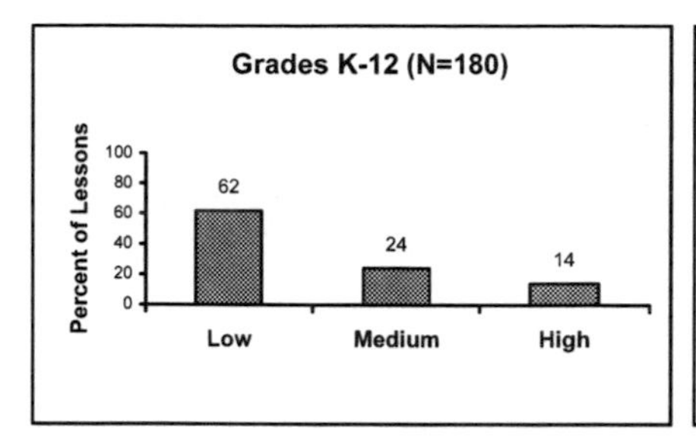

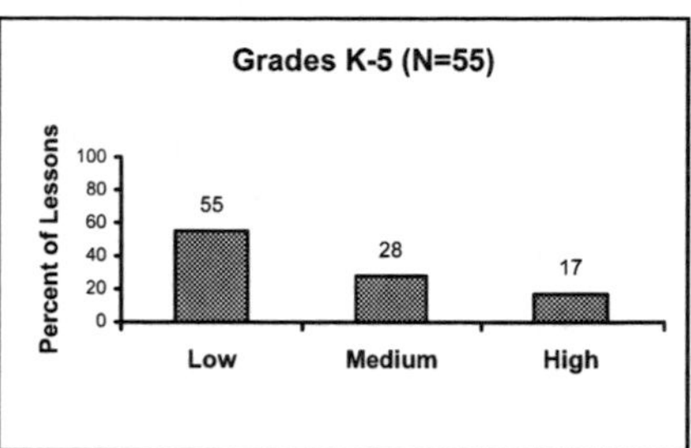

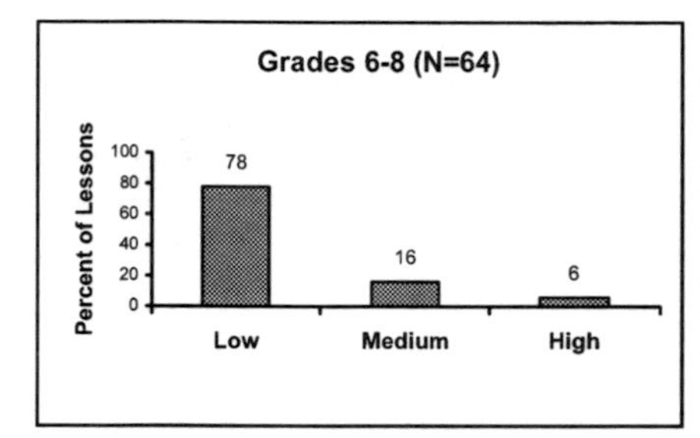

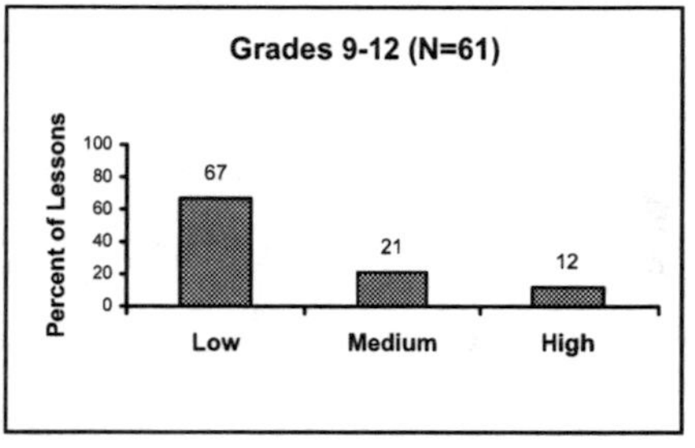

Figure 4.1. Overall rating of the quality of the lesson.

purposefully engaged in deepening their understanding of important science concepts. Some of these lessons were *traditional* in nature, including lectures and worksheets; others were *reform* in nature, involving students in more open inquiries. Observers saw other lessons, some traditional and some reform-oriented, that were far lower in quality, where learning science would have been difficult, if not impossible. To determine which characteristics were most important in determining quality, HRI did an in-depth analysis of lesson descriptions for lessons judged very effective and decidedly ineffective. The factors that seem to distinguish effective lessons from ineffective ones are their ability to:

- Engage students with the science content
- Create an environment conducive to learning
- Ensure access for all students
- Help students make sense of the science content.

These results are presented in the following sections, using excerpts from lesson descriptions to illustrate the findings. All quantitative data provided are weighted to represent all science lessons in the United States, grades K–12.

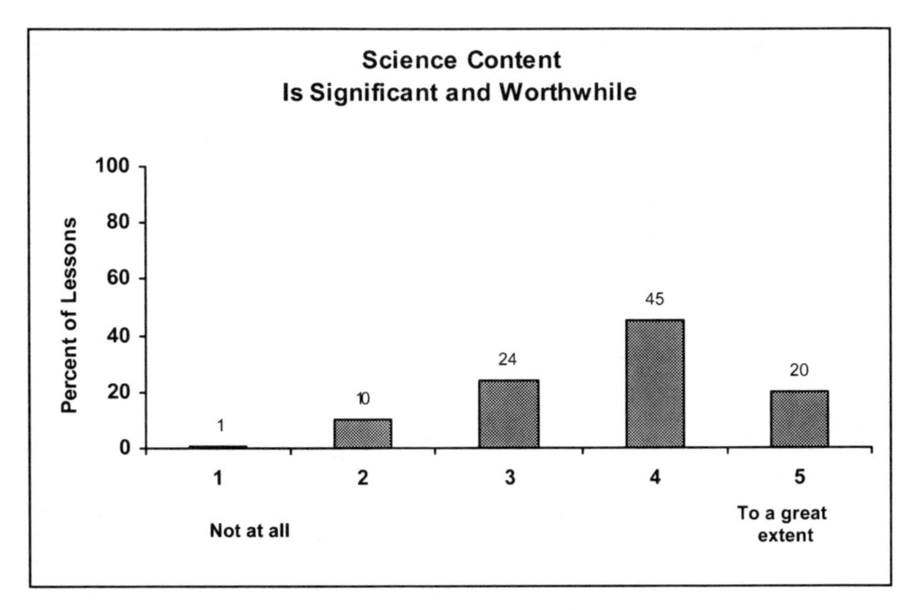

Figure 4.2. Science content is significant and worthwhile.

Engaging Students with Science Content

Certainly one of the most important aspects of teaching for understanding, if not the most important, is that instruction address content that is both significant and worthwhile. Lessons using a multitude of innovative instructional strategies would not be productive unless they were implemented in the service of teaching students important disciplinary and inquiry content. Based on the lessons observed in this study, science lessons in the United States are relatively strong in this area, with the majority of lessons including significant and worthwhile content. (See Figure 4.2.)

Although lessons generally include important content, most lessons are nevertheless low in overall quality. Clearly, while the inclusion of important content is necessary for high quality science lessons, it is not sufficient. We return to this point later in the chapter.

"Inviting" the Learners into Purposeful Interaction with the Science Content

The hallmark of lessons judged to be effective is that they include meaningful experiences that engage students intellectually with science content. These lessons make use of various strategies to interest and engage students and to build on their previous knowledge. Effective les-

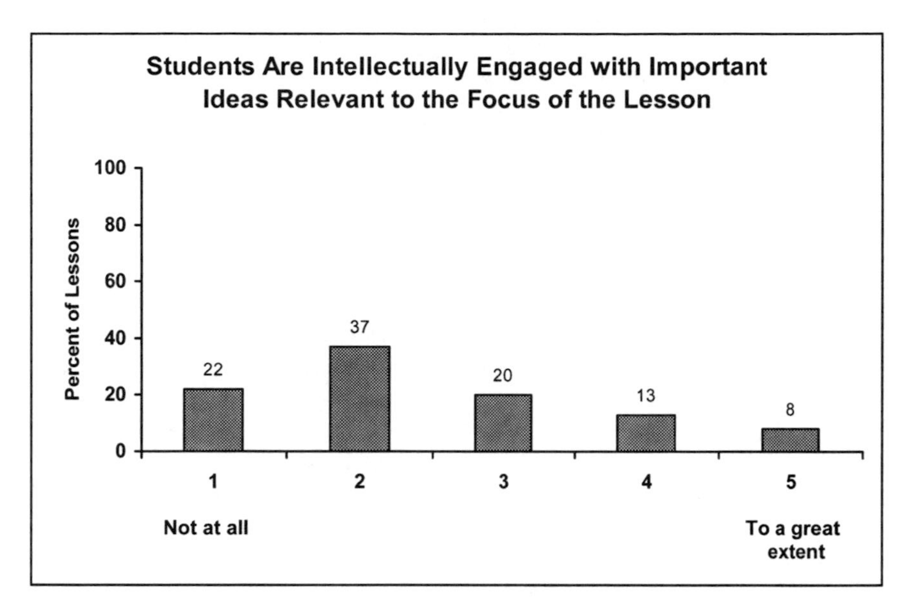

Figure 4.3. Students are intellectually engaged.

sons often provide multiple pathways that are likely to facilitate learning and include opportunities for sense-making. As can be seen in Figure 4.3, few lessons in the nation appear to engage students intellectually with important science content.

Earlier we noted the importance of lessons being *purposeful* in relation to important learning goals, with teachers having a clear understanding of the purpose of each lesson in terms of those goals. In addition, *students* need to see a purpose to the instruction, not necessarily the disciplinary learning goals the teacher has in mind, but some purpose that will motivate their engagement. Lessons need to *hook* students by addressing something they have wondered about, or can be induced to wonder about, possibly but not necessarily in a real-world context. A similar argument was made by Kesidou and Roseman (2002) in their analysis of middle school science programs, citing research support for the idea that, "if students are to derive the intended learning benefits from engaging in an activity, their interest in or recognition of the value of the activity needs to be motivated" (p. 530). The observation protocol used in this study did not specifically ask researchers about the strategies that teachers used to engage students in the lesson, but observers often commented on the presence or absence of this feature in their lesson descriptions.

Observers noted that many lessons *just started*, with no element of student motivation. For example, a teacher began a third grade lesson simply by having the students open their textbooks to the designated chapter, while she handed them a review worksheet. Similarly, a high school lesson began with the teacher distributing a packet of questions and saying, "All right now, these pages should be very easy if you've been paying attention in class. We talked about all of this stuff."

Just starting is not restricted to review lessons. For example, a high school teacher announced that, "Today we're going to talk about Roman Numeral III.H.," referring to a lengthy outline he had given the students previously. In some cases students did not engage in lessons such as these and in other cases they were attentive, typically with an upcoming test rather than interest in the problem being posed as the apparent motivation.

Among lessons that did include some attempt to motivate students or spark their interest, teachers used a variety of strategies. Some lessons *invited* the learners in by engaging them in first-hand experiences with the concepts or phenomena. For example, in a fourth grade science lesson about the basic needs of animals and how different body parts help animals meet these needs, the teacher handed out a tail feather and a magnifying glass to each pair of students, and asked them to examine the feather, pull the barbs apart, and look for the hooks. They then pulled the feather between their fingers, making the barbs stick back together. The teacher then handed out a down feather and they repeated their investigations.

Some teachers, instead of providing students with first-hand experience, invite the students in by using real-world examples to vividly illustrate the concept. For example, the teacher of a high school earth science class began the study of the water cycle by telling students that their state held the dubious honor of being the second driest state in the union.

Teachers sometimes used stories and other fictional contexts to engage students with the content of the lessons. For example, in a first grade science lesson, the teacher read a story about a girl who discovers an arrowhead in her backyard. The class then engaged in an excavation activity in pairs, where one child was the *archeologist* who found the *hidden treasures* in their *midden [refuse heap]* and the other was a *curator* who put their hidden treasures in a *museum*.

Portraying Science as a Dynamic Body of Knowledge

In addition to motivating students to engage with science content, another characteristic of lessons judged to be effective is the manner in which they represent the discipline of science. Lessons can engage students with concepts so they come away with the understanding that science is a dynamic body of knowledge generated and enriched by investigation. Alternatively, lessons can portray science as a body of facts

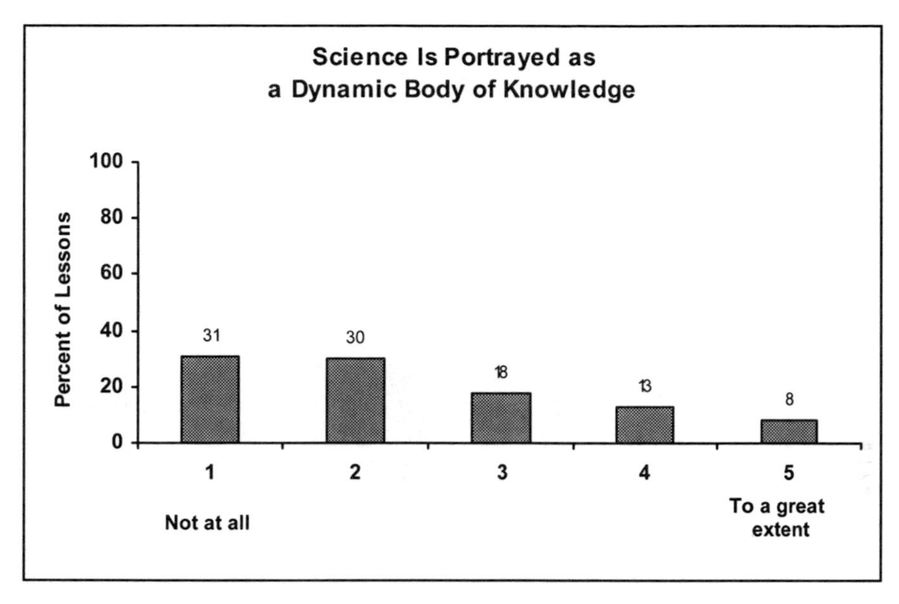

Figure 4.4. Science is portrayed as a dynamic body of knowledge.

and procedures to be memorized. Based on *Inside the Classroom* observations, only 21% of science lessons nationally provide experiences for students that clearly depict science as investigative in nature (rated 4 or 5 on a five-point scale). (See Figure 4.4.)

As an illustration, a sixth grade science lesson consisted of a teacher-led discussion of the process of sedimentary rock formation. By drawing upon the experiences and prior knowledge of the students, the teacher helped the students devise a model of how sedimentary rock is formed. For example, the teacher asked students if they broke a vase what they would need to fix it. The students decided that not only would they need glue, but they would also need something to push the pieces together. The teacher then asked the students, "Where might the force come from [to push sand together to make sandstone]?" The teacher probed students until they considered possible sources of the pressure. This lesson emulated the scientific process of using observable data and knowledge of basic scientific principles to create a model of an unobservable process.

In contrast, in many lessons science was presented as a static body of knowledge, focusing only on vocabulary and facts. Observers of these lessons said things like "the teacher did the thinking throughout the lesson—there was no investigative spirit. The teacher had knowledge, which he attempted to transmit to students." For example, an eighth grade science lesson was designed to give the students a great deal of factual information

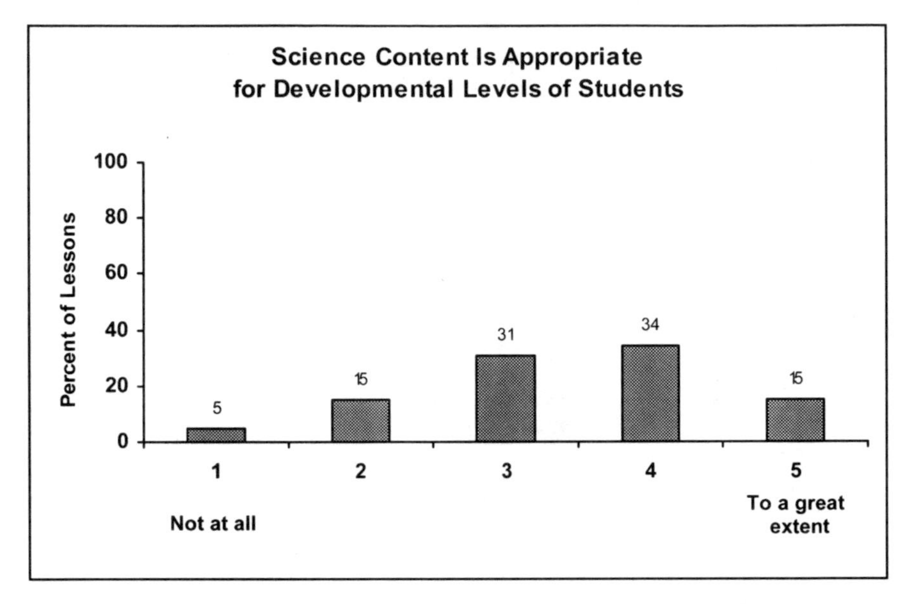

Figure 4.5. Science content is appropriate for developmental level students.

on Newton's Third Law of Motion. The students copied notes from the blackboard for half of the lesson, and the remainder of the lesson was spent with the teacher asking them to recall information from the notes.

Taking Students from Where They Are and Moving Them Forward

Although it is unlikely students are learning if they are not engaged, engagement is not enough. To enable learning, lessons need to be at the appropriate level for students while taking into account what they already know and can do, and challenging them to learn more. As can be seen in Figure 4.5, approximately half of all science lessons are rated high for the extent to which the content is appropriate for the developmental level of the students in the class.

The estimated 20% of lessons nationally that are not developmentally appropriate are only occasionally too difficult for the students. Sometimes students lack the prerequisite knowledge/skills, and the content seems inaccessible to them. At other times, the vocabulary is at far too high a level for the students. More often lessons are pitched at too low a level for some or all of the students.

Some lessons go further than simply providing content at a level that is appropriate for the students. These lessons use multiple representations of concepts to facilitate learning, both to give greater access to students with varying experiences and prior knowledge, and to help reinforce emerging understanding. Many lessons judged to be effective include a variety of experiences where students would be likely to *tap into* one or more of the pathways in developing or reinforcing a concept. For example, beginning with a review of the main facts about fossilization that students had been studying, the teacher in a seventh grade science class provided information about how fossils can be dated and went on to explain radiocarbon dating techniques. She then led the class in constructing standard radiocarbon dating curves, which the students used to date their own *fossils* (plastic bags of pennies). The *heads* represented *C*-14 atoms, which the students then replaced by paper clips, representing *N*-14 atoms. By counting the number of *C*-14 atoms in their fossil, students were able to determine its age. Students who finished this task were then asked to create an *N*-14 standard curve. The observer noted that the lecture was effective, and that the use of the small group, hands-on activity "helped make this rather abstract concept more concrete and interesting."

Creating an Environment Conducive to Learning

Important content and well-designed tasks at an appropriate developmental level are essential in order for students to have an opportunity to learn. So, too, is a classroom culture conducive to learning, one which is both rigorous and respectful. As can be seen in Figure 4.6, 46% of science

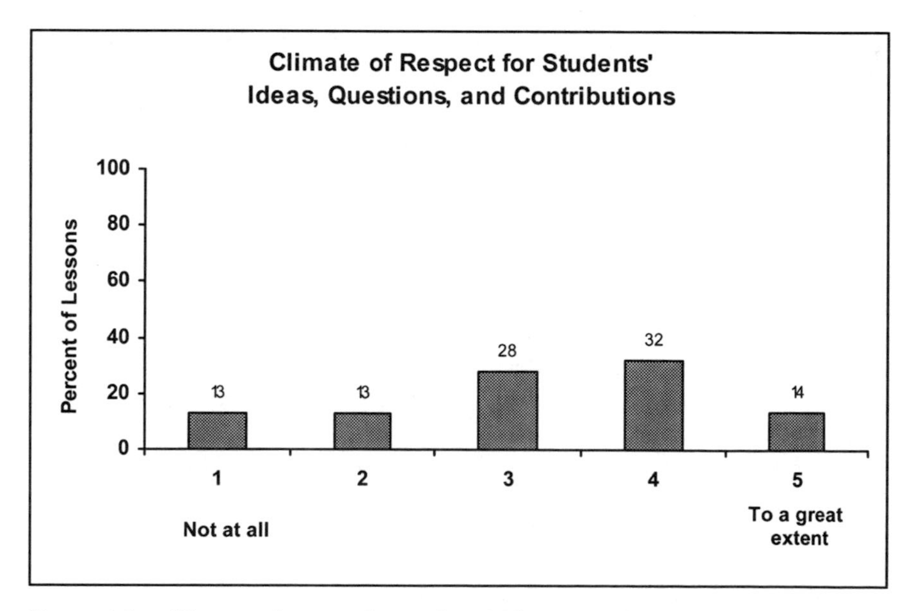

Figure 4.6. Climate of respect for students' ideas, questions, and contributions.

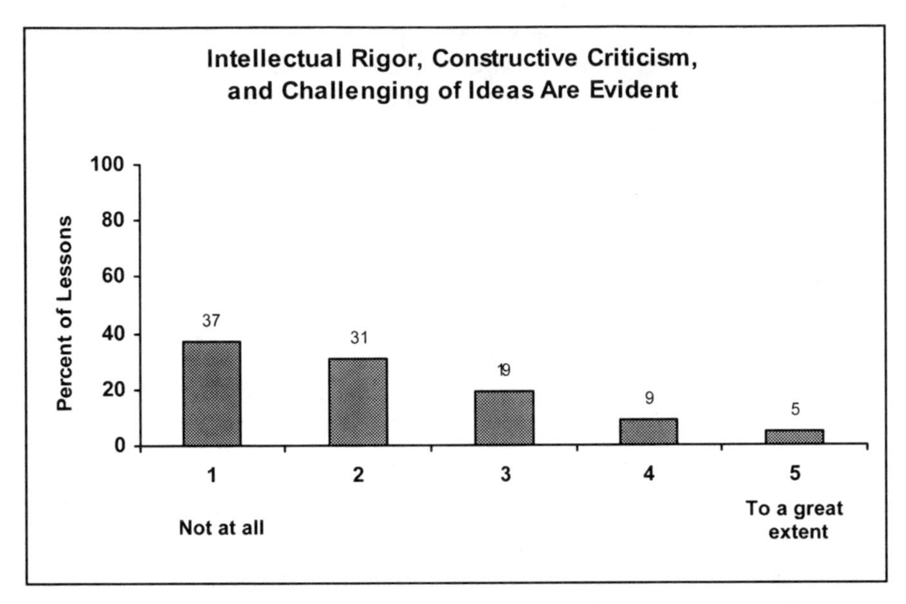

Figure 4.7. Intellectual rigor, constructive criticism, and challenging of ideas are evident.

lessons nationally receive high ratings for having a climate of respect for students' ideas, questions and contributions. Twenty-six percent receive low ratings in this area and the remaining 28% are somewhat respectful. Ratings for rigor are much lower, with only 14% of lessons nationally having a climate of intellectual rigor, including constructive criticism and the challenging of ideas. Sixty-eight percent of lessons receive low ratings in this area, and 19% are somewhat rigorous. (See Figure 4.7.)

Table 4.3 shows a cross tabulation of the two variables. Note that only 12% of lessons nationally are strong in both respect and rigor, and 26% are low in both areas.

Researchers observed highly respectful, highly rigorous lessons in science at the elementary, middle, and high school levels. For example, students in a fifth grade science class worked extremely well in pairs, offering constructive criticism of each other's findings. The observer described an example where one student concluded that a rubber band conducted electricity, but her teammate pointed out that she had accidentally touched the wire to one of the clips, completing the circuit. The pair of students then tried the experiment again, taking care to touch only the rubber band, and found that the rubber band was not a conductor.

> The teacher eagerly answered questions, and encouraged exploration. There was—pardon the pun—an air of electricity and excitement in the

Table 4.3. Cross Tabulation of Climate of Respect and Intellectual Rigor

	Percent of Lessons		
	Intellectual Rigor, Constructive Criticism, and Challenging of Ideas Are Evident		
Climate of Respect for Students' Ideas, Questions, and Contributions	*Low*	*Medium*	*High*
Low	26	0	0
Medium	26	1	2
High	16	17	12

room, and the students had to be shooed away from their activities for recess. It would be hard to imagine a classroom more conducive to learning.

Sixteen percent of lessons nationally could be categorized as respectful but lacking in rigor. *Inside the Classroom* observers used phrases like "pleasant, but not challenging" to describe such lessons. An observer described one such lesson this way:

> emotionally, the culture of this ninth grade science class was good. The teacher had a warm relationship with the students, and it seemed clear that there was great deal of mutual respect. Intellectually, however, the culture in this classroom was very weak. Science was presented as facts and formulas to memorize, with no requirement that things make sense or even be internally consistent. Students were asked to respond to the teacher's questions but did not interact with each other, or propose new ideas for the class to discuss.

Roughly one in four lessons nationally are lacking in respect, in some cases even hostile and demeaning to students, and nearly all of these are also very low in rigor.

Ensuring Access for All Students

Part of the teacher's role is to ensure that all students are in fact accessing the science content, and that no students are left behind. Accordingly, researchers were asked to rate the extent to which lessons encouraged active participation of all students. They also described cases where some students were *left out* of the lesson and cases where the teacher was particularly successful at engaging learners with special needs. As can be seen in Figure 4.8, only 47% of lessons nationally would be rated high in terms of encouraging active participation of all students.

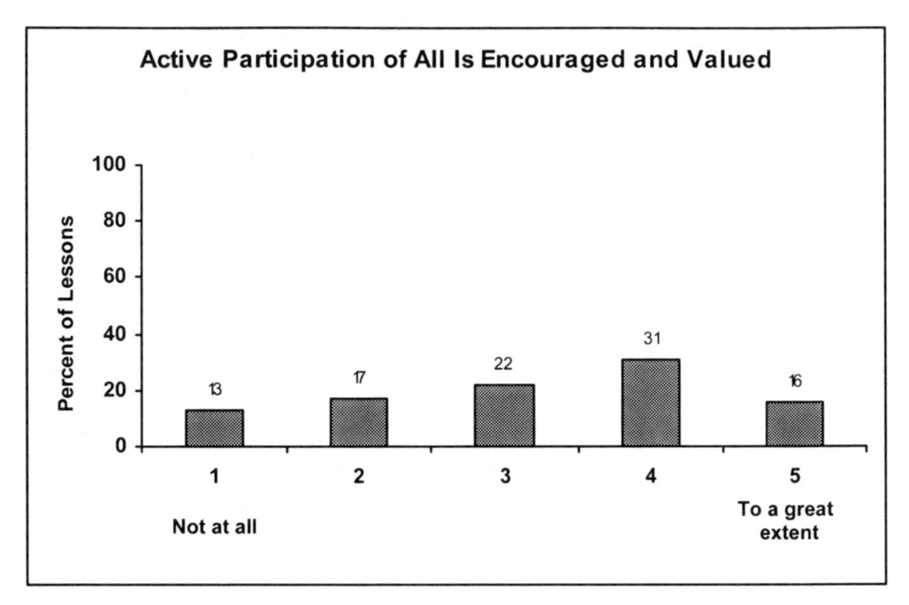

Figure 4.8. Active participation of all is encouraged and valued.

In most cases, low ratings on the active participation indicator reflect overall low levels of student engagement. In a few instances, observers noted differential patterns of participation by gender and/or race. Other observers described lessons where extensive efforts were made to ensure that all students had access to the lesson. A third grade teacher altered her lesson plan to accommodate the varying levels of her students. She required that all students depict what they had observed in the experiment that they had conducted during the class. The more able students could do this in a six-part, step-by-step, illustrated description of the experiment. Other children, who had more difficulty with writing, were allowed to express their understanding through a cartoon or other drawing.

Help Students Make Sense of the Science Content

Focusing on important science content, engaging students, and having an appropriate, accessible learning environment set the stage for learning, but these elements of instruction do not guarantee learning. It is up to the teacher to help students develop understanding of the science they are studying. The teacher's effectiveness in asking questions, providing explanations, and otherwise helping to push student thinking forward as

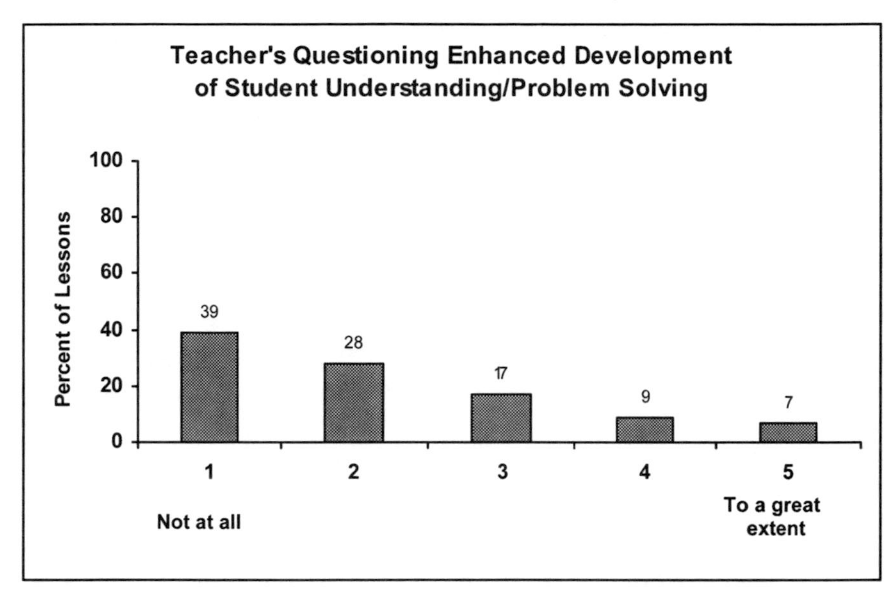

Figure 4.9. Teachers questions enhanced development of student understanding/problem solving.

the lesson unfolds, often appears to determine students' opportunity to learn.

A particularly important strategy for helping students make sense of content is teacher questioning. Researchers observed some lessons in which the questioning was extremely skillful. In these lessons, the teacher was able to use questions to assess where students were in their understanding, and was also able to get them to think more deeply about the science content. There were many more instances where the teacher asked a series of low-level questions in rapid-fire sequence, with the focus primarily on the correct answer, rather than on understanding.

Questioning is among the weakest elements of science instruction, with only 16% of lessons nationally incorporating questioning that is likely to move student understanding forward. (See Figure 4.9.) Lessons that are otherwise well-designed and well-implemented often fall down in this area.

Some teachers are able to use questioning skillfully, both to find out what students already knew and to provoke deeper thinking in helping them make sense of science ideas. For example, as the students in a 10th grade science class were examining the results of their experiment, the teacher asked questions that pushed them to examine their results further and to provide evidence for their conclusions. Examples of questions asked by the teacher are: "How could we test if there is still sugar in the reservoir?" "Why didn't it [the iodine indicator] reach equilibrium?" and "How do you know?"

When teachers ask questions, and individual students respond correctly, it is often difficult to tell if others in the class have a similar level of understanding. Some teachers were able to overcome this difficulty by asking for a show of hands, having established a culture where it was okay to be wrong in the process of working toward understanding. More often observers noted that the teachers moved quickly through the lessons, without checking to make sure that the majority of students were *getting it*. As soon as the few most verbal students indicated some level of understanding, the teacher went on, leaving other students' understanding uncertain.

By far, the most prevalent pattern in science lessons is one of low-level fill-in-the-blank questions, asked in rapid-fire, staccato fashion, with an emphasis on getting the right answer and moving on, rather than helping the students make sense of the science concepts. The following example illustrates this pattern as it played out in a sixth grade science lesson on weather and the atmosphere.

Teacher: The first layer is the what?
Students: Troposphere
Teacher: How many layers are there?
Students: Four
Teacher: What happens in the troposphere?
Student: It rains
Teacher: What happens in that layer?
[Students unsure]
Teacher: w, w, w ...
Student: Water?
Teacher: What have we been studying?
Student: Weather.
Teacher: What are four forms of precipitation?
Students: Rain, snow, sleet, hail.

Teacher questioning is one way, but not the only way to help students understand the science at hand. The important consideration is that lessons engage students in doing the intellectual work, with the teacher helping to ensure that they are in fact making sense of the key science concepts being addressed, as illustrated in the following lesson:

The teacher in a high school human anatomy and physiology class began a lecture by drawing a diagram of a nerve receptor, connected by a nerve

> fiber to (eventually) the brain. He explained the concept of a threshold for a receptor, noting that stimuli could be either subthreshold, threshold, or superthreshold and stressing that only after the threshold is reached does the receptor respond to the stimulus and send a signal to the brain. He spent most of the remainder of the lesson explaining that receptors vary in threshold and, "Your brain recognizes the highest threshold receptor stimulated.
>
> Using the hand as the point of reference, the teacher differentiated among different stimuli—touch, pressure, poke, punch, hammer, excruciating pain. He gave the example of an instance where if "punch" receptors were stimulated, the brain would not register "touch," only "punch." A student asked, "Does it work that way with taste, hearing, and sight?" The teacher responded that it does, and the student asked "How does it work with sight?" The teacher gave the example of caution signs being made of certain colors because the receptors for those stimuli have the lowest threshold, and of an artist using certain colors to create light and draw a person to a particular part of a painting.
>
> The teacher summarized this portion of the lecture, reiterating the all-or-nothing principle and the differentiation of nerve receptors by threshold. He spent the last few minutes of the class moving on to the next portion of his outline, in which he drew and labeled the parts of a synapse. The observer said, "this lecture was extremely engaging, accessible, and focused on worthwhile content. The teacher emphasized sense-making throughout the lesson, using examples familiar to the students and connecting the content to their lives. The students appeared to be very engaged.

Although researchers observed some lessons where students were helped to make sense of the science content as the lesson progressed or at its conclusion, most lessons lack adequate *sense-making*. As can be seen in Figure 4.10, only 13% of lessons in the nation would receive high ratings in this area. Teachers seem to assume that the students will be able on their own to distinguish the big ideas from the supporting details in their lectures, and to understand the underlying science ideas. For example, one researcher described a lesson this way:

> each of the physical science topics demonstrated in this lesson was appropriate to the ninth grade curriculum (mechanical waves, sound and light waves, mixing colors), and could be grasped by these students at some level. Moreover, each of the demonstrations was in itself interesting and motivational for the students, and for the most part kept their attention. However, the teacher presented all of these demonstrations in rapid succession, without providing appropriate ties to the material studied in class. As a result, the overall effect was more show than substance. No attempt was made to anchor the demonstrations into any conceptual framework.

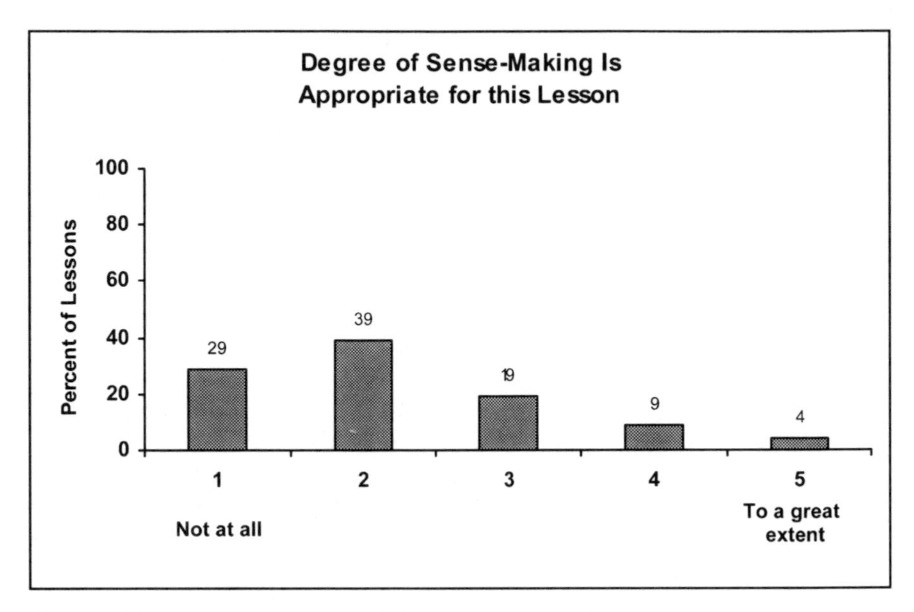

Figure 4.10. Degree of sense making is appropriate for this lesson.

The *NSES* describe a vision of teaching for understanding. However, when two-thirds of the lessons include inadequate attention to sense-making (see Figure 4.10), it is clear that the prevalence of standards-based instruction is quite low. Other data presented earlier in the chapter support this claim. Viewed from a framework of teaching for understanding, only 38% of lessons are considered to be medium or better in quality (see Figure 4.1). We turn now to other images of standards-based instruction, which although incomplete, are commonly held.

IMAGES OF STANDARDS-BASED INSTRUCTION

The term *standards-based* is often used to describe science instruction, but there appear to be different definitions of what standards-based instruction would look like. Two meanings of the term seem the most common and might be thought of as distinct dimensions of instruction. The first relates to the science content in a lesson. Can the content be found in the *NSES*, and is there a grade-level match? A lesson for kindergartners on content from the grade 9–12 standards would not be considered standards-based. This is not to say that conceptual precursors to 12th grade content should not be taught to kindergartners. The second meaning relates to instructional strategies. Was the lesson taught using strategies generally thought of as being consistent with the *NSES*? According to the *NSES*

> In the same way that scientists develop their knowledge and understanding as they seek answers to questions about the natural world, students develop an understanding of the natural world when they are actively engaged in scientific inquiry—alone and with others. (NRC, 1996)

In this sense, standards-based instruction has often been thought of as synonymous with instruction that engages students in scientific inquiry. Below we consider the prevalence of each type of standards-based instruction and how such instruction relates to observers' ratings of quality.

Alignment of Lesson Content with the National Science Education Standards

As noted earlier in this chapter, *Inside the Classroom* researchers took abundant field notes and wrote detailed descriptions of the observed lessons. Project staff read the lesson descriptions and coded the content of the lesson into one of five categories.

1. At least half of the lesson was focused on a content standard, and the content was within the correct grade range according to the *NSES*.
2. At least half of the lesson was focused on a standard outside of life, earth, or physical science (e.g., a standard from unifying concepts and processes).
3. At least half of the lesson was focused on a content standard, but the content was not within the appropriate grade range according to the *NSES*.
4. Less than half of the lesson was focused on a standard, or the lesson content was loosely related to a standard, but did not represent a specific one.
5. The content of the lesson was obviously not from the *NSES*.

Lessons were coded in two stages. First, a pair of researchers read all the lessons and coded the content as either earth science, life science, physical science, or *other science*. The lesson content was coded as other science only if it included no explicit content from the earth, life, or physical science standards. For example, a kindergarten lesson in which children created a classification scheme for a collection of buttons would have been considered other science. Those lessons coded as earth, life, or phys-

Table 4.4. Alignment of Science Lessons with *NSES* Content

The lesson included:	*Percent of Lessons*
1. Substantial focus on grade appropriate content standard	53
2. Substantial focus on "other science"	7
3. Substantial focus on grade inappropriate content standard	17
4. Small focus on content standard in any grade range	18
5. No *NSES* content	5

ical science were assigned to a second pair of coders with relevant content expertise. Each person in the pair coded the lessons independently into one of the five categories described above, then met to resolve any differences. If a physical science lesson, for instance, included no physical science from the *NSES*, but did include a focus on other science (e.g., controlling variables), it was coded a five. In all cases, researchers were able to agree on a single code for each lesson.

Table 4.4 shows the percentage of grade K–12 science lessons in the United States falling in each category. Only 5% of lessons appear to contain no standards-related content, while six in 10 might be considered standards-based; that is a substantial portion of the lesson focuses on content from the *NSES* (either disciplinary or other science content), and there is a grade-level match.

A logical question is whether there is a relationship between a lesson being standards-based (determined by content alignment) and ratings of lesson quality. Table 4.5 shows the percentage of lessons with and without a substantial focus on a grade-appropriate content standard (either disciplinary or other science content), by the quality rating for the lesson. Clearly, lessons thought of as standards-based in terms of content are no more likely than other lessons to be rated as high quality (2-tailed z-test, $p = 0.3356$). Almost two thirds of lessons, regardless of standards alignment

Table 4.5. Standards Alignment of Lessons, by Lesson Quality

	Percent of Lessons		
The lesson included:	*Low*	*Medium*	*High*
Substantial focus on grade appropriate content standard or "other science"	62	22	16
Grade inappropriate content, small focus on content from *NSES*, or no *NSES* content	62	28	11

were considered low in quality, and only slightly more than 10% high in quality.

Following is a description of a lesson that is aligned with the *NSES* in its content—properties and states of matter—but received a low rating for quality. The lesson illustrates several of the areas noted earlier where lessons fall short, including not engaging students with the content, not asking higher order questions, and not providing any opportunities for students to make sense of the content.

> This third grade lesson occurred towards the end of a unit of study on the properties of matter, with the focus of the lesson being to review the basic ideas and definitions of terms before the unit test which was to be given a couple of days after this lesson. Prior to this lesson, students had read and answered questions about matter from the textbook and had observed water as a solid, liquid, and a gas. The content being reviewed in this lesson included knowing that matter is made up of atoms, three physical properties of matter, three states of matter (plasma was not introduced to these students), changes in states of matter, and the concepts of measuring matter as volume and mass.
>
> The teacher began the class by having the students open their textbooks to the chapter on the properties of matter, while she handed them a review worksheet from an accompanying workbook. The teacher then systematically guided the class through the completion of the worksheet by referring the class to a particular question on the worksheet, telling them to turn to a specific page in their textbook and look for the answer, asking one student volunteer to read the answer from the book, then writing the answer on an overhead transparency copy of their worksheet.
>
> An example conversation is included:
>
> Teacher: Let's look at lesson two. Turn to page E16. Fill in the blank. Look on the page. Matter is made of … what?
> Student 1: Atoms.
> Teacher: Adding heat changes a solid to a what?
> Student 2: Liquid.
> Teacher: Good. Now read number three.
>
> At the completion of the worksheet, the teacher then reread the questions and answers to each question to summarize the content in the lesson. The lesson came to a close with the teacher instructing students to keep their worksheets for tomorrow's lesson.

Clearly, this lesson fell well short of the vision of teaching for understanding embodied in the *NSES*, despite being aligned with *NSES* content standards. Again, almost two thirds of lessons that were aligned in terms of content received similarly low ratings.

Table 4.6. Quality Ratings for K–12 Science Lessons, by Instructional Strategy

		Percent of Lessons		
		Distribution of Ratings		
	Number of Lessons	*Low*	*Medium*	*High*
Lecture/discussion	133	56	25	18
Working in small groups	70	51	29	20
Hands-on/laboratory activities	68	38	35	27
Worksheet/textbook problems	52	67	26	7
Reading about science	35	77	18	5

Alignment of Instructional Strategies with the *NSES*

Another image of standards-based instruction involves the instructional strategies teachers use. Table 4.6 shows the frequency of use and the distribution of quality ratings for lessons utilizing each of several instructional strategies. Regardless of the strategy, few lessons were rated as high in quality. In fact, the distributions of ratings for lecture/discussion, hands-on/laboratory activities, and working in small groups are statistically equivalent (Kolmogorov-Smirnov test, with Holm-Bonferroni adjustment for multiple comparisons, $\alpha = 0.05$), though again, the lack of significance may be partly attributable to the sample size of the study. Still, these data suggest, within the framework of teaching for understanding described above, that lessons including any one of these strategies are no more likely to be rated as effectives than lessons using another of the strategies. For example, lessons including lecture/discussion were generally rated as low quality instruction, as were lessons including hands-on/laboratory activities.

The *NSES* call for an inquiry-oriented instructional approach. As noted earlier in this chapter, it is not uncommon for teachers to equate hands-on/laboratory activities with inquiry-oriented instruction, and therefore with standards-based instruction. Results from *Inside the Classroom*, however, suggest that lessons including hands-on/laboratory activities are no more likely than others to reflect the teaching for understanding envisioned in the *NSES*. The lesson descriptions provided by observers give insight into where lessons with hands-on/laboratory activities frequently fall down in quality. Below we present three lessons (one each from grades K–5, 6–8, and 9–12) that illustrate common shortcomings.

1. The second half of this fifth grade lesson began with the teacher saying, “Today as a little treat, we’ll put an element together.” The teacher pulled out gumdrops, candy orange slices, marshmallows, and toothpicks. There was a buzz in the room as the teacher explained that an orange slice was to serve as a base, gumdrops as protons and neutrons, and marshmallows as electrons. A few student helpers passed out the components, and students got to work creating an element of their choice with atomic number less than or equal to seven (due to limited supplies). Students spent about 20 minutes building their models. The best question of the day came from a student asking how many neutrons to use. The teacher brushed the excellent question aside by saying, “Just use about the same number as you have protons.” The only questions asked by the teacher during the activity were “What element are you doing?” and “How many electrons does it have?” It is important to note that the physical model of an atom built from gumdrops and toothpicks is flawed in terms of particle sizes and distances, but these limitations were never addressed by the teacher. At the end of the lesson, each student was asked to hold up his or her model, say which element it was, and say how many electrons it had. Students paid very little attention to their classmates during this portion of the lesson. They may well have been fixated on eating their creations, which they were allowed to do during clean-up.
2. This eighth grade lesson was in a unit on motion, forces, and energy and dealt specifically with projectile motion. It began with the teacher reviewing lab procedures and using the overhead projector to remind students how to properly use a ruler and a protractor. Next, the students went outside to the quad to conduct the experiment in groups of three. While the students were outside, the teacher remained inside, and thus was not aware of, and was unable to correct, the mistakes students were making in their implementation of the lab activity. Students made errors in measuring both distances and angles (e.g., when they launched a marshmallow that went over a two foot wall, they draped the measuring tape over the wall instead of just measuring the horizontal distance. The students were not using the protractor correctly as they did not understand how to line up the base of their catapult with the crosshairs at the base of the protractor). Some groups were not really doing the lab and were just trying to launch their marshmallow at certain targets (“Let’s try to get it over the bush.”).

   Near the end of the period, the teacher called the students back into the classroom and had them work individually on writing their lab reports. Most students spent the time writing out the lab procedure and copying a data table from the textbook and then filling it in. As the students worked, the teacher circulated through the room asking students various questions about their data, which generally led to confusion since their data were of low quality, and students covered by trying to guess the answer the teacher wanted them to give. The teacher's questioning did not engage students with important concepts related to the purpose of the lesson and did not help students make the connection between angle of launch and distance traveled.

3. In this 10th grade class, the teacher began the activity by asking the kids to get out their handout describing the lab activity and spent approximately 5 minutes going over the instructions in an extremely haphazard fashion. During her explanation there was so much background noise that many of the students were likely not able to hear the teacher. In fact, when she asked a few of the students what they were to do, they did not seem to understand their assignment. The teacher gave the students only basic instructions/comments: (1) She asked them what the different pieces of background paper represented. (2) She told them that she "had given them three types of beaks (chopsticks, tweezers and a clothespin) and wanted to talk about those that work the best." (3) She told them to work in pairs; and (4) She told them to complete the questions and the graphing activity on the handout. The teacher did not discuss how many trials they were to do with each background, or beak, or how they were going to record data, and so forth.

   The students worked on the activity in pairs for the remaining 45 minutes of the class period with varying degrees of success. There were at least six pairs where only one of the partners did the work and several pairs where one of the partners was extremely disruptive to the others' learning. The teacher did nothing to help move these pairs along. Throughout the period the students had many questions about the procedures for the activity. They asked each other questions such as "Do I put both kinds of dots on the paper?" and "Which beak do I use first?" They also had trouble with the level of the language on the handout. They made comments to each other such as "What the heck is this? Percent available prey recovered!? Oh well, I don't know, let's go on." Throughout the time that the students spent working on the handout the teacher was very hands-off and uninvolved. Her biggest role was to go around to about half of the groups and explain that "How does the number of each type of disk captured compare with the number of each type remaining on the paper?" meant, in her words "Which dot was easier to pick up and why?" The few questions she asked of the pairs were superficial. For example, "What's that called when something blends in with its environment?" and "Dark, light, neutral moths? You'd better re-read that section" or the mysterious "I don't know; what did I talk about yesterday?" She rarely stuck around for the answer.

Many of the lessons including a hands-on/laboratory activity succeeded in involving students in the *hands-on*, but failed to engage students in thinking about the content. In these lessons, the students were not helped to make the connection between the activity and the learning goal. Instead of helping students make sense of their observations, the teachers' questions tended to be very low-level. Questions typically asked students to report out their observations without any emphasis on what the data meant. Because the connections between the activity and the learning goals were never explored, there was no opportunity for students to intellectually engage in these lessons, resulting in lessons that were essentially *activity for activity's sake*.

At the other end of the spectrum, *Inside the Classroom* researchers observed lessons utilizing hands-on/laboratory activities that were very likely to lead to student learning, though this type of lesson occurred much less frequently.

1. The teacher began this fifth grade lesson by having the students review as a whole class what they had learned the day before when exploring circuits. Responding to eager students who raised their hands, he elicited what it took to make a complete circuit (wires, battery, motor), and what stopped a circuit. The team leaders from each cooperative group then collected and distributed the materials for today's experiment (a battery holder with clips, a battery, a motor with wires, and a plastic bag with assorted items—paper clip, rubber band, wooden stick, sandpaper). Students were told to work with their partners and test the items in the plastic bag while recording their observations on an observation sheet. After they had finished this task, they were free to try out different objects in the room to see which ones conducted electricity. The students instantly became completely engrossed in the exploration, so involved that when the bell rang for recess, the children had to be chased out of the room!

   After recess, the students continued with their explorations. Everything in the room was fair game in their eager investigation of conductors. The students were activity involved in purposeful investigation that was definitely "minds-on" as well as "hands-on."

For example, one student had found that a rubber band conducted electricity, but her team-mate pointed out that she had accidentally touched the wire to one of the clips, completing the circuit. The pair of students then tried the experiment again, taking care to touch only the rubber band, and found that the rubber was not a conductor. The teacher moved around the room to answer questions, encourage investigation, and remind students to record their observations in words and pictures.

After giving students time to finish their investigations and return the equipment, the teacher prompted them to complete their observation sheets and to record their findings in their science journal. Students were encouraged to write down what they had observed using words and pictures, which they did with varying degrees of detail. This served as a very good culmination for the day's activities, and the teacher gladly gave them the extra five minutes students requested to finish this work before they began their next lesson. Final sense making and wrap-up would occur the next day when students pooled their observations and drew general conclusions about conductors and insulators.

2. This 10th grade biology class was completing a laboratory exercise begun the previous day. This lesson began with the teacher having the students, in their lab groups, predict what they expected to have happened with their lab (i.e., whether the starch or sugar diffused across the membrane) and try to explain why (dealing with particle size). After they had made a prediction, the groups got their lab materials and examined them. The teacher then had the students discuss whether their prediction was right or wrong. As a class, they then talked about what had happened in the experiment (what material(s) crossed the membrane, did all of it cross, etc.). As students made hypotheses, the class discussed methods for testing them. As needed, the teacher chimed in with suggestions (e.g., using test tape to measure sugar content), but his role was limited to providing lab techniques that would enable the students to test their ideas, and prodding the groups to make sure they did enough tests to fully explain what had happened. This segment of the lesson worked extremely well, with the students in charge of their investigations and doing the majority of the intellectual work. The teacher kept to his role of facilitator, questioning students and giving them suggestions for lab tests.

   The teacher skillfully guided the students as they finished the lab activity (making observations and analyzing the data), asking questions that pushed students to examine their results and to provide evidence for their conclusions. Examples of questions asked by the teacher are: "How could we test if there is still sugar in the reservoir?" "Why did it [the iodine indicator] not reach equilibrium?" and "How do you know?" The teacher also introduced new words to the class as appropriate. For example, as the students were trying to explain what had happened to the sugar in their experiment, the teacher interjected to the whole class "I hear you discussing, let me introduce a term: equilibrium." Thus, the teacher was able to ease new content into the discussion in the context of the investigation.

After the groups had finished all of their tests, the teacher gave them an assignment to write a story about a paramecium that lived in the local fresh water river who decided to go see his girlfriend who lived in the ocean. The groups were instructed to write about his trip and what he would experience. This activity provided a good opportunity for the students to bring together what they knew about transport across a membrane and apply it to organisms living in their local river.

One key feature that sets these last two examples apart from the previous three is that the hands-on/laboratory activity was purposeful and students knew why they were conducting the activity and by the end of the activity they understood how the activity connected to the learning goals. The students were clearly engaged in the activity, not just physically but intellectually as well. Further, the teachers' questions typically required students to analyze, apply, synthesize, and evaluate what they were experiencing. In each of these lessons, the teachers provided opportunities for sense-making.

Interestingly, high-quality lessons featuring a hands-on/laboratory activity often incorporated a lecture/discussion as well. In many of these lessons, the hands-on/laboratory activity served to motivate the students and to engage them with phenomena, while the lecture/discussion portion helped students interpret, apply, and make sense of their observations. In some lessons, teachers took advantage of teachable moments, which were sometimes planned and sometimes unplanned, to introduce new content. In other lessons, the teacher began with a relatively short lecture/discussion before moving into an activity that would allow students to further investigate the ideas touched upon by the lecture. The common thread throughout these lessons is intellectual engagement of students with the important ideas of science and giving them opportunities to wrestle with those ideas (test them, apply them to different situations, etc.), and then providing students with opportunities to pull it all together and make sense of the ideas.

Table 4.7. Quality Ratings, by Use of Hands-On/Laboratory Activities and Lecture/Discussion

	Number of Lessons	*Percent of Lessons*	*Distribution of Ratings*		
			Low	*Medium*	*High*
Neither	18	10	93	7	0
Hands-on/laboratory activity only	22	15	68	29	2
Lecture/discussion only	108	52	74	19	7
Both*	32	23	17	39	44

*Distribution of ratings significantly different from all other groups, $p < 0.05$ (Kolmogorov-Smirnov test, with Holm-Bonferroni adjustment for multiple comparisons).

Lessons that utilized a hands-on/laboratory activity but not lecture/discussion (or vice-versa) were very unlikely to be rated as high in quality. As can be seen in Table 4.7, only 7% of K–12 science lessons including a lecture/discussion, but not including a hands-on/laboratory activity received high ratings. Similarly, only 2% of lessons containing a hands-on/laboratory activity but no lecture/discussion were rated highly. However, 44% of lessons utilizing both strategies received a high rating. The sample size of the study does not support further disaggregation of these data by grade range. Further, fewer than one in five lessons employing both strategies received low ratings, compared to roughly seven in 10 lessons using lecture/discussion or hands-on/laboratory activities only.

Table 4.8 shows the percentage of lessons that received high ratings on a set of key indicators (i.e., those considered most important within the teaching for understanding framework underlying the quality ratings), and the results of statistical significance testing comparing the four groups on each indicator. Most lessons, regardless of the instructional approach, included content that was significant and worthwhile, reflecting the content alignment of the lessons. In addition, the science content tended to be developmentally appropriate in lessons that included either hands-on/laboratory activities or lecture/discussion. However, consistent with the examples shown above are the ratings related to the level of intellectual engagement and rigor, the extent of sense-making, and the quality of teacher questioning. Lessons that included both hands-on/laboratory activities and lecture/discussion were much more likely to be rated highly on each of these dimensions than lessons that included neither or only one of these strategies.

Table 4.8. Lessons Receiving High Ratings§ on Key Indicators, by Use of Hands-On/Laboratory Activities and Lecture/Discussion

	Percent of Lessons			
	Neither	*Hands-On/ Laboratory Only*	*Lecture/ Discussion Only*	*Both*
The science content was significant and worthwhile.	50	59	59	89†
There was a climate of respect for students' ideas, questions, and contributions.	21	39	37	80†
The science content was appropriate for the developmental needs of the students in this class.	14‡	65	45	62
Students were intellectually engaged with important ideas relevant to the focus of the lesson.	4	2	11	61†
The degree of sense-making of science content within this lesson was appropriate for the developmental levels/needs of the students and the purposes of the lesson.	4	2	8	37†
Intellectual rigor, constructive criticism, and the challenging of ideas were evident.	0	9	6	40†
The teacher's questioning strategies were likely to enhance the development of student conceptual understanding/problem solving (e.g., emphasized higher order questions, appropriately used "wait time,", identified prior conceptions and misconceptions).	0	0	7	55†

§Rated a 4 or 5 on a five-point scale, where 1 was "Not at all" and 5 was "To a great extent."
†"Both" group significantly different from all other groups in that row, $p < 0.05$ (*z*-test with Holm-Bonferroni adjustment for multiple comparisons).
‡"Neither" group significantly different from "Hands-on/laboratory only" and "Both" groups, $p < 0.05$ (*z*-test with Holm-Bonferroni adjustment for multiple comparisons).

CONCLUSIONS AND IMPLICATIONS

This chapter presented the status of standards-based science instruction from three frameworks. From a teaching-for-understanding framework, the picture is not encouraging. Almost two-thirds of grade K–12 science lessons are low in quality, falling short in key areas such as intellectual engagement, sense-making, teacher questioning, and classroom culture. In terms of alignment of lesson content with the *NSES*, the data are more positive. Six in 10 lessons focus on science content that can be found in the *NSES*, and at the appropriate grade level. Further analysis, however, shows that there is no relationship between content alignment and lesson quality (as judged from the teaching-for-understanding framework). Lessons that are standards-based in terms of content are no more likely to receive high ratings for quality than those that are not standards-based.

The third framework centered on instructional strategies. *Inside the Classroom* data confirm findings from survey studies showing that lecture/discussion is by far the most frequently used instructional strategy, with small group work and hands-on/laboratory activities far less frequent. However, there is no relationship between use of these three strategies and ratings of lesson quality. The distribution of quality ratings for lessons including hands-on/laboratory activities is no different than the distribution for lessons including lecture/discussion. A particularly interesting finding is that almost half of the lessons including both hands-on/laboratory instruction and lecture/discussion are high in quality, while less than 10% of lessons including only one of these strategies receive high ratings.

While the data presented in this chapter are less than positive, we have no way of knowing if they represent a change compared to 1996, when the *NSES* was published. No comparable studies of the quality of instruction on a national scale exist. One implication of *Inside the Classroom* findings is the need to collect quality of instruction data on a regular basis; perhaps every 5 to 10 years.

It is clear from *Inside the Classroom* data that teachers need a vision of effective instruction to guide the design and implementation of their lessons. Findings from this study suggest that rather than advocating one type of pedagogy over another, the vision of high quality instruction should emphasize the need for important and developmentally-appropriate science learning goals, instructional activities that engage students with the science content, a learning environment that is simultaneously supportive of, and challenging to students and vitally, attention to appropriate questioning and helping students make sense of the science concepts they are studying.

A number of interventions would likely be helpful to teachers in understanding this overall vision, and in improving instructional practice in their particular contexts. First, teachers need opportunities to analyze a variety of lessons in relation to these key elements of high quality instruction, particularly teacher questioning and sense-making focused on conceptual understanding. For example, starting with group discussions of videos of other teachers' practice, and moving toward examining their own practice, lesson study conducted with skilled, knowledgeable facilitators could provide teachers with helpful learning opportunities in this area.

Second, the support materials accompanying textbooks and other student instructional materials need to provide more targeted assistance for teachers—clearly identifying the key learning goals for each suggested activity, sharing the research on student thinking in each content area, suggesting questions/tasks that teachers can use to monitor student understanding, and outlining the key points to be emphasized in helping students make sense of the science concepts.

Third, workshops and other teacher professional development activities need to themselves reflect the elements of high quality instruction with clear, explicit learning goals such as a supportive but challenging learning environment and a means to ensure that teachers are developing understanding. Without question, teachers need to have sufficient knowledge of the science content they are responsible for teaching. However, teacher content knowledge is clearly not sufficient preparation for high quality instruction. Based on the *Inside the Classroom* observations, teachers also need expertise in helping students develop an understanding of that content, including knowing how students typically think about particular concepts, how to determine what a particular student or group of students is thinking about those ideas, and how the available instructional materials (and possibly other examples, investigations, and explanations) can be used to help students deepen their understanding.

Finally, administrators and policymakers need to ensure that teachers are getting a coherent set of messages. Assessments that measure the most important knowledge and skills will have a positive influence on instruction, as well as providing opportunities and incentives for teachers to deepen their understanding of the science content they are expected to teach, and how to teach it. Only if preservice preparation, curriculum, student assessment, professional development, and teacher evaluation policies at the state, district, and school levels are aligned with one another, and in support of the same vision of high quality instruction, can we expect to achieve the goal of excellence and equity for all students.

ACKNOWLEGMENT

The preparation of the original manuscript was assisted by Rebecca A. Crawford, Lacey R. Dean, Brent A. Ford, Sherri L. Fulp, Elizabeth S. Shimkus, & Kimberley D. Wood at HRI. *Looking Inside the Classroom: A Study of K–12 Mathematics and Science Education in the United States*, was conducted with support from the NSF under grant number REC-9910967. These writings do not necessarily reflect the views of the NSF.

APPENDIX: DEFINITIONS OF LEVELS OF QUALITY

In this final rating of the lesson, consider all available information about the lesson, its context and the teacher's purpose, and your own judgment of the relative importance of the ratings you have made. Select the capsule description that best characterizes the lesson you observed. Keep in mind that this rating is *not* intended to be an average of all the previous ratings, but should encapsulate your overall assessment of the quality and likely impact of the lesson.

- **Level 1**: **Ineffective Instruction**
  There is little or no evidence of student thinking or engagement with important ideas of science. Instruction is *highly unlikely* to enhance students' understanding of the discipline or to develop their capacity to successfully *do* science. Lesson was characterized by either (select one below):

  O **Passive "Learning"**
  Instruction is pedantic and uninspiring. Students are passive recipients of information from the teacher or textbook; material is presented in a way that is inaccessible to many of the students.

  O **Activity for Activity's Sake**
  Students are involved in hands-on activities or other individual or group work, but it appears to be activity for activity's sake. Lesson lacks a clear sense of purpose and/or a clear link to conceptual development.

- **Level 2: Elements of Effective Instruction**
  Instruction contains some elements of effective practice, but there are *serious problems* in the design, implementation, content, and/or appropriateness for many students in the class. For example, the content may lack importance and/or appropriateness; instruction may not successfully address the difficulties that many students are experiencing, and so forth. Overall, the lesson is *very limited* in its likelihood to enhance students' understanding of the discipline or to develop their capacity to successfully do science.

- **Level 3: Beginning Stages of Effective Instruction**
  (Select one below.)

  O Low 3 O Solid 3 O High 3

  Instruction is purposeful and characterized by quite a few elements of effective practice. Students are, at times, engaged in meaningful work, but there are *weaknesses*, ranging from substantial to fairly minor, in the design, implementation, or content of instruction. For example, the teacher may short-circuit a planned exploration by telling students what they *should have found*; instruction may not adequately address the needs of a number of students; or the classroom culture may limit the accessibility or effectiveness of the lesson. Overall, the lesson is *somewhat limited* in its likelihood to enhance students' understanding of the discipline or to develop their capacity to successfully do science.

- **Level 4: Accomplished, Effective Instruction**
  Instruction is purposeful and engaging for most students. Students actively participate in meaningful work (e.g., investigations, teacher presentations, discussions with each other or the teacher, reading). The lesson is well-designed and the teacher implements it well, but adaptation of content or pedagogy in response to student needs and interests is limited. Instruction is quite likely to enhance most students' understanding of the discipline and to develop their capacity to successfully do science.

- **Level 5: Exemplary Instruction**
  Instruction is purposeful and all students are highly engaged most or all of the time in meaningful work (e.g., investigation, teacher presentations, discussions with each other or the teacher, reading). The lesson is well-designed and artfully implemented, with flexibility and responsiveness to students' needs and interests. Instruction is highly likely to enhance most students' understanding of the discipline and to develop their capacity to successfully do science.

REFERENCES

Ausubel, D. (1967a). *Learning theory and classroom practice* (pp. 1–31). Toronto, Ontario, Canada: Institute for Studies in Education.

Ausubel, D. (1967b). A cognitive structure theory of school learning. In L. Seigel (Ed.), *Instruction: Some contemporary viewpoints* (pp. 207–257). San Francisco: Chandler.

Boyd, S .E., Banilower, E. R., Pasley, J .D., & Weiss, I. R. (2003). *Progress and pitfalls: A cross-site look at local systemic change through teacher enhancement*. Chapel Hill, NC: Horizon Research.

Bransford, J. D., Brown, A. L., & Cocking, R. R. (2003). *How people learn: Brain, mind, experience, and school.* Washington, DC: National Academy Press.

Burstein, L., McDonnell, L., Van Winkle, J., Ormseth, T., Mirocha, J., & Guiton, G. (1995). *Validating national curriculum indicators*. Santa Monica, CA: Rand.

Cohen, D. K., McLaughlin, M. W., & Talbert, J. E. (1993). *Teaching for understanding: Challenges for policy and practice.* San Francisco: Jossey-Bass.

Gallagher, J. J. (2000). Teaching for understanding and application of science knowledge. *School Science and Mathematics, 100*(6).

Gardner, H., & Boix-Mansilla, V. (1994). Teaching for understanding in the disciplines—and beyond. *Teachers College Record, 96*(2).

Horizon Research. (2003). *Special tabulations of the 2000–01 lsc teacher questionnaire and classroom observation data.* Chapel Hill, NC: Horizon Research.

Kesidou, S., & Roseman, J. E. (2002). How well do middle school science programs measure up? Findings from Project 2061's curriculum review. *Journal of Research in Science Teaching, 39*(6), 522–549.

Mayer, D. P. (1999). Measuring instructional practice: Can policymakers trust survey data? *Educational Evaluation and Policy Analysis, 21*(1), 29–45.

National Research Council. (1996). *National science education standards.* Washington, DC: National Academy Press.

O'Sullivan, C. Y., Lauko, M. A., Grigg, W. S., Qian, J., & Zhang, J. (2003). *The nation's report card: Science 2000.* NCES 2003-453. Washington, DC: U.S. Department of Education. Institute of Education Sciences. National Center for Education Statistics.

Perkins, D., & Blythe, T. (1994). Putting understanding up front. (Teaching for understanding). *Educational Leadership, 51*(5), 4.

Porter, A. C., Kirst, M. W., Osthoff, E. J., Smithson, J. S., & Schneider, S. A. (1993). *Reform up close: An analysis of high school mathematics and science classrooms* (Final Report to the National Science Foundation on Grant No. SPA-8953446 to the Consortium for Policy Research in Education). Madison, WI: University of Wisconsin-Madison, Wisconsin Center for Education Research.

Spillane, J. P., & Zeuli, J. S. (1999). Reform and teaching: Exploring patterns of practice in the context of national and state mathematics reforms. *Educational Evaluation and Policy Analysis, 21*(1), 1–27.

Stake, R. E., & Easley, J. (1978a). *Case studies in science education, Vol. I: The case reports.* Champaign University of Illinois at Urbana.

Stake, R. E., & Easley, J. (1978b). *Case Studies in Science Education, Vol. II: Design, overview and general findings.* Champaign: University of Illinois at Urbana.

Tyler, R. (2002). Teaching for understanding in science: Constructivist/conceptual change teaching approaches. *Australian Science Teachers' Journal, 48*(4).

Weiss, I. R., Banilower, E. R., McMahon, K .C., & Smith, P. S. (2001). *Report of the 2000 national survey of science and mathematics education.* Chapel Hill, NC: Horizon Research.

Weiss, I. R., Pasley, J. D., Smith, P S., Banilower, E. R., & Heck, D. J. (2003). *Looking inside the classroom: A Study of K–12 mathematics and science education in the United States.* Chapel Hill, NC: Horizon Research.

Wong, D., Pugh, K., & the Dewey Ideas Group at Michigan State University. (2001). Learning science: A Deweyan perspective. *Journal of Research in Science Teaching, 38*(3), 317–336.

CHAPTER 5

TEACHER PERCEPTIONS OF SCIENCE STANDARDS IN K-12 CLASSROOMS

An Alabama Case Study

Dennis Sunal and Emmett Wright

Teachers are aware of, and reflect on, the purpose of education set forth by leaders of their time. Since the publication of a plan for moving the United States toward national education goals and national standards, 15 years ago, what impacts have the national and state science standards had on teachers? An investigation was conducted using open-ended essay e-mail surveys of teachers in the state of Alabama to determine perceptions and expectations of the national and state science standards using similar-sized samples of elementary, middle school, and high school teachers. The results demonstrated that, while the majority of teachers were aware of the science standards, sizable groups lacked knowledge of purpose, use, or importance of science standards in classroom practice.

The Impact of State and National Standards on K-12 Science Teaching, 123–151

INTRODUCTION

Education standards and accountability are recognized facets of education today for educational stakeholders in the United States. The decisions made and daily events experienced by students, parents, teachers, and administrators are impacted in some way by the standards. Stakeholders, in turn, are held accountable for student achievement of the standards.

As described in Chapter 2, *History of the Science Standards Movement in the United States*, by DeBoer, the science standards for K-12 students are neither new nor unique. Many of the basic ideas found in today's standards have long histories. Examples include; a common curriculum for all high school students suggested by the College Entrance Examination Board in 1990 (Bracey, 2002) and the Committee of Ten (National Education Association, NEA, 1893), inquiry teaching and learning described in the *National Science Education Standards (NSES)* (National Research Council, NRC, 1996) and student learning in the late nineteenth century by Herbert Spencer (Spencer, 1864, p. 124), teaching for conceptual understanding suggested in *Benchmarks for Science Literacy* (American Association for the Advancement of Science, AAAS, 1993) and by the Geography Conference of the Committee of Ten (NEA, 1893). A short time after the Committee of Ten report, the NEA's reports of 1911 and 1918 provided an alternative view of the purpose and organization of education, substituting some very different guiding principles. The 1918 NEA report, *Cardinal Principles of Secondary Education*, identified life goals, appropriate student behaviors, health, and vocational training as the central goals of education rather than mental development. This identification was a forerunner of proposals in recent times. As an example, The *Cardinal Principles* can be compared with the focus on life goals in *AMERICA 2000: An Education Strategy* (U.S. Department of Education, 1991) on character education, school choice, lifelong learning, and the cultivation of communities of lifelong learning. The pendulum swings back and forth. But, as it swings, each time there is a different color, twist, or depth of meaning. So, perhaps a spiral is a better description than is a pendulum of the slow forward/upward motion of evolutionary educational practice. The swings have oscillated between equity and excellence over the past 40 years. The present focus is on standards of excellence and accountability. Currently there are no mandated national science standards in the United States. However, individual states have now mandated science standards for schools that are similar to, or closely model published national science standards, *NSES* and *Benchmarks*. In most cases the state standards are a direct result of the No Child Left Behind Act of 2001 (U.S. Department of Education, 2001).

Many teachers at any period are aware of, reflect on, and adopt the purposes and goals of education set forth by educational and political leaders of the time. However, change is incremental and slow (Cuban, 1993). Typically, during any major educational reform movement, 10% to 25% of the teaching force will resist change. Another 10% to 25% of the teachers embrace change, develop new professional knowledge and skill for their classrooms, and create faithful reform. However, the majority of teachers attempt change but ultimately do not incorporate meaningful and long lasting reforms in their classrooms. At best they create poor and usually ineffective versions of the reform, referred to as reform hybrids (Cuban, 1993).

Teachers find change difficult because the organization and expectations of the educational institution and its social responsibilities inhibits risk taking, creates ambiguity, and discourages the professional inquiry required for change to occur (Cohen, 1988; Cuban, 1993). Hypotheses have been suggested for the lack of change in elementary and secondary instruction over the past century. These hypotheses include: (a) the culture at large creates strong forces inhibiting change; (b) ongoing staff development, follow-up, and monitoring are inadequate; (c) the organizational context and structure of the institution is resistant to change, shaping instructors' practice; (d) the perceived realities of the classroom influence a teacher to institute ineffective incremental changes rather than the major ones needed; and (e) instructors' beliefs and expectations about teaching and learning limit change (Cuban, 1993).

BACKGROUND

National Reform

During the current reform movement beginning in the 1990s, most states in the United States have adopted education standards closely reflecting the content and purpose of current national standards, *NSES* (NRC, 1996) and *Benchmarks* (AAAS, 1993) (Blank, Langesen, Sardina, Pechman, & Goldstein, 1997). These standards serve as goals for change in districts, schools and classrooms. As stated in the *NSES* (NRC, 1996) the purpose of science standards is to "present a vision of a scientifically literate populace" (p. 2). "The intent of the Standards can be expressed in a singe phrase: Science standards for all students" (p. 2) and to "provide criteria that people at the local, state, and national levels can use to judge whether particular actions will serve the vision of a scientifically literate society" (p. 3). The "*Standards* will require major changes in much of this country's science education" (p. 2) in both the way it is learned and taught and in the organization and content of the curriculum.

Alabama State Reform

In 1995 the state of Alabama adopted the mission of science literacy for all students (Alabama State Department of Education, ALSDE, 1995). The 1995 *Alabama Course of Study: Science* (*ACOS: Science*) reflected the goals and purpose of the national science reform movement of the 1990s and the early drafts of the NRC's *NSES* sent out for discussion and comment. Greater emphasis was placed on providing higher-level content standards. The Alabama science standards became a mandated set of standards for teaching science in K-12 classrooms.

> In the state of Alabama "The *Alabama Course of Study: Science* (Bulletin 2001, No. 20) provides the framework for the K-12 science education in Alabama's public schools. Content standards in the documents are minimum and required (Alabama Code #16-35-4). They are fundamental and specific but not exhaustive. When developing a local curriculum, each school system may include additional Content Standards to address specific needs or focus on local resources. Implementation guidelines, resources, and/or activities may be added. (ALSDE, 2001. p. iv)

ALSDE stressed that Alabama's high school graduates must be able to compete on both national and international levels during the first half of the twenty-first century (ALSDE, 2001). This recognition, reflected in both the 1995 and 2001 state documents, was directly influenced by both national (NRC, 1996) and international (Trends in International Mathematics and Science Study, TIMSS, 1999) reform documents. State science education reform incorporated "recommendations for a more rigorous, integrated, hands-on, minds-on instructional approach to teaching fundamental science concepts" (ALSDE, 2001. p. ii).

During the time of the investigation reported in this chapter, 2004, Alabama K-12 education experienced wide, but not complete change, adopting only some of the basic elements of national science reform. ALSDE reported that the majority of science content contained in the state standards reflected the national reform documents.

> The 2000-2001 Science State Course of Study Committee made extensive use of the following documents in developing the minimum required content: *Alabama Course of Study: Science* (Bulletin 1995, No. 4): *National Science Education Standards* published by the National Research Council (NRC); Project 2061's *Science for All Americans*; and *Benchmarks for Science Literacy*

> published by the American Association for the Advancement of Science (AAAS); and *Pathways to the Science Standards* published by the National Science Teachers Association (NSTA). (ALSDE, 2001. p. iv)

Analysis of examples of content and teaching standards in the *ALCOS: Science*, however, over this time period illustrate that the changes made were hybrids, and not close replications of the process and content of the reform movement taking place on the national level.

At the fourth-grade level the national science content standards developed by the NRC (1996) listed the fundamental or key concepts and principles for all students that underlie life science to include; "the characteristics of organisms," "life cycle of organisms,' and "organisms and their environment." Under the first fundamental idea "the characteristics of organisms", three principles are given,

- Organisms have basic needs.
- Each plant or animal has different structures that serve different functions in growth, survival, and reproduction
- The behavior of individual organisms is influenced by internal cues (such as hunger) and external cues (such as change in the environment). Humans and other organisms have senses that help them detect internal and external cues (p. 129).

These statements of principle or explanation were to be developed by teacher in an inquiry strategy to provide experiences where students create a meaningful hierarchy of facts and concepts to develop the final principle. "From the earliest grades, students should experience science in a form that engages them in the active construction of ideas and explanations" (p. 121).

The *ACOS: Science* also covers similar science topics. The evolution over several editions of these topics at the fourth-grade level is illustrated in Table 5.1. Statements of standards that were influenced by national reforms as well as state lobby groups are evident over the 17 years of changes, 1988 to 2004, made in the *ACOS: Science*. The 1988 version of *ACOS: Science* (ALSDE, 1988) described key standards in fourth-grade life science in general terms. The standard statements were not described as principles or concepts. As an example, the key idea of "characteristics of living things" was defined by the statements "common characteristics" and "categories" (p. 64). By the time of the 1995 edition of the *ACOS: Science* (ALSDE, 1995), this same fourth-grade content standard was grouped under the idea of "diversity." This idea was defined by two student actions that produce facts "Classify living things using various characteristics" and "Examine fossil evidence for change in organisms over

Table 5.1. Evolution of the Alabama Course of Study: Science 1988-2004: Sample Science Content Objectives for Fourth-Grade Life Science, Diversity, and Adaptations Standard

1988 Alabama Course of Study: Science

I. Life Science

A. Characteristics of living things

1. Common characteristics
2. Categories

B. Groups of Organisms

1. Characteristics of organisms
2. Helpful and harmful organisms
3. Physical characteristics of organisms
4. Helpful and harmful microorganisms

(Alabama State Department of Education, 1988, p. 64)

1995 Alabama Course of Study: Science

Diversity

41. Classify living things using various characteristics.
42. Examine fossil evidence for change in organisms over time.
    Examples: dinosaurs became extinct, some plant species are extinct or have changed, horseshoe crabs have remained relatively unchanged.

(Alabama State Department of Education, 1995, p. 52)

2001 Alabama Course of Study: Science

Diversity and Adaptations

24. Classify animals into groups according to specific characteristics.
    - Vertebrates
    - Invertebrates
    - Warm-blooded
    - Cold-blooded
    - Body coverings
    - Locomotion

(Alabama State Department of Education, 2001, p. 35)

2004 Alabama Course of Study: Science

Diversity and Adaptation

8. Classify animals into groups according to specific characteristics, including vertebrates or invertebrates, endotherms or ectotherms, and methods of locomotion.
    - Describing the organization of organisms, including cells, tissues, organs, organ systems, organisms, population, community, and ecosystem.
    - Classifying organisms into kingdoms, including Animalia, Plantae, Monera, Protista, and Fungi.

(Alabama State Department of Education, 2004, p. 24)

time" (p. 52). However, there was little or no emphasis on key principles or understanding of the concepts and the relationships that demonstrate how the principle works.

The 2001 edition of *ACOS: Science* (ALSDE, 2001) was revised again and the fourth-grade life science standard was given the title "Diversity and Adaptations." This idea was defined by student actions and concept terminology (names), "Classify animals into groups according to specific characteristics. Vertebrates, Invertebrates, Warm-blooded" (p. 35). There was now an emphasis on facts and concept definitions, but again little or no effort to develop simple key principles or explanations.

This same trend continued and was extended to more abstract concept terminology in the 2004 edition of the *ACOS: Science* (ALSDE, 2004). Using the same title "Diversity and Adaptations" the science topic was defined by "Classify animals into groups according to specific characteristics, including vertebrates or invertebrates, endotherms or ectotherms, and methods of locomotion. [Subtopic] Classifying organisms into kingdoms, including Animalia, Plantae, Monera, Protista, and Fungi" (p. 35). This standard focused on abstract concepts leaving the teacher to develop the facts and examples necessary for distinguishing between the concepts. Again,this standard does not result in a key principle or explanation appropriate for a fourth-grade student.

Over 17 years and four editions of *ACOS: Science*, the description of the science learning process standards for students has evolved in a similar fashion. The various editions of the *ACOS: Science* described the science learning process standard as follows.

- 1988, "Hands on activities selected to promote process skill development" (ALSDE, 1988, p. 8).
- 1995, "Learning that requires that students reflect upon and make sense of their activities. An effective teaching cycle (the Five E Learning Cycle of guided inquiry was suggested]) includes evaluation of student abilities to use, extend, and/or apply what is learned" (ALSDE, 1988, p. 8).
- 2001, "Effective teaching ensures active students engagement, interaction with the environment, and reflection upon the learning activity. It is the teacher's role to guide, focus, challenge, and encourage student learning at all levels of inquiry" (ALSDE, 1988, p. 7). Again, the Five E Learning Cycle of guided inquiry was suggested for use.
- 2004, "The goal of scientific literacy is best achieved through an inquiry-based K12 science program that incorporates scientific knowledge and skills with opportunities to apply both in practical

> ways" (ALSDE, 2004, p. 1). Other than the use of the word "inquiry" in this sentence, there is no mention of a learning/teaching model to be used for inquiry in the *ACOS: Science*. However, several paragraphs discuss the meaning of "theory" as related to evolution for teachers of science (p. iii).

Standards for the student science learning process described in the 1988 edition of *ACOS: Science* changed from the 1960's and 1970's concept of hands-on science to greater emphasis on a guided inquiry focused learning content through process and application skills in the 1995 and 2001 editions. This suggested a movement to hands-on and minds-on science. The latest 2004 edition of *ACOS: Science* has stepped back from this emphasis as described in the document and has increased emphasis on the learning of content.

The examples of standards provided for the fourth-grade in the *ACOS: Science* illustrate the evolutionary changes of the state science standards for all grades, K-12. Over this time period the changes made were hybrids of the intent of the national reform movement. The *ACOS: Science* did not replicate in a faithful manner, the national reforms in content or process as represented by *NSES*.

After more than 15 years since the publication of *AMERICA 2000*, which described a plan for moving the United States toward national education goals and national standards, what impact have the science national and state standards had on our teachers? The national and Alabama state efforts in developing and disseminating science standards have produced hard copy documents, Websites, news media publicity, and professional training. Larry Cuban (1993) states that change occurs more slowly than expected, and teachers find change is difficult or not possible in education. Is this view applicable to science education reform in Alabama in recent years? How do teachers view science standards? What expectations do teachers have for the science standards? Does teachers' understanding of the purpose expressed in the standards match their classroom use of the science standards in practice? What are the perceived problems in implementing the science standards? What are teachers' attitudes about classroom use of the science standards and what perceptions do they express to peers? Do these perceptions and expectations follow the intent of the science standards and are they compatible with that intent?

PROCEDURE

To better understand the impact of science standards on Alabama teacher practice, a survey was conducted in 2004. The survey was designed to assess the awareness of and attitudes toward, and personal views of teachers regarding the extent to which Alabama science standards play a part in the teaching of science. The survey was e-mailed to a random sample of practicing classroom teachers; stratified by elementary, middle, and high school in order to obtain a matched description of responses by grade level. The e-mail survey was sent to 174 teachers in 16 northern Alabama school systems, out of a state total of 67 county and 57 city school systems. After reminders, the return rate was 81%, or 141 teachers. Among those responding were 86% of the elementary teachers, 80% of the middle school teachers, and 78% of the high school teachers. The return sample surveys closely matched the population with a small difference bias to the lower grades. The returned sample was judged appropriate to represent the population surveyed.

The survey consisted of three open response essay questionnaires. All potential respondents were instructed that they would receive three questionnaires, one every 2 weeks relating to the science standards. Each consecutive questionnaire was e-mailed on receipt of responses to the previous survey. The questionnaires involved three levels of reflection. The issues covered in the Teacher Reflection on Science Standards questionnaires (TRSS) were (1) TRSS 1 (Level 1): Understanding and Use of Science Standards in School, (2) TRSS 2 (Level 2): Feelings Toward Science Standards, and (3) TRSS 3 (Level 3): Personal View of Teaching Science. TRSS 1 used four questions to ask teachers to reflect on their use and understanding of science standards as a classroom teacher (Table 5.2). TRSS 2 used four questions to ask teachers to reflect on their attitudes toward using science standards as a classroom teacher (Table 5.5). TRSS 3 used three questions to ask teachers to reflect on their view of teaching science in their own classrooms (Table 5.8). Content validity and clarity were determined using a focus group of science education experts consisting of two science educators and six science teachers enrolled in a science education doctoral program at the University of Alabama. The questionnaires were field tested with 17 K-12 teachers of science. Category themes and frequencies were determined for each questionnaire item and are reported in the results section.

RESULTS

Teacher's Understanding and Use of Science Standards (TRSS 1)

All teachers of science in Alabama are guided by, and required to, address the minimal standards in planning and teaching science of the *ACOS: Science* (ALSDE, 2001). Because of the *ACOS: Science* stated overlap with the *NSES* (NRC, 1996) and *Benchmarks* (AAAS, 1993) there was a

Table 5.2. Teacher Reflection on TRSS 1: Use and Understanding of Science Standards

TRSS 1: What is your understanding and use of science standards in school—state and national?

This survey relates to your use and understanding of "science standards." Refer to your classroom planning and teaching in the responses below. Return this as an email as soon as you finish it.

1. Name and identify the various versions of state and national "science" standards?
2. What is the purpose of the science standards for teachers?
3. How do you use the science standards in your teaching? Tell a story here and include which science standards you are using and a sequence of how they are used by you in planning and teaching.
4. What professional training have you had specifically in the nature or use of the science standards?

Table 5.3. Identify the Science Standards for Your Classroom—TRSS 1 Item 1

Science Standard Cited	*Teachers Responding, N & %*
ALCOS and NSES	72/51%
ALCOS	100/71%
NSES	83/59%
Benchmarks	42/30%
Unknown	35/25%
AHSGE	25/18%
County Guide	25/18%
NSTA Pathways to the Science Standards	14/10%
Science Textbook	11/08%

built-in expectation that teachers also would be addressing these standards. Teachers' incorporation of the standards were found to be varied as evidenced by their responses to items in TRSS 1 related to their understanding and use of the science standards (Table 5.2). For TRSS item one, 71% (100 teachers) identified the *ACOS: Science* and 59% (83 teachers) identified the *NSES* as a science standards document that should guide instruction (Table 5.3). About half of the teachers (51%, $N = 72$) identified both sources as a guide for instruction. A representative response for this group was,

> The science standards are contained in the *National Science Education Standards* created in 1996 by the National Academy of Sciences and the *Bench-*

> *marks for Scientific Literacy*, Project 2016 established by the American Association of Advancement of Science. Alabama also has within its state course of study, a science section outlining not only science content but also a section on scientific process skills. (fifth-grade teacher)

Other responses cited as science standards were *Benchmarks* (AAAS, 1993) and *NSES* (NRC, 1996) with 30% of teachers citing both. However, 10% of teachers citing *NSTA Pathways to the Science Standards* (Texley & Wild, 1996) but did not also cite *NSES*.

Two additional groups of teacher responses should be noted. One significant group of teachers (25%) indicated they did not know about the science standards. Statements frequently made by these teachers included, "Science standards, I am not really sure, Life Science, Physical Science, not really sure" (sixth-grade teacher) and "I know there are science standards just like there are math and reading standards. However, I do not use them. I usually just go through the textbook trying to cover all that I can" (fifth-grade teacher).

Another group, 18%, cited the standardized examinations taken by students during grades K-8, the Stanford Achievement Test (SAT) (Harcourt Assessments, 2005), and in high school, the Alabama high school graduation exam (AHSGE) (ALSDE, 2003). See Table 5.4, which lists AHSGE objectives. Teachers sometimes referred to these objectives as *test standards*. Both the textbooks and standardized exams used in Alabama are poorly coordinated with the state or national science standards. For example, the AHSGE covered science as "70% biological science and 30% other science areas" (ALSDE, 2003, p. 13). However, before students take the AHSGE in 10th grade, the *ACOS: Science* equally represents the sciences in the K-8 grade levels with physical science taught in 9th and biology in 10th grades. In response to the question, TRSS 1 - Item 2, teachers stated that science content covered in grades leading up to the 10th grade becomes predominately biological science. The state and national standards as well as the textbooks used in all Alabama schools in grades K-8 are general science and in ninth grade, physical science. This effect substantially reduces learning in nonbiological science areas required by the state and national science standards for those grade levels as demonstrated by this teacher's comment,

> I am trying to recall actual science standards at the state and national levels, however, since the AHSGE covers mostly biology, my 9th grade classes in physical science are mostly biology topics and I spend little time on physics and chemistry. (ninth-grade teacher)

In TRSS 1, Item 2, teachers described their understanding of the purpose of science standards. All respondents stated that the purpose is to

Table 5.4. ALHSGE Student Performance Record Demonstrating Objectives

ALABAMA HIGH SCHOOL GRADUATION EXAM
High School Scores
SCIENCE – Date ____________

Objective #	*Objectives*	*# of Items*	*Biology Item*	*Physical Sci. Item*	*9-2003 Sys. Scores*	*Ranked Worst To Best*
I-1	Analyze the methods of science	7	X	X		
II-1	Trace matter and energy transfer	8	X	X		
II-2	Relate particle motion to matter states	4		X		
II-3	Apply the periodic table	4		X		
II-4	Identify physical and chemical changes	4		X		
III-1	Distinguish among taxonomic groups	4	X			
III-2	Differentiate characteristics of plants	8	X			
III-3	Differentiate characteristics of animals	10	X			
IV-1	Recognize genetic characteristics	4	X			
IV-2	Define the function of DNA	6	X			
V-1	Distinguish cell structures and functions	14	X			
V-2	Differentiate between mitosis and meiosis	4	X			
VI-1	Define the components of an ecosystem	7	X			
VII-1	Relate energy conservation to transformation	4	X	X		
VII-2	Relate waves to energy transfer	4		X		
VIII-1	Apply Newton's three laws of motion	4		X		
VIII-2	Relate force to pressure in fluids	4	X	X		
	TOTAL	100				

BLANK FORM TO BE USED TO ANALYZE SCHOOL OR STUDENT SCORES BY SUBJECT AREA provide "guidelines" or a "framework" for teachers in planning and teaching science lessons as indicated by these representative statements,

> I believe that the science standards set for teachers are necessary and have a certain purpose for the teacher. I believe the main purpose is to serve as guideline to explain what each teacher is expected to teach at a certain grade level. By doing this, teachers do not overlap and reteach what has already been taught. It gives the teacher a way to build upon the knowledge the students have gained in previous years (fourth-grade teacher). The standards provide teachers with a framework of what to teach. Standards should be used as a guide when planning lessons. (third-grade teacher)

These guidelines were seen as mandatory by 50% of these teachers, who usually stated that the standards were "To ensure that all students are receiving the same materials in the science curriculum" (12th-grade science teacher) and "they insure consistency in content" (seventh-grade science teacher) or "explain what each teacher is expected to teach" (first-grade teacher). Of these teachers, 19% included a statement similar to the response, "Standards relate to what students will be tested on in standardized tests" (third-grade teacher).

In addition, the science standards were understood by 70% of the teachers (N = 119) as guidelines for science content, as a list of content objectives and not a guide for other aspects of a teacher's role related to pedagogy and assessment, or as a plan for systemic change in schools. Typical statements included, "The purpose of the science standards in my teaching is an outline or guide of what is to be taught in science" (fourth-grade teacher) and "The purpose of the science standards for teachers is to let teachers know what educators and the public expect students to be learning at each level of science study" (10th-grade science teacher). The remaining teachers (30%) stated that the standards provided a framework for science curriculum, pedagogy, literacy, equity, assessment, and systemic change. For example, one teacher stated that standards "provide guidance for assessment, professional development, planning, and implementation of science curricula" (third-grade teacher).

In TRSS 1, Item 3, teachers described through a narrative story their understanding of how to use standards to plan for and teach their students. Overall, 62% (87 teachers) of the responses included classroom stories demonstrating the application of science standards in their daily professional lives. The viewed purpose of the science standards most often related closely to a teacher's personal practice in teaching science. For 37% (32 teachers) of the response stories, the standards become a listing of mandatory topics to be covered in science classes. A 10th grade science teacher's story reported, "In my teaching, I use science standards to guide my planning of units of instruction. I include both the state course of study objectives and also the AHSGE standards and objectives in each weekly lesson plan." Another 10th-grade science teacher's story recalled,

> The first thing that comes to mind is teaching the cell's structure and the functions of the respective cell parts. The Graduation Exam outlines that cell structure and function are the most heavily tested facets of science. Seventy percent of the test is biology, and the other thirty covers various components of physical science. As a teacher, pressure is placed on us to deliver these objectives in such a way that students will have the best opportunity in passing these standardized tests.

18% (15 teachers) of the story responses described the science standards as a philosophy and guide to develop science literacy in students. "To me, the standards provide an overall view of how my classroom should look and what should be going on in the heart of it. With the science standards in mind, I use these materials to plan for science in my classroom" (sixth-grade teacher). In the science textbook this teacher finds that,

> Science content is more difficult to bring to the surface and requires more thought as to how to do this. Within these materials there are frequent opportunities to assess and several activities that require an application of learning, all those goals outlined in the national standards. The part of the science that needs the most orchestrating is the concept building that is continually churning in the student's minds.... In truth, the science standards beautifully guide us to the most authentic scientific setting for our students; unfortunately, there are many concepts to teach, many other accountability issues to meet, and not enough time to truly devote to the endeavor of such science. It is my own story that also falls into conflict with these problems and one that I can only continue to strive to work toward that utopian science classroom, one young scientist at a time. (fifth-grade teacher)

Fewer teachers, 20% or 17 teachers, reported using materials and standards based curricula, modules, or kits. These teachers considered themselves fortunate to use science materials coordinated in content, pedagogy, and assessment with the national science standards. An elementary teacher expressed her ability to access NRC (NRC, 2005) materials developed by the Smithsonian Museum as being fortunate, compared to most other teachers.

> We are lucky enough to have modules that fulfill most of the science standards. The national standards that are not covered through the module are easily supplemented into the lessons. When beginning a module we look at the objectives for each week that we are going to cover and match them with the standards we are to cover. Any standards that are not taught in the module's objectives we implement into either an already developed lesson or

> make a lesson on our own, usually using the *AIMS* (*Activities Integrating Mathematics and Science* [*AIMS*], 2004) resources. (fourth-grade teacher)

Still other teachers, 25% or 23 teachers, rarely use or do not use state or national standards to inform their planning or teaching of science. For some in this group change has been occurring.

> In my first years of teaching, I just went through the textbook from lesson to lesson hoping I was covering everything; and sometimes, I still do. I really didn't understand the importance and necessity of using standards and using the standards to write objectives. I credit my ... (graduate education) classes for better preparing me to be a highly qualified teacher. (fifth-grade teacher)

Finally, to a small but significant percentage of teachers (17%) who overlap with other groups, expressed that science teaching is a difficult task and meeting the standards receives little encouragement or support from the school system. One teacher described her experiences as follows:

> I have taught kindergarten for 12 years. I have never been provided with any type of science materials and resources. When we adopted a new science program two years ago I thought we would, for the first time, receive some science materials, but we haven't. The other kindergarten teacher and I have always had to rely on materials that we've purchased ourselves. This year we are stressed over DIBELS Assessment and have even less time to spend on teaching science. Much of the science that I teach is integrated into reading in the form of units. For example, we are studying about the letter 'B,' and in science we are studying bears and their habitats, which fall under life science. Next week we'll study weather in earth and space science. For physical science, we will make ice cream. I realize that I fall short in teaching process and applications. (kindergarten teacher)

TRSS 1, Item 4 asked for information about the extent of professional training in the nature or use of the national and/or state science standards. Half of the teachers reported that they had not had organized professional training on the state or national science standards. Such training would include philosophy, content, pedagogy, or assessment. To the extent teachers developed awareness, it was done on their own. A number of teachers reported experiences similar to that of one fourth-grade teacher, "The only training I have had on the science standards is mainly what I have figured out on my own. I don't recall ever attending a workshop or seminar dedicated solely to science." Some teachers (30%) stated that they received training in the nature and use of the science standards in graduate and continuing education college courses. Just over 25% reported that their experience resulted from meetings in school where peer teachers were charged to coordinate content covered with the *ACOS: Science*, textbooks, and/or standardized tests to be given, for example the SAT and AHSGE. One fifth-grade teacher reported "Grade-level meeting discussions on implementation of the timeline for covering the state course of study in preparation for the SATS are one way we approach preparation for teaching the science standards." One third of this last group, (08% of the total, 11 teachers) reported that they attended a workshop on science standards, most often more than 5 years ago.

Teachers Attitudes Toward Science Standards

Teachers' attitudes toward the science standards were described in questionnaire TRSS 2, Feelings Toward Science Standards (Table 5.5). In TRSS 2, Item 1 32% of the teachers described a positive attitude toward the usefulness of the science standards. The standards were "essential," provided "equitable science education," and "standards give teachers a foundation for providing students with more inquiry-based experiences in science." These statements were illustrated with stories from their classroom teaching.

> The state standards for science provide cooperative guidelines to implement activities that are grade-level appropriate.... This provides the teacher with criteria to formulate lesson plans. I feel the state standards are a necessary link and complements inquiry-based instruction promoting scientific literacy in students. The majority of the teachers at our school simply use the textbook to guide their instruction. So I feel the standards are very useful if they are used. (fifth-grade teacher)

Table 5.5. TRSS 2: Feelings Toward Science Standards

Survey Topic 2: What are your feelings toward science standards for schools—state and national?

This activity relates to your use and understanding of "science standards." Refer to your classroom planning and teaching in the responses below. Return this as an email as soon as you finish it.

1. Describe your feelings in another story about the usefulness of the science standards. Identify which ones you are referring to.
2. What were the most important reasons why you use or were not able to use the science standards in your teaching?
3. What were the significant barriers in using the science standards?
4. What type of advice would you give another teacher in using the science standards in teaching? Tell a story about what you would say to this teacher.

About an equal number of teachers, 30%, expressed neutral or qualified feelings toward the science standards. Although a reason was seen for the standards, it was different from those provided in the standards. One teacher stated,

> I use the Alabama Course of Study every day with my lesson planning. I started this semester to put the actual ACOS number on each lesson plan. I feel that will help me if I am ever questioned as to why I am teaching a particular topic. We are mandated to teach evolution to our 8th grade students and in my community, this is a great source of contention. I find it to be of great importance to be able to quote national standards and ACOS standards in dealing with that situation. (eighth-grade science teacher)

A larger group, 38% of the teachers, described a negative attitude toward the use of the science standards in teaching. Again, there were a variety of reasons given. Many in this group responded that they perceived themselves as overloaded.

> In dealing with the science standards set by both the state and national guidelines, I often wonder if they consider what we actually deal with in the classroom. For example, I must remediate each of my students who have not passed the AHSGE in Science. Well, if I go by Course of Study standards alone, then I miss the standards set by the Graduation exam. If I go by national standards, I also miss the specifics as required by the AHSGE. There needs to be consistency between them all. (11th-grade science teacher)

> There are too many generalized areas that should be addressed specifically in order for students to be scientifically literate. Likewise, there are some specifics from each set of standards that don't contribute much to scientific processes and behaviors. I think that standards are necessary, but they should also be realistic. (10th-grade science teacher)

> I have often harbored negative feelings about standards-based educational reform, especially when those standards have been heavily tied to accountability issues and standardized testing. In those cases, I feel the testing was driving the establishment of standards instead of the other way around. (second-grade teacher)

TRSS 2, Item 2 asked teachers to describe important reasons why they use or were unable to use the science standards in their teaching (Table 5.6). Science standards were described as useful as a guide to planning and teaching, by 18% of the teachers. About three-fourths of these respondents were elementary teachers.

> I also have found that they help me to actually be a science teacher. They take me away from the textbook and help me create experiences for my stu-

Table 5.6. Important Reasons Why Teachers Use or Were Unable to Use the Science Standards in their Teaching—TRSS 2 Item 2

Reason for Using Science Standards	*Teachers Responding, N & %*
Do use—they are useful	25/18%
Do use—they are required	42/30%
Do not use or use less than expected	73/52%

dents that are real. They provide opportunities for students to actually do science in a way that one might do science as a career. People that have jobs in the field of the sciences do not just read and listen; they investigate, experiment, and search for new ideas and understandings. (fifth-grade teacher)

A second group of respondents, 30%, reported using the science standards as criteria for teaching science. The standards were useful because they described what was required of them in teaching science. The standards make teaching more predictable and decisions easier. "I always find which standards a unit addresses before I begin to teach it. Furthermore, if I can't find any place where the material is called for, I don't teach it" (eighth-grade science teacher). "One of the most important reasons why I use the standards in my teaching is because it is required. It is a guide to what has to be taught in science and I rely on these standards to help me teach science concepts" (10th-grade science teacher). The great majority of these teachers, about 75%, were secondary teachers whose students would be taking the AHSGE.

Many teachers (52%) reported that there were important reasons for not using the standards, using them part of the time, or using them less than they perceived was expected of them. The reasons were described in the next question, TRSS, Item 3.

TRSS 2, Item 3 asked teachers to describe significant barriers in using the science standards in their teaching. Descriptions often were strongly worded and detailed. Reasons given can be summarized by five descriptors; pressure from other subjects to be taught, lack of training, lack of time and pressure of testing, lack of resources, and miscellaneous (Table 5.7). The teachers were represented equally across the grade levels for each reason except for pressure from other subjects to be taught, which was expressed exclusively by elementary teachers.

Pressure from other subjects was due primarily to pressure to teach reading. This was understood to mean that teachers would not teach science, teach little science, or teach it with less emphasis than required in the Alabama state course of study. One fifth-grade teacher stated,

Table 5.7. Significant Barriers in Using the Science Standards—TRSS 2 Item 3

Barriers to Using Science Standards	*Teachers Responding, N & %*
Pressure from reading, math	16/11% or 32% of K-6 teachers
Lack of Professional Development	45/32%
Lack of time	38/27%
Lack of resources	31/22%
Miscellaneous	11/08%

> We receive pressure to constantly produce greater achievements in students' reading scores. In addition to that, I teach fifth grade which means I receive twice the pressure as the other grades because the results from the Alabama Writing Assessment are published and used as an indicator of which schools are placed on academic alerts. It is made quite clear to us that math is the next area of focus. While we are cautioned to follow the state guidelines for all areas, it is also made clear where our greatest efforts are to be placed.

Others reported,

> I am acquainted with the national standards, but do not fully implement them in my classroom. Because our school's focus is on reading, writing, and math, there is very inadequate time to teach or prepare for science or social studies. (third-grade teacher)

> The ACOS is all I have used up until now. It seems that we have to focus most of our time on reading and math, and there is hardly any time left for anything else. Since I do not have much time, I have got to make the time to make it more important. (second-grade teacher)

Lack of opportunities for inservice professional development was an important barrier for 32% of the teachers. Lack of training decreased awareness, created confusion, and resulted in a lack of emphasis on meeting the science standards. Two examples of barriers reported on the survey were,

> Teachers do not receive enough professional development in using standards. For them to be of real value to teachers, we need sufficient training. At my school, very few, if any, teachers know that there are national standards in all content areas. (eighth-grade teacher)

> In working with another teacher I would explain that it is important to understand the standards as a way of teaching science and not as a curriculum. (fifth-grade teacher)

> For me, the significant barriers in using the science standards were simply not being knowledgeable of them. I had no prior knowledge of actually implementing standards into my lessons. I am sure that somewhere along my BS degree, I was exposed to the term "standards", but the concept didn't stick with me. (fifth-grade teacher)

Another reported group of barriers, reported by 27% of teachers, to using the science standards was lack of time and pressure of testing. The overfilled curriculum, time taken by standardized test preparation, and lack of instruction time that tends to get filled with school administrative concerns and other distractions all reduce the time on needed to address the science standards. Two examples include,

> There are too many generalized areas that should be addressed specifically in order for students to be scientifically literate. Likewise, there are some specifics from each set of standards that don't contribute much to scientific processes and behaviors. I think that standards are necessary, but they should also be realistic. There is not enough time to teach every standard. (10th-grade science teacher)

> I teach a unit on the Properties of Matter. These objectives are specifically included in the AHSGE Science Standards, but are conveniently left out of the ninth-grade Biology standards. Well, if students take the test after they have biology, those objectives have not been met. Those are physical science standards which the students get after they take the test the first time. (ninth-grade science teacher)

The reality of daily life in the classroom causing difficulty in addressing the standards was reflected in the lack of resources, by 22% of the teachers' responses. Examples include,

> Occasionally, I cannot meet certain objectives due to lack of resources. Lab equipment cost money, lab supplies cost money, a class of 34 cannot use a computer lab with only 15 computers. Therefore, most objectives that are not utilized thoroughly tend to be the ones where scheduling, space, or resources are limited. (seventh-grade science teacher)

> I still do not have the materials I need/want to teach science the way I want to, but I will after teaching several more years. (second-grade teacher)

The remaining important barrier cited was a set of responses by 8% of the teachers on a variety of ideas. These responses included textbooks required in the course and extreme range of students in classes as barriers to addressing the science standards

When asked to give advice explained in a narrative story to other teachers in using the science standards, TRSS Item 4, teachers' responses fell into three categories. The largest group, 62%, cautioned other teachers to take accountability seriously. One 11th-grade science teacher provided the following advice:

> Advice for other teachers using the standards would be to make sure you meet the AHSGE standards and objectives first. Also, follow the course of study when planning the lesson. You want to be sure that if some child says you didn't teach something and that was why he did NOT pass the AHSGE, then you have your proof in your lesson plans. Always keep a copy for yourself and document everything. Also, if a student does not EARN his grade, it would not be wise to pass them on because they probably won't pass the graduation exam and that could become a huge problem.

The second largest group, 23% of respondents, encouraged teachers to reflect on them as goals that foster effective science learning; connecting teachers to appropriate decisions about curriculum, pedagogy, and assessment. Advice from a fourth-grade teacher included,

> I would strongly attempt to show him/her that the standards provide true science learning for the students. I would discuss the issue of the amount of planning it takes to have the activities ready and available for the students. I would show the teacher how involved it gets the students and how much like an actual career choice science investigations are for the students.

An eighth grade science teacher ended her advice with the encouraging statement, "It is very much worth the journey and a journey that we must all continue taking." The third group, the remaining 15% of the teachers reported that they could not provide any useful advice due to their lack of knowledge and use of the science standards. A fourth-grade teacher said,

> I do not feel that I could advise another teacher on using the standards because I have had inadequate training. Even though I have picked and chosen parts and pieces I felt comfortable using, I have not incorporated the standards as a complete document.

Teachers Views of Teaching Science

The remaining questions were included in items on questionnaire TRSS 3 and related to Personal View of Teaching Science (Table 5.8). Teachers were asked to write a story of a science lesson that recently occurred in their own classroom that reflected how they viewed science in their classroom teaching. One hundred and thirty-one teachers responded with a story of a science lesson. Their final submitted stories were grouped into categories.

Table 5.8. TRSS 3: Personal View of Science Teaching

Survey Topic 3: What is your personal view of teaching science?

This activity relates to your view of teaching science. Refer to your own classroom planning and teaching in the responses below. Return this as an email as soon as you finish it.

1. Identify the specific grade level (and course for secondary teachers) and the science lesson topic.
2. Tell a story (anecdote) of a lesson that occurs in your classroom. The story should be told in a first person narrative starting with a description of the setting. Use a real classroom situation. You may choose to enhance a real situation with rich detail. However, the story should represent how you view science in your classroom teaching. Your story may include some or all of the following components and you may arrange them in any order.
   - Description of the teacher
   - Teacher's background and/or experiences
   - Description of classroom, school, or community
   - Description of students
   - Teacher's feelings and intentions
   - Students' feeling and intentions
   - Dialogue
   - Description of other relevant parties (e.g., parents, principals, other teachers)
   - Outcome(s)
3. Lessons or morals that can be drawn from the story that would be advice would you give another teacher about teaching science using the science standards?

The first category was stories that demonstrated the national and state science standards. These were called Inquiry Lessons (IL). Thirty-six percent of the lessons were grouped into this category. They demonstrated interactivity between students, use of hands on materials, a reflective approach, concepts developmentally appropriate, and an investigative strategy in student learning of new concepts.

> My classroom is grouped cooperatively for the purpose of peer tutoring and promotion of learning.... I initiate interest in the upcoming lesson to be taught on micro-worlds by asking some students to volunteer to bring in jars of pond water.... I am still intrigued by the curiosity of the students as the days pass by while the infusion process of the pond water and straw is taking place.... Once the infusion is at a ripe state and after the students have been introduced to the appropriate use of the microscope, I have set the stage for the finale. When the students view the pond water and straw infusion and observe the microorganisms found in the water it is interesting to hear their remarks. One child remarked, "You mean when I go swimming in my pond, I'm swimming in that?" (fifth-grade teacher)

A second category, Transmission Lessons (TL), represented by 54% of the lessons, demonstrated an authoritarian approach by the teachers that relied on presentation, step-by-step instructions, and an incidental approach to science as opposed to conceptual or thematic unit approach.

> The activity dealt with using marshmallows and toothpicks to represent how molecules like Water, Carbon Dioxide, and other basic molecules are made from different elements joined together. I knew this lesson would take some extensive planning.... My feelings were uplifted at the thought of it being a hands-on activity with active participation from the students.... The lesson was a huge success with both my students and me. They enjoyed it and I feel learned much with this process. It allowed them to not only learn the molecular formulas, but to take it a step further and actually have a representation that proved a visual on how different molecules join together to make different elements and compounds. (fifth-grade teacher)

> My lesson begins by assessing prior knowledge. I ask the students such questions as "What seasons do we experience in our area?" and "How is the weather different in each season?" The next part of the lesson will include an investigation using a flashlight to represent the sun, and shining the flashlight onto paper straight down. I will have a student circle the beam on paper. We will observe the brightness of the light. Next, the student will tilt the book and paper. Keeping the flashlight in the same position, we will draw another circle to represent the beam at this angle. We will compare this information and come to the conclusion that the light looked brighter inside the circle from the straight rays. After the investigation, I will use a graphic organizer to write down vocabulary words and meanings. Upon completion of these activities, we will now read and discuss the information provided in the text. I give my students study guides that provide them with the information that will be tested and at the end of the chapter they will be tested. (third-grade teacher)

The third category, 8% of the total, represented narratives that could not be grouped into the first two and were titled Non-Lesson (NL). These narratives described their ideas about science teaching or just a list of topics taught.

The final question, TRSS 3, Item 3, asked each teacher to describe meaning that can be drawn from the story in Item 2 that would serve as advice for other teachers instructing standards-based science. Analyzing only the questionnaires of teachers narrating a story of a classroom lesson, found in TRSS Item 2, 38% provided meaning that supported the goals of the national and state science standards. A second-grade teacher described the stories' meaning as,

> Hand or spoon-feeding our kids does not help them in the long run. It truly hinders them in their development of their minds. They will not know how to think for themselves, therefore not do so. Use hands-on materials as

much as you can. Let the students make up their own minds about questions and answers.

Sixty-two percent of the teacher responses provided a meaning that did not support the standards. A kindergarten teacher stated "My students and I enjoy hands on activities in science but most of the time I don't have the materials and run out of time."

Relationships between responses were analyzed comparing type of science lesson story, the teachers' view of classroom science provided in TRSS 3, Item 2, and responses on other questions. Chi-square analysis was used to determine the significance of the relationship. The first comparison examined the lesson story and the type of professional development training experienced in the science standards, TRSS 1 Item 4. The total number of teachers responding appropriately to both items was 131. Professional training in the science standards was found to be related to describing a story representing a science IL, chi-square statistic = 62.40 with $p < .001$. The group of teachers with little or no training in science standards provided views of science lessons that were more often classified as TL.

A second comparison was made between the type of lesson story, TRSS 3, Item 2, and the type of meaning that can be drawn from it that would be advice given to another teacher about teaching science using the science standards, TRSS 3, Item 3. Advice that was negative or different from the science standards was related to lesson stories told by teachers that were TL, chi-square statistic = 65.87 with $p < .001$. Teachers providing advice to others that was supportive of the science standards were more likely to describe IL stories about their own classrooms.

The third comparison was between the science lesson stories and most important reasons why the teachers use of the science standards in their teaching, TRSS 2, Item 2. Teachers reported use of the science standards as a guide or criteria for planning and teaching were found to be related to related to describing science lesson stories as IL, chi-square statistic = 37.67 with $p < .001$. Teachers who stated that they did not use science standards were found to be more likely to describe teaching as a TL story.

Other analyses for which nonsignificant differences were found were completed comparing science lesson stories with (1) the stated purpose of the science standards for teachers, TRSS 1 Item 2, chi-square statistic = 1.01 with $p = 0.30$; 2) significant barriers in using the science standards TRSS 2 Item 3, chi-square statistic = 0.46 with $p = .0.50$; and 3) grade level of science lesson topic, TRSS 3 Item 1, chi-square statistic = 2.71 with $p = 0.10$.

CONCLUSIONS AND IMPLICATIONS

The current science standards movement was acknowledged in the literature more than 15 years ago. Publication of the national science standards began more than a decade ago with state governments following up with developing their versions of the science standards. Where do we stand in reaching the goals of these policy documents for science learning and teaching in today's classrooms? Based on data from the schools and teachers in northern Alabama, the goals have not been met. Many teachers are confused about the science standards, what they are, their goals and purposes, and how they should be applied in their science planning, teaching and assessment activities. Although about half of the teachers surveyed were aware of appropriate sources for the science standards, with 51% identifying state and national documents, 25% did not know of the science standards or their purposes, and an additional 18% identified a standardized test as being the source.

How do teachers view the science standards? Even though half of the teachers identified standards documents, 70% of all of the teachers viewed the science standards as relating only to required content objectives and not as a guide for the teacher's role related to pedagogy and assessment or as a plan for systemic change in schools.

Teachers' understanding of the purpose expressed in the science standards matched, to a great extent, their classroom use of the standards in practice. The lack of adequate professional development for a majority of the teachers provides a possible explanation for the often-minimal implementation of the standards.

What are the perceived problems in implementing the science standards? The results indicate a neutral or negative attitude was expressed toward the science standards by almost 70% of these teachers. Although elementary teachers reported using the science standards more often than secondary teachers, the standards failed to be implemented by the majority of the teachers. Five major reasons were given for failure to implement the science standards: (1) pressure from other subject areas, (2) lack of professional development, (3) pressures of time for instruction and for testing, (4) lack of resources, and (5) the constraints of daily life in the classroom such as lack of enough computers. Given the opportunity to give advice to other teachers about using the science standards, the majority cautioned other teachers to take accountability seriously and to use the standards as a content guide to lesson planning. The perceived problems of implementing the science standards appear to be unrelated to the original goals and purposes of those standards related to *science literacy and equity.*

Do these perceptions and expectations follow the intent of the science standards and are they compatible with that intent? In providing narrative stories of classroom lessons that recently occurred in their classrooms, the majority of these teachers described lessons that did not fit the intent of the science standards. Of the lessons described, 54% were TL as compared to IL. When asked to describe the meaning that can be drawn from the narrative lesson story that would serve as advice given to another teacher about using science standards, 62% of the teachers provided meaning that did not support the standards. For most of these teachers, their perceptions and use of the science standards do not support the intent of science standards.

Statistical relationships were found with narratives of recent science lessons and responses to several of the questions asked on the surveys. A significant relationship was found between professional development in the standards and the type of science lesson that recently occurred in their classrooms. The effects of professional development were evident: those teachers having training in the science standards described IL demonstrating the purpose and intent of the standards while those lacking training described TL. Another relationship found was that those giving advice to other teachers supportive of the standards were more likely to describe an IL. A third relationship found was that teachers reporting appropriate use of the science standards as a guide or criteria for planning and teaching described science lesson occurring in their classrooms as IL. Classroom practice appears to be related to these teachers' understanding of the nature, goals, and purpose of the science standards.

Educational change and science reform occurs today more slowly than expected for some teachers. Many other teachers find change is very difficult or not possible in science teaching. George DeBoer stated in Chapter 2, *History of the Science Standards Movement in the United States*, that

> Schools, as organizations, aren't designed as places where people are expected to engage in sustained improvement of their practice, where they are supported in their improvement, or where they are expected to subject their practice to the scrutiny of peers or the discipline of evaluations based on student achievement.

There is much evidence to support parts of this statement. Is this view applicable to science education reform in Alabama in recent years? Although some teachers are required to do inservice professional development or participate in action research programs in the school on science standards, many teachers in Alabama do not do so. Outsiders make the majority of instruction and curriculum decisions for the classroom. This leads to the development of hybrid copies of reforms that do not agree with the intended reform philosophy as embedded in the science standards. The impression is given that teachers are at fault when standards are not implemented. However, Larry Cuban (1993) suggests that there are several competing hypotheses. One relates to teachers' knowledge of subject matter, and their professional and personal beliefs. These were explored in the study presented here. However, Cuban also suggests, alternatively, that the potential cause for little or no change in schools is due to several factors that are outside of the control of teachers. They are

1. Societal cultural beliefs about the nature of knowledge
2. School organization and practices
3. Ineffective execution by policymakers
4. Organizational structure of district, school, and classroom
5. Traditional culture of teaching

The intent of the science standards in the *ACOS: Science* and the perceived goal and problems of using those science standards evinced by Alabama teachers of science are closely related to the concept of *excellence* and have little to do with *equity*. The focus of standards in *ACOS: Science* and of the majority of sampled teachers is on students' passing of mandated achievement testing which is seen as necessary for American international competitiveness. The source of this excellence focus is clearly portrayed by the 2001 No Child Left Behind Act (U.S. Department of Education, 2001), the business community, and other forces outside of the control of teachers and administrators. The science for all, equity focus of the national science standards does not match current trends in the political and economic areas in American society. A sixth item should be added to the list of factors suggested by Cuban related to limited change in teaching over time in American classrooms: *external forces that counteract the goals of ongoing reforms in education*. These forces can stop well-intentioned and successful reform from ever taking hold. We are left with traditional teaching and stagnant or declining achievement in meaningful science learning outcomes as defined by the national science standards and evidenced in international comparative standardized tests such as TIMSS. Lawrence A. Cremin (1989) expressed similar conclusions when he wrote,

> American economic competitiveness with Japan and other nations is to a considerable degree a function of monetary, trade, and industrial policy, and of decisions made by the President and Congress, the Federal Reserve Board, and the Federal Departments of the Treasury, Commerce, and Labor. Therefore, to conclude that problems of international competitiveness can be solved by educational reform, especially educational reform defined solely as school reform, is not merely utopian and millennialist, it is

at best a foolish and at worst a crass effort to direct attention away from those truly responsible for doing something about competitiveness and to lay the burden instead on the schools. It is a device that has been used repeatedly in the history of American education. (Cremin, 1990, pp. 102-103)

SUMMARY

An investigation was conducted using open-ended essay e-mail surveys of teachers in the state of Alabama to determine perceptions and expectations of the national and state science standards using similar-sized samples of elementary, middle school, and high school teachers. The results demonstrated that, while the majority of teachers were aware of the science standards, sizable groups lacked knowledge of purpose, use, or importance of science standards in classroom practice. This study indicates that science education reform efforts as represented by national and state science standards, do not in themselves heavily impact classroom practice unless teachers have access to professional development that will scaffold their understanding and use of those standards. Teachers are aware of, and reflect on, the purpose of education set forth by leaders of their time but implementation can be problematic.

REFERENCES

Activities Integrating Mathematics and Science. (2004). *AIMS activities and publications.* Retrieved January 20, 2005, from http://www.aimsedu.org/

Alabama State Department of Education. (1988). *Alabama course of study: Science, Bulletin 1988, No. 35*. Montgomery, AL: Author

Alabama State Department of Education. (1995). *Alabama course of study: Science, Bulletin 1995, No. 4*. Montgomery, AL: Author

.Alabama State Department of Education. (2001). *Alabama course of study: Science, Bulletin 2001, No. 20*. Montgomery, AL: Author.

Alabama State Department of Education. (2003). *Great expectations: A guide to Alabama's high school graduation exam.* Montgomery, AL: Author. Retrieved January 20, 2005, from http://www.alsde.edu/general/great_expectations.pdf

Alabama State Department of Education. (2004). *Alabama course of study: Science, Bulletin 2004, draft of the Alabama course of study.* Montgomery, AL: Author.

American Association for the Advancement of Science. (1993). *Benchmarks for scientific literacy, Project 2061*. New York: Oxford University Press.

Blank, R. K., Langesen, D., Sardina, S., Pechman, E., & Goldstein, D. (1997). *Mathematics and science content standards and curriculum frameworks: States' progress on development and implementation.* Washington, DC: Council of Chief State School Officers.

Bracey, G. W. (2002). The war against America's public schools: Privatizing schools, commercializing education. Boston: Allyn & Bacon.

Cohen, D. (1988). *Teaching practice: plus a change*. (Issue paper No. 88-3). East Lansing, MI: Michigan State University, The National Center for Research on Teacher Education.

Cremin, L. A. (1990). *Popular education and its discontents*. New York: Harper & Row.

Cuban, L. (1993). *How teachers taught*: *Research on teaching monograph series*. New York: Teachers College Press

Harcourt Assessments. (2005). *Stanford achievement test series*. San Antonio, TX: Harcourt Assessment. Retrieved January 20, 2005, from http://www.stanford10.com

National Education Association. (1893). *Report of the committee on secondary school studies*. Washington, DC: U.S. Government Printing Office.

National Research Council. (1996). *National science education standards*. Washington, DC: National Academy Press.

Spencer, H. (1864). *Education: Intellectual, moral, and physical*. New York: Appleton.

Texley, J., & Wild, A. (1996). *NSTA pathways to the science standards*. Arlington, VA: National Science Teachers Association.

Trends in International Mathematics and Science Study. (1999). Highlights from TIMSS 1999 results. Retrieved January 20, 2005, form http://nces.ed.gov/timss/

U.S. Department of Education. (1991). *America 2000: An education strategy*. Washington, DC: Author.

U.S. Department of Education (2001). *No child left behind*. Retrieved August 8, 2003, from http://www.ed.gov/offices/OESE/esea/nclb/titlepage.html

CHAPTER 6

REALIZING THE NATIONAL SCIENCE EDUCATION STANDARDS

Channels of Influence Using a State Level Perspective—A Kansas Case Study

Stephen Marlette and M. Janice Goldston

In February, 2001, the state of Kansas adopted science standards consistent with the *National Science Education Standards* (*NSES*) (1996). A self-reported questionnaire was mailed to elementary principals and teachers in the spring of 2002 to gain insight into the implementation of these standards. Items on the questionnaire were framed around Fullan's (2001) four characteristics of the change process that affect implementation. These characteristics include need, clarity, complexity, and the quality. The frequency and percent of the principal and teacher responses were determined for the questionnaire using these characteristics. In addition, differences in principals and teachers perspectives were explored using the chi-square statistic. Findings were reported under the three channels of influence—teacher development, curriculum, and assessment and accountability (Weiss, Knapp, Hollweg, & Burrill, 2002). The results lead to six conclusions: (1) As a whole, perspectives of teachers and principals are not substantially differ-

The Impact of State and National Standards on K-12 Science Teaching, 153–184

ent; (2) Principals and teachers believe that science instruction should be aligned with the standards; (3) Principals and teachers do not have a clear understanding of what it means to be standards based; (4) More resources need to be leveraged to support science instruction; (5) Science education is not a priority at the elementary level; (6) While the intent of the *Kansas Science Education Standards* (*KSES*) was to provide educators the criteria "to judge whether current actions serve the vision of a scientifically literate society" (Kansas State Board of Education, KSBE, 2001, p. 3), policies and documents other than these are influencing classroom decisions.

INTRODUCTION

> I came to the conclusion that it is not new materials or national reports that make a difference for most science teachers. Something much deeper and more important is influencing the direction of science teaching. Resistance to change must be related to other factors. One could be that science teachers do not know or do not share the larger purposes of science education stated in so many reports. It could also be that the needs of science teachers are largely met by current practices, or that the school system does not really support any sustained effort to improve science teaching. (Roger Bybee, 1997, p. 44)

The *NSES* "provide criteria to judge progress toward a national vision of learning and teaching science" (NRC, 1996, p. 12). In order for a national vision to be realized, it must first be operationalized and embraced at the local, district, and statewide level. Thus, the course for reform will not be direct because each state has autonomy. This chapter focuses on statewide survey data conducted in Kansas, a state like many others that has used the *NSES* as a guide for the development of state science standards (KSBE, 2001; Leonard, Penick, & Douglas, 2002; Raizen, 1998). In examining the results of the survey the researcher utilized the framework of change theory (Fullan, 2001), to explore characteristics of the *KSES* within the context of the channels of influence (NRC, 2002) to gain insights into *NSES* implementation.

Charged with the task of bringing "greater clarity and specificity to what teachers should teach and students should learn" (KSBE, 2001, p. 3), the science standards committee recommended and the KSBE adopted science education standards in February of 2001. Both the *NSES* and the *KSES* consist of eight domains (KSBE, 2001). In combination, these domains have captured a balance in emphasis of several traditional science goals (Bybee & DeBoer, 1994; Trowbridge, Bybee, & Powell, 2000). These goals include: (1) acquiring scientific knowledge, (2) learning the process or methodologies of science, and (3) understanding the applications of science, especially the relationship between science and

Table 6.1. The Congruency of Traditional Science Instruction Goals with National and Kansas Science Education Standards

Goal of Science Instruction	*Domains in the National Science Education Standards*	*Domains in the Kansas Science Education Standards*
Development of scientific knowledge	Physics and chemistry, life science, Earth and space science, science and technology, history and nature of science and unifying concepts and processes	Physics and chemistry, life science, Earth and space science, science and technology, history and nature of science and unifying concepts and processes
Development of processes or methodologies	Science as inquiry, science and technology	Science as inquiry, science and technology
Understand the applications of science	Science in personal and social perspective	Science in personal and environmental perspective

society and science-technology-society (Bybee & DeBoer, 1994; DeBoer, 1991; Trowbridge, Bybee, & Powell, 2000). The congruency of the domains in the *KSES*, the *NSES*, and some traditional goals of science instruction are outlined in Table 6.1.

While the *KSES* are not dogmatic in that they require a specific local curriculum (KSBE, 2001, p. 3), they do provide clear direction for science instruction within the state. Like the *NSES*, the intent is for Kansas educators to use the standards "to judge whether current actions serve the vision of a scientifically literate society" (KSBE, 2001, p. 3). Given this stance, the Kansas standards state, "Science is much more than a body of information; it is a process of discovery" (KSBE, 2001, p. 91) and both the *KSES* and *NSES* point out that "conducting hands-on science activities does not guarantee inquiry" (KSBE, 2001, p. 93; NRC, 1996, p. 23). This means more than simply providing hands-on science activities. Thus, both sets of standards align and place inquiry as a content standard that is an integral part of all other content standards and central to science learning.

Initially, it appears the *NSES* have had a strong influence within the state and for all practical purposes have become the state standards in Kansas. While this is an important step in realizing standards-based science reform, the *NSES* clearly articulate the necessity for the different components of the science education system to share this common vision (NRC, 1996). This is the more difficult task. As Bybee concluded in a historical review of science education reform, "The past 50 years have witnessed only limited success in improving science education. Science educators have failed to transform purpose into practice, and they have also consistently underestimated the power of school systems and science teachers to maintain status quo" (1997, p. 24). Thus, a more critical examination through the lens of change theory is needed to understand how the principles in the *NSES* are impacting the schools and science practices.

Fullan (2001), a leader in the field of school reform and change theory, identifies three phases within the change process that can be identified in most reform endeavors. These include Phase I-initiation, Phase II-implementation, and Phase III-continuation. While standards-based reform may have been attempted earlier at various localities in Kansas, the adoption of the *KSES* formalized the science reform initiative into statewide policy representing Phase I. Phase II or implementation refers to the first few years of use as educators work to put the ideas of the standards into practice. Phase II provides the context for this study. Phase III –continuation relates to whether the vision of science education presented in the *KSES* becomes institutionalized throughout the state on multiple levels. The ideas in the standards can become integrated into the fabric of the science education system or like many other reform attempts, fall by the way side.

Change efforts that concentrate on the "on-paper" changes, like the adoption of the *KSES*, and ignore how the ideas are being translated into practice, in short, fail (Fullan, 2001). Critical in the change process is considering what is happening at the school level. This study critically examines Phase II of the change process—the implementation of the *KSES*. Using principals' and teachers' perspectives, it provides a view of the implementation phase approximately 1 year after their adoption of the *KSES* in the hope of identifying obstacles and/or barriers that may keep the standards from becoming institutionalized. Each of the three channels of influence explored in the study include: (a) teacher development, (b) curriculum, and (c) assessment and accountability (Weiss et al., 2002). Data are interpreted from a change theory framework that is defined by the characteristics of the *KSES* and that affects implementation (Fullan, 2001). These characteristics include need, clarity, complexity, and the quality.

LITERATURE REVIEW

> I don't think I can have science yet, I'm not a good reader ... I think you get science in Junior High. My brother is in Junior High and he has science. (Julia Thomas & Carol Stuessy, 1998)

The *NSES* "urge teachers to recognize that science is a discipline that must be taught to every student in every elementary school and that it must be taught using an inquiry/thinking approach" (Lowery, 1997, p. 6). Like the *NSES*, the *KSES* challenge the way science has traditionally been taught. For many, the main goal of science teaching has been completing the text or covering the content in the syllabus (Gallagher, 1996). The underlying assumption is that students need to memorize content. DeBoer (1991) identifies this goal as the most dominant focus of classroom instruction for the past 100 years. Despite other reform efforts that emphasized the processes of science and the social applications of science, none have achieved successful long-term implementation (DeBoer, 1991).

According to the *NSES*, the expectation is that science instruction should change what it emphasizes. Teaching in science can no longer be seen as simply telling students what they need to know. The expectations are for students to achieve higher levels of knowledge and understandings (NRC, 1996). This suggests that in light of the nation's reform agenda of increasing teacher effectiveness and student achievement, teachers will have to learn new skills and perspectives while simultaneously unlearning practices and beliefs that have dominated educational practice for years. Rhoton (2001, Observing the Standards-based Science Classroom section, para. 2) states that, "Perhaps there is no greater challenge facing science education reformers than helping teachers move from current practices to strategies [consistent with the *NSES*]." An examination of some of the issues involved in teaching science at the elementary level will help clarify the challenges facing reform and provide a baseline in which to discuss the data in this study.

Too often at the elementary level, science is a low priority subject. While most middle level and high school teachers specialize in one or two content areas, elementary teachers have responsibilities in several content areas. Besides science, the K-6 curriculum includes such areas as mathematics, reading, writing, history, music, geography, and health. The many demands on the time available in the school day create competition regarding what gets taught. Frequently, this leaves science as an end of the day add-on activity or eliminates it from the curriculum altogether. This is more prevalent now with the No Child Left Behind Act influence than at the time of this study. On a national survey comparing time devoted to reading/language arts, mathematics, science, and social studies, Fulp (2002, p. 11) found "grade K–5 self-contained classes spent an average of 25 minutes each day in science instruction, compared to 114 minutes on reading/language arts, 53 minutes in mathematics, and 23 minutes in social studies." Lowery (1997) recommends an average of 30 minutes per day be spent on science instruction grades 1-3 (150 minutes per week) and 45 minutes per day grade 4-6 (225 minutes per week). As the data indicate, at the elementary level reading and mathematics have the highest priority. Fulp's (2000) findings align with a current national survey conducted by, Pasley, Smith, Banilower, and Heck (2003) who in general found that K-12 science teaching had changed little in the past 20 years.

The *NSES* point out that teachers need to "work within a collegial, organizational, and policy context that is supportive of good science teaching" (NRC, 1996, p. 27). Pfannenstiel, Seltzer, Yarnell, and Lanbson, (2000, Qualitive Aspects of Professional Development section, bullet 3) reported that in Kansas both "teachers and principals agree that their schools provide a safe haven for teachers to try out new strategies in their classrooms." The study reported over 90% of teachers and principals agreed that "teachers can practice new skills or strategies in a low-risk environment" (Pfannensteil et al., 2000, Qualitive Aspects of Professional Development section, bullet 3). Likewise, the study found that "90% of teachers and principals agree that school administrators support teachers in applying what they have learned from professional development activities to classroom" (Pfannensteil et al., 2000, Qualitative Aspects of Professional Development section, bullet 3). While this is positive in regards to developing general teaching skills, data suggests that the low priority of science in the curriculum directly affects the opportunities elementary teachers have to develop their skills teaching science.

Fulp (2002) reports that nationally, three quarters of the K-5 science teachers indicated they had 15 or fewer hours of science-related professional development in the preceding 3 years. In the Kansas study, Pfannenstiel et al. (2000, Policy Implications of Professional Development Findings section, bullet 7) made policy recommendations that included expanding "opportunities for elementary teachers to participate in professional development on science content areas." The study found that fewer than half of the elementary teachers in the study participated in professional development on science instructional strategies. The focus of professional development at the elementary level was literacy and mathematics. In other words, opportunities to improve teaching are available, but not if you want to focus on science. Further compounding this situation is the weak science content background of most elementary teachers.

Research studies suggest that elementary teachers do not have adequate content knowledge to teach science (NRC, 2001). Fulp (2002, p. 5) reported nationally "fewer than one-third of elementary teachers reported feeling very well qualified to teach each of the science disciplines." In this same study, 58% "of elementary teachers have not taken a college/university science course since 1990, and 49 percent have not taken either a science course or a course on how to teach science since

1990" (p. 10). It is no wonder nearly "two-thirds of the elementary school science teachers, 70 percent of those in grades K-2 and 58 percent of those in grades 3-5" (p. 6), reported not being familiar with the *NSES* document. "Of the third of elementary school teachers of science who reported they were at least somewhat familiar with the NRC *Standards*, roughly 70 percent said they agreed with them" (p. 6). Fulp (2002, p. 10) concluded with "These data indicate a serious need for retooling a large percentage of the elementary school science teaching force."

Gess-Newsome (2001) briefly summarized research outlining teaching strategies that support the goals of the *NSES*. These include lessons that are inquiry-oriented, problem solving based, and those that provide an accurate portrayal of the knowledge, structure and nature of the discipline. A summary of the self reported data (Pfannenstiel et al., 2000) provide insights into the nature of instruction students receive in Kansas. Elementary teachers did not describe inquiry-based strategies as a daily feature of their classroom practices. About one-fourth of the elementary teachers use project-based learning on at least a weekly basis and about 40% use inquiry/problem-based learning on at least a weekly basis. Approximately two-thirds of both elementary teachers almost never use the research/scientific process in their classrooms. At the conclusion of the report Pfanennstiel et al. (2000, Policy Implications of Teaching Practices Finding section, bullet 1) made recommendations to "Offer professional development opportunities on inquiry-based strategies."

These data have outlined some of the issues elementary science reform in Kansas must address in moving from the initial adoption of the *KSES* to the implementation and eventually continuation phases of the change process. In examining this process, it is important to consider the outcomes that will be utilized to determine the progress of the change efforts. One outcome many will be concerned with is the impact the changes have had on student learning. To this end, the Kansas assessment program administered the science assessments in grades four, seven, and 10. The 2000-2001 administration of the test served as the baseline year and measured both knowledge and process using multiple-choice questions to provide a total score (Kansas State Department of Education, n.d.) While reading and mathematics are measured yearly, science and social studies are assessed every 2 years. Student achievement data, however, is not the focus of this study.

Fullan (2001) explains that another outcome to consider is whether "experiences with change increase subsequent capacity to deal with future changes" (p. 50). In other words, to what extent have changes made a difference in the system in which teaching and learning occur? Certainly seeing a rise in science achievement scores would be a major accomplishment, but to narrowly focus on student achievement numbers misses the larger goals of standards reform. The inclusion of program and system standards in the *NSES* make it clear that in meeting the standards, change will have to go beyond improving test scores (NRC, 1996). Ideally, changes like that those advocated in the *KSES* will be both an individual and a systemic transformation (NRC, 2000).

The educational system can be viewed as a hierarchy of subsystems in which the school is the central institution (NRC, 1996). The school is in turn a component of the local district, which in turn is a component of the state educational system. As the science education reform ideas expressed in the *KSES* enter this system, they will have to reach the various layers of each subsystem. They will do this through various channels of influence. The three channels considered in this study and the ones most relevant in determining the impact of the *NSES* include *teacher development*, *curriculum*, and *assessment and accountability* (Weiss et al., 2002). In focusing on only these three and treating them separately there is risk. It over-simplifies the complexity and interactive nature of the change process and the multidimensional properties operating within the educational system. Those wanting immediate results often fail to understand that implementing the *NSES* "is a large and significant process that will extend over many years" (NRC, 1996, p. 9).

Fullan (2001) provides insights into why the change process poses such a challenge by pointing out that sophisticated nature of the change that must take place. This includes: (1) the possible use of new curriculum materials, (2) the possible adoption of new teaching approaches, and (3) the possible alteration of beliefs and understandings. While important, simply adopting new curriculum and learning new teaching strategies will not create the kind of change that is sustained over time. Reform like that advocated in the standards must also influence what is valued and what is believed (Anderson, 1996; Fullan, 2001; NRC, 2000; Tippins, Nichols, & Weseman, 1998). This implies that school practitioners have had time to develop an understanding of the standards, time to practice and develop new competencies, time to locate and/or develop new resources, and time to reflect on the impact of what they have done so they understand the benefits. In determining the extent or how quickly school based practitioners take to implement the science standards, research suggests the characteristics of the standards themselves will play a significant role (Fullan, 2001; Rogers, 1995).

Fullan (2001) identifies four characteristics of the change process that affect implementation. These include: *need*, *clarity*, *complexity*, and *quality*. Need is related to the extent in which those faced with adoption perceive it to be a needed change. In this case, it includes perceptions of how important implementing the *KSES* is compared to other priorities. This is relevant to the implementation of the science standards when dealing with what Fullan (2001) calls an "overloaded improvement agenda" (p. 76).

Clarity relates to the extent to which those adopting the innovation understand its' essential features. Fullan (2001) comments that trying to make changes that are not clear in the minds of the adopters can lead to frustration. He also cautions against what he calls "false clarity" (p. 77). This could occur if those adopting the *KSES* oversimplify the changes needed to make them work. For example, there may be a false impression that the *KSES* have been implemented by simply adopting the standards, when in fact the teachers have failed to embrace the deeper more fundamental changes related to science teaching and learning that need to occur in the classroom.

Obviously not every teacher, principal, or school system will have to undergo the same level of change in order to implement the *KSES*. Some may already have science programs that are operating in a manner consistent with the vision of the standards while others may have far to go. Fullan (2001) explains, "Complexity refers to the difficulty and the extent of change required of individuals responsible for implementation" (p. 78). This underscores the importance of considering the teaching strategies, curriculum, and beliefs of those implementing the science standards. The higher the levels of complexity the more effort will be required to achieve implementation.

The last factor relates to quality and practicality of the change. Fullan (2001) describes the scenario where adopting the innovation is more important than the actual implementation. Poor quality occurs in cases where "decisions were made on the grounds of political necessity" or "on the grounds of perceived need without time for development" (p. 79). The possibility of adopting the science standards out of necessity rather than actually embracing them is very real. If meaningful change is to occur, the resources and time required for teachers to make the needed changes happen must be provided. Given the aforementioned background, the following sections describe a statewide study that examined the status of science reform from the perspectives of elementary teachers and principals in the state of Kansas.

METHODS

In order to gain a clearer picture regarding the implementation of the *KSES*, principals and teachers in Kansas were asked to complete separate versions of a mailed, self-administered questionnaire during the spring of 2002. One section of the questionnaire contained 12 items written to correspond to the factors identified by Fullan (2001) as affecting implemen-

Table 6.2. Part I of the Kansas Science Instruction Survey (Principal Version)

Item #	
1	The science curricular resources, including texts and other supplemental material available for use by teachers, align well with the Kansas Science Education Standards.
2	The instructional practices of teachers in this school should be closely aligned with the recommendations in the Kansas Science Education Standards.
3	I would have difficulty recognizing whether text activities and other supplemental science materials have been modified so they more closely align with the Kansas Science Education Standards.
4	I have a clear understanding of the learning theory upon which the science standards are based.
5	I would have difficulty recognizing if the way teachers in this school teach science is consistent with the recommendations in the Kansas Science Education Standards.
6	Compared to some other content areas in our school, improvement in science does not have as high of a priority.
7	It is important for students in this school to investigate scientific questions even if it means covering less science content.
8	With regard to implementing the Kansas Science Education Standards, teachers have adequate planning time to discuss, implement, and adjust the science curriculum to best meet the needs of students in our school.
9	My understanding of what is important in science education is different from that advocated in the Kansas Science Education Standards document.
10	The environment in this school encourages and supports teachers to develop a deep understanding of the Kansas Science Education Standards.
11	If teachers in this school were to teach in a manner consistent with the Kansas Science Education Standards, their teaching style would have to change.
12	In this school, obtaining high quality professional development that focuses on the implementation of the Kansas Science Education Standards is a problem.

tation (see Table 6.2). These included the presence of a need (Items 2, 6, & 7), the degree of clarity (Items 3, 4, & 5), the level of complexity (Items 1, 9, & 11), and the quality and practicality associated with the process (Items 8, 10, & 12).

Respondents were asked to rate each item in this section using the following scale:

4 = It is *completely true*.
3 = It is *somewhat true*.
2 = It is *somewhat untrue*.

1 = It is *completely untrue*.
D = I do not have the information needed to respond to the statement.

In addition, an open response section of the survey included two items. The purpose of this section was to allow principals and teachers the opportunity to share their own ideas as to what the goal of science instruction should be.

To ensure that valid inferences could be made about the principals' and teachers' perspectives in this study, a panel of expert judges reviewed the questionnaire for clarity and content validity. A focus group consisting of one science educator and seven graduate students in a science education research course was used to confirm the panel's recommendations and provide additional suggestions. Before mailing the instrument, it was field-tested on seven teachers and principals.

The questionnaires were sent to the principals and two teachers in each of 149 different elementary schools across the state. Of the 149 questionnaires mailed to principals, 76 were returned (51%). Of the 296 questionnaires sent to teachers, 101 were returned (34%). To ensure that the principals and teachers adequately represented rural, urban, and metropolitan perspectives, an attempt was made to select equal numbers of elementary schools from each category. Beale code data (Butler & Beale, 1994) were analyzed to determine the extent to which the sample represented the population. Analysis revealed the percentage of principals in the sample for this study matched the population fairly closely. The study sample of teachers more strongly represented those in nonmetropolitan districts that are not adjacent to metropolitan areas. However, for the purposes of this study it was concluded that the sample of teachers was sufficient enough to adequately represent the variety of perspectives held by the teachers who teach science in Kansas.

The frequency and percent of the principal and teacher responses on each of the 12 items was calculated. In addition, differences in principals and teachers perspectives were explored using the chi-square statistic. The null hypothesis was that the principals and teachers would not have different perspectives on each of the 12 items related to Fullan's (2001) categories of need, clarity, complexity, and quality. To minimize the likelihood of making a Type I error, each item test was conducted at alpha level .004 (.05 /12), thus maintaining an experiment wise alpha level of .05 (Huck, 2000). In this manner, it was possible to identify whether principals' and teachers' perspectives were consistent on the 12 items related to the standards implementation.

RESULTS

Even though the official policy adopted to guide science education in Kansas closely reflects the *NSES*, it is critical for the change process to consider what is happening at the school level. Discussion of the questionnaire results are organized around three channels of influence: teacher development, curriculum, and assessment and accountability. Data related to quality, complexity, clarity, and need are used to more clearly understand what is happening with teacher development and the curriculum. The respondent's qualitative comments regarding the goal of science instruction are organized around themes to provide insights into assessment and accountability.

Teacher Development

The setting in which teachers work can either enhance or constrain their opportunities to engage in professional learning. A rich "infrastructure" to support standards-based science teaching would be an indication that national standards are influencing teacher development (Weiss et al., 2002, p. 54). Fullan (2001) concluded "change will always fail until we find some way of developing infrastructures and processes that engage teachers in developing new understandings" (p. 37).

Quality Issues

In this study, the presence of a rich infrastructure was investigated through items related to the quality of the change process (Fullan, 2001). The Standards for Professional Development for Teachers of Science in the *NSES* suggest providing professional growth opportunities for teachers that revitalizes them intellectually, rather than simply developing specific skills. Teachers need opportunities to develop theoretical and practical understandings of the science discipline (NRC, 1996).

Elementary principals and teachers responded to questionnaire items 10 and 12 related to school environment and access to professional development. The frequency and percent of their responses in each category are provided in Tables 6.3 and 6.4. By combining the true categories in Table 6.3, the data indicate that over 80% of the principals and 70% of the teachers believe school environments encourage teachers to develop deep understanding of the *KSES*. Similar to earlier findings (Pfannenstiel et al., 2000) indicating 90% agreement that teachers could practice new skills in a low-risk environment and 90% agreement that teachers have a safe place to try new strategies. It appears that the majority of principals and teachers in this study have a positive view of their school environ-

Table 6.3. Teachers have a School Environment that Encourages Them to Develop a Deep Understanding of the Kansas Science Education Standards

	Untrue *N (%)*	*Somewhat True* *N (%)*	*Completely True* *N (%)*	*Total* *N (%)*
Principal	12 (15.4)	43 (55.1)	23 (29.5)	78 (100)
Teacher	28 (27.7)	47 (46.5)	26 (25.7)	101 (100)

Chi-square (3) = 3.870; p = .144.

Table 6.4. Obtaining High Quality Professional Development Related to the Kansas Science Education Standards is a Problem

	Completely Untrue *N (%)*	*Somewhat Untrue* *N (%)*	*Somewhat True* *N (%)*	*Completely True* *N (%)*	*Total* *N (%)*
Principal	12 (15.8)	24 (31.6	30 (39.5)	10 (13.2)	76 (100)
Teacher	26 (26.8)	23 (23.7	33 (34.0)	15 (15.5)	97 (100)

Chi-square (3) = 3.829; p = .280,

ments. It is important to note that data in Table 6.3 also indicates over one in four teachers (27.7%) do not think this is true. This suggests that while the environment in a general sense may be encouraging, one that focuses on helping teachers to more deeply understand the *KSES* is still a concern by some individuals. This concern is more pronounced by the results found in 6-4 related to professional development.

Analyzing responses from item 12 and combining true categories as shown in Table 6.4, over half of the principals (52.7%) and almost half of the teachers (49.5%) indicated access to professional development was a problem. This data is consistent with national data (Fulp, 2002) and the recommendation for greater opportunities for professional development *in science* made in the Kansas study (Pfannenstiel et al., 2000). It would seem that access to high quality professional development related to the *KSES* standards would be an important aspect of an encouraging work environment and certainly part of an infrastructure of support. Data from Pfannenstiel et al. (2000) suggest literacy and mathematics receive the lion's share of professional development opportunities.

However, simply acknowledging a lack of professional development opportunities related to science instruction at the elementary level does not go far enough. As policy makers and science leaders work in specific localities to build a rich infrastructure of support for standards-based science, they need to take into account teacher needs. In conjunction, acknowledging certain abilities, predispositions, levels of understanding, and beliefs and values held by principals and teachers that can potentially impose obstacles to implementing standards based approaches to science teaching is critical. The following section related to complexity provides further insights into elements of infrastructure that are influential during this change process.

Complexity Issues

Complexity of the change process related to the implementation of the *KSES* was examined by determining how closely principals and teachers perceived that their teaching practices align with the standards and the extent to which they align philosophically with ideas in them. Elementary principals and teachers responded to the following two questionnaire items, 11 and 9. The frequency and percent of their responses in each category are provided in Tables 6.5 and 6.6. The data in Table 6.5 suggest that both principal (48% untrue /52% true) and teacher (54.3% untrue/ 45.7% true) respondents are fairly equally split over whether teachers' teaching style would have to change to be consistent with the standards. Without observational data, it is not possible to confirm the veracity of their opinions. However, the large number indicating a need to change is consistent with the earlier data in the Pfannenstiel et al. (2000) report

Table 6.5. To be Consistent with the Kansas Science Education Standards, the Teaching Style has to Change

	Completely Untrue *N (%)*	*Somewhat Untrue* *N (%)*	*True* *N (%)*	*Total* *N (%)*
Principal	11 (14.7)	25 (33.3)	39 (52.0)	75 (100)
Teacher	20 (21.3)	31 (33.0)	43 (45.7)	94 (100)

Chi-square (2) = 1.332; $p = .514$.

Table 6.6. My Understanding of What is Important in Science Education is Different from that Advocated in the Kansas Science Education Standards Document

Grade Level Position	*Completely Untrue* *N (%)*	*Somewhat Untrue* *N (%)*	*Somewhat True* *N (%)*	*Completely True* *N (%)*	*Don't Know* *N (%)*	*Total* *N (%)*
Principal	22 (28.2)	29 (37.2)	16 (20.5)	4 (5.1)	7 (9.0)	78 (100)
Teacher	17 (16.7)	33 (32.4)	31 (30.4)	6 (5.9)	15 (14.7)	102 (100)

Chi-square (4) = 5.900; $p = .207$.

concerning the lack of inquiry-based strategies, problem-based learning, and use of research/science processes.

In regards to what is important in science education (Table 6.6), the number of principals (25.6%) and teachers (36.3%) in this study have the perspective that what is advocated in the *KSES* differs from what they think is important. While Fulp (2002) only reported data for those indicating that they were familiar with the standards, the results of those indicating they did not agree with the *NSES* (approximately 30%) are similar. Fullan (2001) points out the difficulty in changing beliefs can often "challenge the core values held by individuals regarding the purposes of education" (p. 44). Whether the differences in the data are perceived or real, it suggests a small segment of this study's participants might offer resistance. Therefore, determining what educators value and what aspect of the standards is causing tensions might prove valuable in designing professional development opportunities.

Another issue brought to light in Table 6.6 was the number of principals (9%) and teachers (14.7%) indicating they did not know whether the Kansas standards align with what is emphasized as important in science education. This issue is worth noting and is connected to issues of clarity—the next topic of discussion.

Clarity Issues

In reference to the *NSES* and professional growth, Loucks-Horsley and Bybee (1998) ask the question, "How will we know when we are there?" Both researchers suggest it is critical for the infrastructure supporting standards based implementation to be designed to help principals and teachers answer this question. Thus, the dimension of clarity in this study is related to the degree respondents indicated they understood the standards and whether they were able to recognize standards-based science instruction.

Elementary principals and teachers responded to the following two questionnaire items, 4 and 5, linked to issues of clarity. The frequency and percent of their responses in each category are provided in Tables 6.7 and 6.8. In regards to learning science, Lowery (1997) provided constructivism as a basic principle in the standards. While not all education practitioners value learning theory, Schunk (1997) explains how learning theory provides a framework to use in making decisions about how to teach for meaningful learning. Data in Table 6.7 indicate a number of principals (22.7%) and teachers (29.2%) think they do not have a clear understanding of the learning theory underpinning the standards. In addition, Table 6.8 indicates over one in three principals (36.4%) and almost half (46%) of the teachers are having difficulty recognizing teaching practices consistent with the *KSES*. Clearly identifying the essential features of the *KSES*

Table 6.7. I have a Clear Understanding of the Learning Theory on which the Science Standards are Based

	Untrue *N (%)*	*Somewhat True* *N (%)*	*Completely True* *N (%)*	*Total* *N (%)*
Principal	17 (22.7)	39 (52.0)	19 (25.3)	75 (100)
Teacher	28 (29.2)	54 (56.3)	14 (14.6)	96 (100)

Chi-square (2) = 3.337; p = .189.

Table 6.8. I have Difficulty Recognizing Science-Teaching Practices as Consistent with the Recommendations in the Kansas Science Education Standards

	Completely Untrue *N (%)*	*Somewhat Untrue* *N (%)*	*True* *N (%)*	*Total* *N (%)*
Principal	22 (28.6)	27 (35.1)	28 (36.4)	77 (100)
Teacher	25 (25.0)	29 (29.0)	46 (46.0)	100 (100)

Chi-square (2) = 1.681; p = .431.

is an important aspect in the implementation process, yet the number of principals and teachers indicating a lack of clarity verify that this is an issue that must to be considered during the implementation phase. However, Kansas teachers fared better when these finding are compared against national data (Fulp, 2002) which indicated that two-thirds of the elementary teachers were not familiar with the *NSES*.

Need Issues

KSES were adopted in the spring of 2001. However, adoption of these standards at the state level does not mean that principals and teachers are motivated to embrace them locally. Need associated with the change process was examined by investigating whether principals and teachers indicated whether their current practices should align with the *KSES*.

Responses by elementary principals and teachers connected to need, items 2 and 7, surrounding the implementation of the standards are seen in the Tables 6-9 and 6-10. The *KSES* identifies several teaching practices as important to quality science instruction. These include allowing students time to conduct investigations. However, spending time on investigations means a pivotal tradeoff of less content coverage. Cited by the NRC (2000), Anderson (1996) identified teachers' commitment to coverage of material as an issue that arose in schools trying to initiate new approaches to science and mathematics instruction. His work suggests teachers perceive a need to prepare students for the next level of schooling. It is not difficult to understand that a teacher might not embrace a new practice that requires them to shortchange what they feel is important for students. Thus, item 7 examined to what extent coverage of material was an issue to respondents in this study. Interestingly, the data in Table 6.9 challenges the contemporary stance with almost 90% of principals (89.3%) and teachers (89.8%) acknowledging the need for students to be involved in investigations even if it means covering less content. This suggests educators indicating a disagreement (approximately 30%) in regards to what is important in science education (see Table 6.6) may not be due to a *coverage* mentality, but other issues yet uncovered.

In addition, the data in Table 6.10 suggests that even though there was a level of disagreement regarding what is important in science education, over 90% of the educators in this study sensed a need for teaching practices to be aligned with the KSES. This may suggest that the portion of the educators in this study not agreeing with the standards feel a certain level of pressure to align with them. Fullan (2001) argues that many complex forces work in tandem to maintain the status quo and pressure to change is positive. However, he warns that pressure without support can lead to resistance. This reinforces why careful consideration of contextual elements in designing an infrastructure to support change is paramount.

Table 6.9. It is Important for Students in this School to Investigate Scientific Questions Even if it Means Covering Less Content

	Untrue N (%)	*Somewhat True* N (%)	*Completely True* N (%)	*Total* N (%)
Principal	8 (10.7)	37 (49.3)	30 (40.0)	75 (100)
Teacher	10 (10.2)	52 (53.1)	36 (36.7)	98 (100)

Chi-square (2) = .242; p = .886

Table 6.10. Instructional Practices Should be Closely Aligned to the KSES

	Untrue N (%)	*Somewhat True* N (%)	*Completely True* N (%)	*Total* N (%)
Principal	2 (2.6)	22 (24.4)	54 (69.2)	78 (100)
Teacher	1 (1.0)	33 (33.7)	64 (65.3)	98 (100)

Note: Because two cells (33.3%) had expected counts of less than 5; the chi-square test for this level of analysis was not valid.

Curriculum

By adopting state science standards that are virtually the same as the content in the *NSES* (see Table 6.1), the state of Kansas made a positive step toward seeing the vision of the *NSES* enacted. As stated earlier, the *KSES* were developed to provide districts with a vision of what constitutes scientific literacy at various grade levels. Interpreting what this means for each specific school district is left to those in charge of making local decisions and is contextually based. If the *NSES* are influencing change, one would expect teachers to have access to curricular resources that are consistent with the goals outlined in the standards. In the event that teachers are provided with more traditional curriculum materials, it would be important for them to be able to use the *KSES* to create or select appropriate learning materials. In other words, one would expect teachers to have the resources to study standards in conjunction with their curriculum materials. The following sections address aspects of curriculum.

Complexity Issues

Fulp (2002) reported that nationally teachers in two-thirds of grade K-5 classrooms use one or more commercially published textbooks/programs. Tobin, Tippins, and Gallard (1994) argued that textbooks often determine the curriculum and thus affect the quality of instruction (which may or may not be standards-based). Textbooks, however, are not the only curricular resources utilized at the elementary level. Tolman, Hardy, and Sudweeks (1998) reported an increase in the use of nontext materials such as commercial modules, kits, and canned units in the elementary classroom, but concluded textbooks "have retained their foothold as major resources in the study of science" (p. 44). Because of the importance of this issue, complexity in the change process for standards-based implementation in relation to the alignment of science curricular resources was examined.

Elementary principals and teachers responded to questionnaire item 1 on curricular resource alignment. Frequency and percent of responses in each category are provided in Table 6.11. The data in Table 6.11 indicates that more than 85% of the educators in this study believe their curricular resources align well with the *KSES*. However, in light of the clarity issues related to their inability to recognize teaching practices, 36.4% of principals and 46% of teachers (see Table 6.8), their ability to discern the alignment of curricular resources is suspect. In other words, do they believe there is alignment because the cover of the book says it is and using the book falls comfortably into their ways of teaching? As the data in Table 6.12 confirms, data across survey questions consistently links to clarity issues.

Table 6.11. The Science Curricular Resources Align Well with the Kansas Science Education Standards

	Untrue *N (%)*	*Somewhat True* *N (%)*	*Completely True* *N (%)*	*Total* *N (%)*
Principal	8 (10.5)	49 (64.5)	19 (25.0)	76 (100)
Teacher	13 (13.4)	58 (59.8)	26 (26.8)	97 (100)

Chi-square (2) = .495; p = .781.

Table 6.12. Teachers: I Am Not Sure How to Modify Text Activities and Other Science Materials; Principals: I Would have Difficulty Recognizing Whether Materials have been Modified

	Completely Untrue *N (%)*	*Somewhat Untrue* *N (%)*	*Somewhat True* *N (%)*	*Completely True* *N (%)*	*Total* *N (%)*
Principal	10 (13.5)	29 (39.2)	29 (39.2)	6 (8.1)	74 (100)
Teacher	37 (37.8)	25 (25.5)	30 (30.6)	6 (6.1)	98 (100)

Chi-square (3) = 12.723; p = .005.

Clarity Issues

Trowbridge, Bybee, and Powell (2000), report that since the 1960s the prevailing view of leaders in science education has been for activity-based programs or nontextbook dominated programs. Efforts of teachers trying to implement inquiry-based techniques like those advocated in the standards are hampered if their materials are text-based (NRC, 2000). Using the standards to determine learning goals, teachers can decide which aspects of their materials to use in helping to reach those goals. Therefore, it may be necessary to modify and adapt materials in order to provide the kind of science instruction advocated in the *KSES*.

Elementary principals and teachers responded to questionnaire item 3 on curriculum resource modification. Frequency and percent of responses in each category are provided in Table 6.12. The combined percentages of the true categories in Table 6.12 indicated that almost one half (47.3%) of principals and a large number of teachers (36.7%) felt they lacked clarity regarding the modification of text activities. These numbers are noteworthy and indicate a lack of clarity for implementation and make clear that learning what and how to modify existing curriculum when designing the change infrastructure should be a priority.

Table 6.13. Teachers have Adequate Planning Time for Science

	Completely Untrue N (%)	*Somewhat Untrue N (%)*	*Somewhat True N (%)*	*Completely True N (%)*	*Total N (%)*
Principal	16 (20.5)	36 (46.2)	18 (23.1)	8 (10.3)	78 (100)
Teacher	45 (44.1)	41 (40.2)	12 (11.8)	4 (3.9)	102 (100)

Chi-square (3) = 13.688; p = .003.

Quality Issues

The writers of the standards ask teachers to go beyond simply telling and teaching facts. In order for students to develop personal meaning from the content and processes, and to apply this knowledge to their daily lives, teachers need to create inquiry-based learning environments for their students (NRC, 1996). This type of teaching requires informed and intentional thought, preparation, and design. To do this, teachers need time.

Questionnaire item 8 was targeted at gaining insights into time available for science planning. Frequency and percent of responses in each category are provided in Table 6.13. By combining the untrue categories, data in Table 6.13 indicates over 84.3% of the teachers and 66.7% of the principals did not think the statement was true. These data suggest a lack of planning time related to science is a major issue for teachers. This was the only item in which differences in principals' and teachers' responses were found to be significantly different (p <.004). The fact more elementary teachers (44.1%) indicated this was completely untrue than principals (20.5%) suggests the difference in perspective is not so much a difference concerning whether each think this is a problem, rather it reflects a difference in the degree to which each perceives it as a problem. Furthermore, it may be reflective of each of the survey group's position within the school system itself. In other words, the principals may not understand the magnitude of the problem because they do not deal with planning on a daily basis, as do teachers.

Need Issues

In spite of the large majority of principals and teachers (over 90%) in this study indicating the need to be more in line with the standards (see Table 6.10), the complex nature of change makes it apparent that this is no simple matter and requires thoughtful multifaceted consideration of needs. Fullan (2001) identified factors that complicate the issue of need. At times the realization of one need is overshadowed by other more important needs. Need in the context of this work becomes a matter of

Table 6.14. Improvement in Science Does Not have as High of a Priority as Other Content Areas

	Completely Untrue N (%)	*Somewhat Untrue N (%)*	*Somewhat True N (%)*	*Completely True N (%)*	*Total N (%)*
Principal	15 (19.2)	14 (17.9)	40 (51.3)	9 (11.5)	78 (100)
Teacher	10 (9.8)	12 (11.8)	50 (49.0)	30 (29.4)	102 (100)

Elementary: Chi-square (3) = 10.560, p = .014.

priorities. This is relevant to the implementation of the science standards when dealing with what Fullan (2001) calls an "overloaded improvement agenda" (p. 76). In other words, how important is improvement in science compared to other improvement priorities.

Elementary principals and teachers responded to the following item. Frequency and percent of responses in each category are provided Table 6.14.

Data in Table 6.14 indicate most elementary principals (62.8%) and elementary teachers (78.4%) marked somewhat true or completely true. This is consistent with the work by Pfannenstiel et al. (2001), and respondents' views on the lack of professional development opportunities (see Table 6.4). Lastly, the data are consistent with national findings regarding the minimal time given to science in the school day (Fulp, 2002). Combined, these data suggest science is not a priority subject within the elementary school curriculum. Furthermore, in light of overloaded improvement agendas facing schools, questions arise regarding the resolve the educators will have to continue with their change efforts in science.

Assessment and Accountability

> The methods used to collect educational data define in measurable terms what teachers should teach and what students should learn. (NRC, 1996, p. 76)

State assessments, like the one in Kansas, extract from the standards certain academic expectations. The information generated from them is used by teachers, school officials, and policy makers to make decisions. Accountability, on the other hand, involves "using some of this information to generate incentives to validate or change the behaviors of students and educators" (NRC, 2001, p. 59). The utilization of assessments like state exams and accountability reports are used to judge school quality. It is not unreasonable to visualize the pressure this can place on teachers to adjust their practices accordingly. Because of this, Jorgensen and Shymansky (1997) describe assessment as both the frustration and hope of lasting change in science education. They describe it as frustration because of the "lack of congruence between the educational goals for science and the nature of many science tests" (p. 108). The hope, they proclaim, is the emergence of authentic assessments that "document what students think, understand, and can do in science" (p. 108).

To this end, *NSES* created assessment standards. These standards were developed to serve as a guides "for developing assessment tasks, practices, and policies" (NRC, 1996, p. 75). In the view of the standards, assessment and learning are inseparable. The standards advocate for assessments to focus on what is most important for students to learn in science. It also includes utilizing multiple measures to gain insights into all aspects of science achievement. This does not rule out the use of paper-and-pencil tests, but they do call for an increased emphasis on "authentic assessment" where assessment activities require students to, "apply scientific knowledge and reasoning to situations similar to those they will encounter in the world outside the classroom, as well as to situations that approximate how scientists do their work" (NRC, 1996, p. 78).

Like other states, Kansas has developed an assessment and accountability system. As was stated previously, the *KSES* were developed to "provide criteria for Kansas educators and stakeholders to judge whether particular actions will realize the vision of a scientifically literate society" (KSBE, 2001, p. 3). Following the development of these standards, the Kansas state board of education contracted a second party to develop the assessment using the standards as a guide. The assessment measures both knowledge and process using multiple-choice questions to obtain a total score. The baseline year for administration of this science test occurred during the spring 2001; 1 year prior to the administration of this questionnaire. The results of the assessment are published in an annual assessment report to provide a review of statewide performance. Information from this annual assessment is also included in an annual accountability report that has been mandated by state legislation. Data in both of these reports are on a statewide level. District and building data are reported in school building report cards as mandated by the federal government.

With inquiry-based learning experiences as the center piece of science reform, it is difficult to imagine how utilizing only a multiple choice assessment format on the state exam would be adequate to measure inquiry abilities or understandings. The standards recommend that, "Assessment practices and policies provide operational definitions of what is important. For example, the use of an extended inquiry for an assessment task signals what students are to learn, how teachers are to teach, and where the resources are to be allocated" (NRC, 1996, p. 76). If this is the case, what operational definition do practitioners receive when states like Kansas base assessment and accountability reports solely on multiple-choice data? The *KSES* sends a strong message concerning what is valued in science education. Might principals and teachers be tempted to settle for raised scores on an exam that by its very nature has limited value in assessing inquiry understandings or inquiry skills rather than pursuing the deeper vision of science education provided in the *KSES*?

Practitioners responded to two open items. First, *What should be the primary goal of science instruction for students at your grade level?* Second, *Explain what this means for students by providing one or two specific examples.* While respondents were not asked to comment specifically about the standards or the assessment, five principals and eleven teachers included references to them in their comments. Data from this analysis raise questions about the effectiveness of the state assessment in promoting the kind of science reform advocated in the standards.

Comments from principals and teachers made it clear both the standards and the assessment are having an impact on science instruction. For example, it was reported that the *KSES* are being used to assist aligning district science programs, and teachers are adding/deleting topics of study. In addition, the effectiveness of the state assessment in pressuring change was made evident. One teacher wrote:

> At my grade level, it is difficult to find time for science. I try to cover the topics in the Kansas Science Education Standards needed so 4th graders can do well on the state science assessment. Right now, that is my primary science instruction goal. I have taught some concepts I have never taught before, such as sound and light. I have had to quit teaching other concepts, or not put as much emphasis on them, such as dinosaurs.

Likewise a principal pointed out:

> The primary grades should have science integrated into the language arts and math curriculum. Beginning with 3rd grade science should be taught with more of a formal lesson. The state assessments are the driving force in beginning science formally in third grade.

These comments indicate the inclusion of science on the state assessment has provided a much-needed confirmation that science is an important subject that should be an integral part the elementary curriculum. However, the comments also make it clear for some, the assessment, rather than the standard document has become the goal for science instruction.

If the state assessment was designed to reflect what was valued in the *KSES* this would not be an issue. However, multiple-choice questions are limited in the kind of data they supply. "Multiple choice and short answer responses are convenient for assessing the things that students should know at the 'drop of a hat' or 'cold.'" Many things valued in the standards, however, require at least the time for reflection, more than a couple of minutes (NRC, 2000). By choosing to measure science understandings and abilities with only multiple-choice questions the danger to quality instruction becomes evident. As stated earlier, frequently teachers adjust their instruction to reflect what is valued on the assessment.

Evidence regarding the type of instruction this encourages is provided in one elementary teacher's comment. She wrote, "We have a daily science question that is aligned with the state assessment. Our district just rewrote curriculum for all grade levels. I worked on preparing the 4th grade curriculum." It would be unfair to draw too many conclusions about this teacher's instructional practices, but her words do illustrate how science instruction could be reduced to a series of test preparation exercises. This is not the type of science instruction envisioned by the writers of the standards. Change may have occurred, but what about the quality of the change?

SUMMARY AND CONCLUSIONS

The adoption of the *KSES* as state policy is great news for advocates of *NSES* reform. Science education in the state, including the elementary level, has been provided with a vision for reform that has the potential to play a key role in the transformation of science teaching and learning. However, written policy must be translated into practice. Data related to teacher development, curriculum, and assessment and accountability provided insight into the extent the message of reform is being communicated to school based practitioners through each of these channels. Examination of the data yielded six conclusions. The discussion that follows is focused on the implications of each of these conclusions.

1. As a whole, perspectives of teachers and principals are not substantially different.

Surprisingly, there was much continuity between principals' and teachers' perspectives. Differences were found to exist in only one of the 12 chi-square tests that were conducted. This is contrary to initial expectations. It is possible that by maintaining an experiment wise alpha of .05, some differences were not detected. However, this level of significance was warranted, given the error a less conservative approach would have been introduced.

Being in touch with the classroom is important for principals since Fullan (2001) describes this role as one that ensures school programs for staff and students are focused and coherent. This apparent coherence should also make professional development easier, since staff developers will be working with individuals who appear to share similar perspectives.

2. Principals and teachers believe that science instruction should be aligned with the standards.

The adoption of the state standards consistent with the *NSES* was a big step towards standards based reform. Even though a percentage of principals (25.6%) and teachers (36.3%) indicated they might have philosophical differences with the *KSES*, it appears over 90% of the educators in this study believe science instruction should align with the standards. This represents a strong foundation for change.

3. Principals and teachers do not have a clear understanding of what it means to be standards based.

Fullan (2001) identified clarity as an essential element if implementation of an innovation like the standards is to occur. He discusses that lack of clarity is expected and may not be bad as long as those needing it get support. Without this, trying to adopt changes will only lead to frustration. The study identified large numbers admitting to having doubts about their ability related to modifying text activities (principals 47.3% and teachers 36.7%) and in recognizing teaching practices (principals 36.4%, teachers 46%). It is clear that support for these people is necessary if reform is to move forward.

The large number of respondents indicating a lack of clarity raises important questions. What criteria do these educators utilize to make instructional decisions? What criteria do they use to make judgments regarding quality science instruction? This illustrates why too often the adopted textbook or other adopted materials become the curriculum. It also suggests why the state assessment, not the standards, is influencing the planning and evaluation of science instruction for many educators.

Related to clarity is the idea of "false" clarity. Fullan (2001) mentions that it is common for people to oversimplify change and not realize all it entails. In essence, they think they understand or are making changes, but in reality are not. One of the limitations of self-reported data is its inability to detect whether respondents have a false sense of clarity. However, the large number of principals and teachers (over 85%) indicating their science curricular resources align well with the standards raises this as a possible issue. Why did not more of them admit they did not know about the alignment of their curricular resources when so many were willing to admit they were not sure about being able to recognize teaching practices and revise curricular resources? Is it a matter of trust and reliance in the work of the publishing companies? Whether or not the curricular resources actually align well cannot be determined from the data, but respondents' perspectives of reality remains important. If principals and teachers believe their curricular resources align well with the standards, they will not see a need to improve in this area. Thus, this may be an area of needed improvement that is overlooked. In a much broader sense, if practitioners do not object to the quality of their curricular materials, why should suppliers change?

4. More resources need to be leveraged to support science instruction.

Change in science education will not happen if the resources are not made available that will allow it to happen. With the number of principals and teachers indicating lack of clarity, it is clear there is a need for professional development. Unfortunately, almost half of the respondents indicated access to high quality professional development related to science is a problem. Bybee and Loucks - Horsley (2001) elaborate on why adopting standards alone will not be enough to help translate policy into programs and into practice. Another constraint on science teaching at the elementary level is lack of planning time (Teters & Gabel, 1984). Both principals (66.7%) and teachers (84.3%) confirmed this as an issue.

Both professional development and planning time are integral elements needed in creating a culture where teachers can work in developing science programs that are standards based. This study did not probe other resource related issues like lack of supplies, insufficient lab space, quality of textbooks (Schmidt, McKnight, & Raizen, 1997), and outdated facilities and equipment (Lewis, Snow, Farris, Smerdon, Cronen, Kaplan, & Greene, 2000). Loucks-Horsley, Love, Stiles, Mundry, and Hewson (2003) provides an overview regarding what is needed for capacity building and "scaling up", but for the purpose of this study it is enough to acknowledge that investing financially in resources without simultaneously investing in those elements needed to ensure sustainability would be a mistake.

In spite of all this, over 70% of the all teachers and over 80% of all principals reported that school environments encouraged teachers to develop a deep understanding of the standards. When combined with the high numbers of principals and teachers indicating lack of planning time, these percentages are perplexing. It would seem access to professional development and having planning time to discuss the science curriculum should be more closely related to developing a deep understanding of the standards. This raises the question, what principals and teachers consider deep understanding? This is an area requiring further inquiry.

5. Science education is not a priority at the elementary level.

Evidence from several sources indicates many educators do not place a priority on improving science. The most direct evidence for this lies in the fact that almost 80% of the elementary teachers and 60% of the elementary principals indicated improvement in other content areas have priority. Comments from respondents suggest mathematics and reading concerns surpass any concerns practitioners may have about science. One elementary principal stated, "Science instruction in the primary grades is less important than establishing effective readers." With the new mandate to test elementary in reading and math, the time for science may be even further eroded. A fourth grade teacher explained, "I don't believe science should be emphasized in a K-4th grade building. More time should be spent in reading and math." Fullan (2001) reports that many schools are faced with innovation overload. There are limits to the number of improvement programs in which a school can be involved. This implies that more pressing needs may overshadow the need for science instruction to be aligned with the standards. This low priority of science leaves doubts about how quickly change in science instruction will occur.

6. While the intent of the KSES was to provide educators the criteria "to judge whether current actions serve the vision of a scientifically literate society" (KSBE, 2001, p. 3), policies and documents other than these are influencing classroom decisions.

Science education does not occur in isolation. As a discipline it is one of several competing for a place in the elementary school curriculum. In a complex system like that of education, it is not always clear how decisions in one area will influence other areas. For example, adopting a state assessment policy that requires yearly data on mathematics and reading and data on science only biyearly sends a clear message to practitioners that these subjects take priority—not only in terms of instructional time, but with resources like professional development opportunities. Both of which were identified as issues in this study. The intent of this example is not to argue the merits of scientific literacy in relation to other disciplines. All are important and vital aspects of the curriculum. Rather it points out the complex nature in which the system operates and how high stakes policies determine curriculum by legitimizing or marginalizing content.

Along these same lines, qualitative comments also suggest that even though science scores are assessed every other year, it is motivating some to give science instructional time. However, the utility of the multiple-choice assessment in promoting standards-based change is questionable. Accountability policies might influence instruction by pressuring teachers to align content with the assessment, but there are serious concerns about whether this assessment is measuring what is most valued. For some, the assessment is having a *greater* influence on science instruction than the *KSES* document. The extent this is an issue is unclear. Data in this study was not conclusive, but further exploration into this area is warranted.

CONCLUDING THOUGHTS

> *Well, the hard work is done. We have the policy passed; now all you have to do is implement it.* Outgoing deputy minister of education to colleague. (Fullan, 2001, p. 69)

The data and discussion provided in this study emphasizes how the elements of the system including, teacher development, curriculum, and assessment and accountability has worked but must change if we are to move toward the vision of the *NSES*. Under the umbrella of this vision, the push for change must be a long-term, ongoing and multifaceted effort reaching all levels of the system. Fullan (2001) outlines three roles state leaders in Kansas can play in facilitating educational change like the one started with the adoption of the *KSES*.

One push state leadership can provide is an accountability system. Kansas has one in place and to a certain degree there is evidence that it is having an impact. Scores in science have increased between the 2001 and 2003 test years (see Table 6.15) (Kansas State Department of Education, n.d.). Also, as has been previously mentioned, qualitative data indicate some teachers not normally including science in their daily routine are doing so now.

Table 6.15. Kansas Science Assessment Grade 4: Race/Ethnicity Desegregation for All Students on Total Score, 2001 and 2003

Performance Level	*% White*		*% African American*		*% Hispanic*	
	2001	*2003*	*2001*	*2003*	*2001*	*2003*
Exemplary	15	18.2	2.3	3.2	2.2	4.2
Advanced	29.8	32.1	9.8	12.0	11.5	14.1
Proficient	25.4	25.3	16.1	20.4	21.7	22.6
Basic	22.5	19.0	36.2	38.8	39.8	39.2
Unsatisfactory	7.4	5.4	35.6	25.5	24.9	19.9

However, as has been stated previously, an issue needing closer examination is the impact adopting this multiple choice only instrument is having on classroom instruction. In addition, selecting to only emphasize accountability fails to change beliefs, enhance professional skills, or provide clarity of purpose. All of which were identified as issues in this study. This is why Fullan (2001) identifies the need for state policy makers to provide incentives like professional training in conjunction with accountability policies. While not showing up on the test scores above, data in this study suggest that the investment in this kind of support is desperately needed. Not the least being a commitment to bring clarity to the public, policy makers, and practitioners regarding what it means to provide a standards-based science program. Without this, judgments regarding science education will be based on the results of test scores like the one in Table 6.15. Which leads directly into the last recommendation, state leadership must also invest in capacity building.

Capacity building implies not only investing in the development of individuals, but also investing in the system in which teachers work so that it supports change. Teachers need an environment that encourages them to apply their knowledge and expertise in local contexts focused on continuous improvement. The implication here is clear—teachers need to be empowered to respond to issues and changes in a purposeful and reasoned manner. To do this, they need support, resources and opportunities for this to happen. Fullan (2001) identifies five aspects of capacity as being “the individual skill and knowledge of the teachers, professional learning communities, program coherence, availability of resources, and a school principal who helped develop the previous four factors” (p. 224). Investing state resources into each of these would do much in moving science reform or any such reform forward.

An examination of Table 6.15 does demonstrate growth in scores between years and is a tribute to those working hard to make science education meaningful. However, it also provides evidence that an achievement gap between ethnicities still exists. Certainly, developing teachers’ knowledge and skills related to the science standards and expanding the assessment and accountability system to include measures that more accurately reflect what is valued in the science standards is a needed step. However, it is doubtful that these alone would make these differences go away. Investing in capacity is a key.

Realizing the changes advocated in the KSES promises to be a lengthy process that will require an investment of support in order to succeed. A teacher provided insights into her personal struggle.

> I believe that a love for science should come *first* because students dislike the subject. Science should allow the students to love and engage in science

> activities that are student-centered *without* the horrible pressures of getting through all the multi-million state objectives about which many youngsters have not a care in the world. Students remember from by-gone years when they made life-sized whales in the snow with paint. Reptiles and amphibians, bird studies, and spiders are all subjects that I have developed really fun & exciting things and are the things my students remember about science and about life. I try *very hard* to make the standards interesting and meaningful-but sometimes it is tough.

History of past reforms indicates that for meaningful change to occur, reform surrounding the standards must become more than implementation. Those practitioners most affected by the standards must be provided with the means to embrace them at a deeper level. Teachers, principals, and educational policy makers must move beyond simple awareness of the standards and have the opportunity to wrestle with the difficult questions related to how to negotiate comprehensive change that is consistent with the standards within the channels of influences affecting each local school culture. This is no easy task, and skepticism regarding the impact and permanence of the science standards and reform may well be expressed by an elementary teacher who stated,

> Implementing the standards, addressing fewer topics rather than acquiring a broader knowledge base may/may not prove the end we want. Inquiry was much the way to go in the early 70s—it went away—now it is back. Such are the cycles in education.

Clearly, if advocates of the standards wish to avoid becoming just another chapter in the history of science education, it is imperative that the change process must occur at all levels; individual, district, and statewide; while concurrently recognizing practitioners' voices, providing resources, and allotting time for professional growth and transformation of teaching.

REFERENCES

Anderson, R. D. (1996). *Study of curriculum reform* (Volume 1 of the final report of research conducted under contract no. RR91182002 with OERI. U.S. Department of Education). Washington DC: U.S. Government Printing Office. (ISBN 0-16-048865-6)

Butler, M. A., & Beale, C. L. (1994, September). *Rural-urban continuum codes for metro and nonmetro counties, 1993* (Staff report No. 9425). Beltsville, MD: Agriculture and Rural Economy Division, Economic Research Service, U.S. Department of Agriculture.

Bybee, R. (1997). *Achieving scientific literacy: from purpose to practices*. Portsmouth, NH: Heinemann.

Bybee, R., & DeBoer, G. (1994). Research on goals for the science curriculum. In D. Gabel (Ed.), *Handbook of research on science teaching and learning* (pp. 357-387). New York: Macmillan.

Bybee, R., & Loucks - Horsley, S. (2001). National science education standards as a catalyst for change: The essential role of professional development. In J. Rhoton & P. Bowers (Eds.), *Professional development planning and design*. (pp. 1-12). Arlington, VA: National Science Teacher's Association.

DeBoer, G. (1991). *A history of ideas in science education: implications for practice*. New York: Teachers College Press.

Fullan, M. (2001). *The new meaning of educational change*. New York: Teachers College Press.

Fulp, S. (2002). *2000 national survey of science and mathematics education: Status of elementary school science teaching*. Chapel Hill, NC: Horizon Research.

Gallagher, J. (1996). Implementing teacher change at the school level. In D. Treagust, R. Duit, & B. Fraser (Eds.), *Improving teaching and learning in science and mathematics* (pp. 222-231). New York: Teachers College Press.

Gess-Newsome, J. (2001). The professional development of science teacher for science education reform: A review of the literature. In J. Rhoton & P. Bowers (Eds.), *Issues in science education: Professional development planning and design*. (pp. 91-100). Arlington, VA: National Science Teacher's Association.

Huck, S. (2000). *Reading statistics and research*. New York: Longman.

Jorgensen, M., & Shymansky, J. (1997). Assessment in science: A tool to transform teaching and learning. In J. Rhoton & P. Bowers (Eds.), *Issues in science education* (pp. 107-113). Arlington, VA: National Science Teacher's Association.

Kansas State Board of Education. (2001). *Science education standards*. Topeka, KS: Kansas State Department of Education.

Kansas State Department of Education. (n.d.). *2001 assessment report*. Retrieved April 1, 2004, from http://www.ksde.org/assessment/2001_assess_results.html

Kansas State Department of Education. (n.d.). *2003 assessment report*. Retrieved April 1, 2004, from http://www.ksde.org/pre/assessment2003.htm

Leonard, W., Penick, J., & Douglas, R. (2002). What does it mean to be standards-based? *Science Teacher, 69*(4), 36-39.

Lewis, L., Snow, K., Farris, E., Smerdon, B., Cronen, S., Kaplan, J., & Greene, B. (2000). *Condition of America's public school facilities: 1999*. Washington, DC: U.S. Department of Education, National Center For Educational Statistics.

Loucks-Horsley, S., Love, N., Stiles, K., Mundry, S., & Hewson, P. (2003). *Designing professional development for science and math* (2nd ed.). Thousand Oaks, CA: Corwin Press.

Loucks-Horsley, S., & Bybee, R. (1998). Implementing the national science education standards. *Science Teacher, 65*(7), 22.

Lowery, L. (Ed.). (1997). *NSTA pathways to the science standards: Guidelines for moving the vision into practice, elementary school edition*. Arlington, VA: National Science Teacher's Association.

National Research Council. (1996). *National science education standards*. Washington, DC: National Academy Press.

National Research Council. (2000). *Inquiry and the national science education standards*. Washington, DC: National Academy Press.

National Research Council. (2001). *Educating teachers of science, mathematics, and technology: New practices for a new millennium*. Washington, DC: National Academy Press.

Pfannenstiel, J., Seltzer, D., Yarnell, V., & Lanbson, T. (2000). *Study of professional development practices and early childhood education in Kansas*. A report prepared by the Research & Training Associates, Overland Park, Kansas, for the Kansas State Department of Education. Retrieved April 1, 2004, from http://www.ksde.org/pre/prodev_study2000.html

Raizen, S. A. (1998). Standards for science education. *Teachers College Record. 100*(1), 66-121.

Rhoton, J. (2001). School science reform: an overview and implications for the secondary principal. *NASSP Bulletin, 85*(623). Retrieved October 14, 2001, from http://www.nassp.org/news/bltn_sch_sci_rfrm301.htm on 10/14/2001

Rogers, E. (1995). *Diffusion of innovations*. New York: Free Press.

Schmidt, W., McKnight, C., & Raizen, S. (1997). *A splintered vision: An investigation of U.S. science and mathematics education*. Dordrecht, The Netherlands: Kluwer Academic Publishers.

Schunk, D. (1997). *Learning theories*. Englewood Cliffs, New Jersey: Merrill.

Teters, P., & Gabel, D. (1984). *Results of the NSTA survey of the needs of elementary teachers regarding the teaching of science.* Washington, DC: National Science Teachers Association. (ERIC Document Reproduction Services No. ED 253 398)

Thomas, J. A., & Stuessy, C. L. (1998). *Directing change: Guiding reform in elementary science education in Texas.* Paper presented at the annual meeting of the Association for the Education of Teachers of Science, Minneapolis, MN.

Tippins, D., Nichols, S. E., & Weseman, K. (1998). Contemplating criteria for science education reform: The case for Olympia school district. *School Science and Mathematics, 98*(7), 389-396.

Tobin, K., Tippins, D., & Gallard, A. J. (1994). Research on instructional strategies for science teaching. In D. Gabel (Ed.), *Handbook of research on science teaching and learning* (pp. 45-93). New York: Macmillan.

Tolman, M., Hardy, G., & Sudweeks, R. (1998). Current science textbook use in the United States. *Science and Children 35*(8), 22-25.

Trowbridge, L., Bybee, R., & Powell, J. (2000). *Teaching secondary school science.* Upper Saddle River, NJ: Merrill.

Weiss, I. R., Knapp, M. S., Hollweg, K. S., & Burrill, G. (2002). *Investigating the influence of standards: A framework for research in mathematics, science, and technology education.* Washington, DC: National Academy Press.

Weiss, I., Pasley, J., Smith, S., Banilower, E., & Heck, D. (2003). *Looking inside the classroom: A study of K-12 mathematics and science education in the U.S.* Retrieved April 1, 2004, from http://www/horizon-research.com/insidetheclassroom/reports/looking/complete.pdf

CHAPTER 7

SCIENCE STANDARDS

Cause and Object of Influence—A Kansas Case Study

John R. Staver

The author describes how national science standards—an example of policy—have influenced other policies, programs, and practices. Using Kansas as an example, the author documents the principal influence of national standards as increasing the alignment of curriculum, instruction, and assessment with state standards, which were developed from national science standards. The author then describes how the No Child Left Behind Act (NCLB) and controversy over teaching evolution have influenced Kansas' standards. State level policy makers implemented NCLB because it is the law, and NCLB became the broad umbrella under which the state framework for accrediting schools operates. The author describes and examines events in 1999-2001 when the Kansas State Board of Education (KSBE) removed evolution, deleted ancient earth and universe concepts, and fundamentally altered the nature of science in the state science standards. Following the 2000 elections, these changes were rescinded by the KSBE in 2001. The author then brings his description and analysis forward to 2005 as a new KSBE led by a Christian fundamentalist majority considers substantial revisions to the state's current science standards.

The Impact of State and National Standards on K-12 Science Teaching, 185–211

INTRODUCTION

Describing a framework for reform in science education, Bybee (1997) sets forth four fundamental categories of reform initiatives: Purpose, policy, program, and practice. Regarding each category, he explains,

> Statements of purpose are universal and abstract, and apply to all concerned with reforming science education.... Policies are more specific statements of standards, benchmarks, state frameworks, school syllabi, and curriculum designs based on the purpose.... Programs are the actual materials, textbooks, software, and equipment that are based on policies and developed to achieve the stated purpose.... Practice is the specific actions of the science educators. (p. 29)

In a perfect environment of reform, purposes, policies, programs, and practices would be connected tightly, and each component would support the other components. Moreover, the components taken together would become a highly coherent, elegant, and internally consistent system. But, the United States of America is not perfect; it is a democracy. Observers of and participants in school reform throughout the United States frequently witness a great deal of inconsistent, even counterproductive activity as stakeholders clash over reform initiatives and as one reform initiative conflicts with others in the same category as well as those in other categories.

PURPOSE

My purpose herein is twofold. First, I will describe how national science standards—an example of policy—have influenced other policies as well as purposes, programs, and practices. Second, I will describe how standards themselves have been influenced by broad reform initiatives beyond science education and by values and belief systems—those concepts in the minds of stakeholders that determine the correctness of the reform vision expressed in national science standards.

SOME PERSPECTIVE

Before embarking on the task of achieving these two purposes, I want to offer a brief perspective on the term *standard*. The dictionary (Merriam-Webster, 1970) on my office shelf contains eight definitions of standard:

> 1: A conspicuous object (as a banner) formerly used at the top of a pole to mark a rallying point esp. in battle or to serve as an emblem 2 a: a long narrow tapering flag that is personal to an individual or corporation and bears heraldic devices b: the personal flag of the head of a state or of a member of a royal family c: an organization flag carried by a mounted or motorized military unit d: BANNER 3: something established by authority, custom, or general consent as a model or example: CRITERION 4: something set up and established by authority a rule for the measure of quantity, weight, extent, value 5a: the fineness and legally fixed weight of the metal used in coins b: the basis of value in a monetary system 6: a structure built for serving as a base or support 7: a plant grown with an erect main stem so that it forms or resembles a tree; *also*: a fruit tree grafted on a stock that does not induce dwarfing 8a: the large upper posterior petal of some flowers b: one of the three inner usu. erect and incurved petals of an iris. (p. 853)

The third and fourth definitions are most pertinent to the meaning of standard as the term is used in educational reform. National standards in science education, as envisioned by their developers, represent both a voluntary model and guidelines for science content, teaching, teacher preparation, professional development assessment, programs, and systems.

Standards as a voluntary model and guidelines emerged from numerous calls for reform throughout the 1980s as well as international comparisons of achievement in the 1980s and 1990s. These calls for reform cited a myriad of ills concerning K-12 education in the United States. Based on U.S. students' poor performance on international comparisons, one central theme portrayed school reform in the service of improved American economic productivity (Atkin & Black, 1997). Another central theme reflected the vision of an informed citizenry in the service of a democracy, a vision first expressed more than 200 years ago by our nation's founding fathers: "The whole people must take upon themselves the education of the whole people and be willing to bear the expense if it" (John Adams, 1758, cited in Center on National Educational Policy, 1996, p. 17), and "Above all things, I hope the education of the common people will be attended to; convinced that on this good sense we may rely with the most security for the preservation of a due degree of liberty" (Thomas Jefferson, 1787, cited in Center on National Educational Policy, 1996, cover page). In current terms, a healthy, vibrant United States of America—beyond its economic productivity—needs an entire citizenry that is literate in science, mathematics, and technology in today's and tomorrow's highly scientific and technological society (Klopfer & Champagne, 1990). National science standards reflect both visions in the minds of many, but many others see standards as representing improved economic productivity or higher levels of citizen's science literacy but not both.

National science standards moved quickly to a national level center stage with the publication of *Curriculum and Evaluation Standards for School Mathematics* by the National Council of Teachers of Mathematics (NCTM) in 1989. Whereas NCTM had labored for a decade to achieve consensus in the mathematics education community and to prepare standards, this work had taken place largely beyond the glare of national public attention and focus. The publication of the NCTM standards, as they came to be known, provided much of the required activation energy for the broad science education community to commence production of national standards. Another key component that helped push the activation energy over the top was the initiation of Project 2061 by the American Association for the Advancement of Science (AAAS), and the subsequent publication of *Science for All Americans* in 1990 by James Rutherford and Andrew Ahlgren, Executive Director and Deputy Director of Project 2061, respectively.

An attempt to develop national standards began hastily within the National Science Teachers Association (NSTA) in 1990. Unable to develop a consensus across the diverse science discipline and science education communities, the NSTA leadership dismantled the entire effort and handed it to the National Academy of Sciences in 1991, where it was reconsidered and reconstructed from the ground upwards, given the necessary credibility, and provided ample nourishment. The *National Science Education Standards (NSES)* were published 5 years later (National Research Council [NRC], 1996). Rutherford, Ahlgren, and their colleagues at Project 2061 preceded the *NSES* in 1993 with *Benchmarks for Science Literacy*, a document that delineated steps for increasing science literacy in grades K-12. Science literacy for all students, of course, is the universal, abstract purpose or vision of this reform (NRC, 1997), and standards as a concept and as expressed in the *NSES* and *Benchmarks* are specific statements of policy based on the vision or purpose of higher levels of science literacy for all.

INFLUENCE OF STANDARDS

On State Policy

In what ways have national science standards influenced the purpose, policies, programs, and practices of the current reform movement in science education? Let me sharpen the initial interrogative by focusing on a question about policy: How have national standards influenced state policy in the form of state science standards? My initial Google search on November 29, 2004 used three keywords, *standards*, *science*, and *align*; it produced approximately 433,000 hits. The 10 hits on page one of the search offer users a variety of resources for aligning science curriculum, key concepts, activities, texts, and other materials to state and national science standards. This electronic search opens the door to understanding a major influence of national standards.

In the years immediately following the release of the *NSES*, 49 of 50 states in the union—Iowa is the exception—commissioned, developed, and adopted state science standards (Lerner, 2000). In virtually all states, state standards were themselves modeled on the shoulders of the *NSES*, *Benchmarks*, or both documents. Moreover, state standards represent statements of what students should know and be able to do at various levels of their K-12 education. Such standards focus on science content. Although national standards were conceived and published as a voluntary model and guidelines, they have been institutionalized as requirements at the state level as states worked to develop their own standards using the *NSES* and/or *Benchmarks* as models. It is noteworthy to point out that virtually all states developed content standards but not standards for instruction, assessment, program, and system, even though these additional categories of standards are in the *NSES*.

I will discuss Kansas as an example because I was centrally involved in the development of Kansas' science standards. Except for the controversy over evolution, I will discuss this issue in terms of how values and belief systems influence standards. I, also, will hypothesize that the standards development process in Kansas shares much in common with the development of science standards in other states. I also expect that Kansas is unique in many aspects. Given this, readers must decide for themselves how much, if any, of what I describe generalizes to their states and local school districts.

Dr. Loren Lutes, then superintendent of Unified School District (USD) 318 Elkhart, Kansas and now superintendent of USD 341 Oskaloosa, Kansas, and I served as cochairs of a 27-member team of K-12 teachers of science, university science educators, and scientists that developed science standards for the KSBE. Our team began its task in June, 1998 and worked steadily over the next 14 months. In all of its work, the writing team relied substantially on the *National Science Education Standards* (NRC, 1996) and *Benchmarks* (AAAS, 1993). During the first 6 months, the team met at least monthly to set directions, give members assignments for future work, discuss its progress, and hear from outside experts. For example, in the summer of 1998 Dr. Rodger Bybee, then Executive Director of the Center for Science, Mathematics, and Engineering Education at the NRC and Chair of the *NSES* Content Standards Working Group and now Executive Director of Biological Sciences Curriculum Study (BSCS), was invited to talk to the committee. The team delivered an initial draft to the KSBE in January, 1999. From January through August, 1999, the committee held monthly meetings to discuss feedback as well as a variety of issues, revise its work, and set plans. Simultaneously, the committee and/or its individual members attended several public hearings around the state, received an external review of an early draft from the council for basic education, produced subsequent revisions of its initial draft, and engaged in numerous interactions with the board at monthly KSBE meetings, with board members in small groups, and in one-on-one conversations. The writing team presented its fifth and final draft to the board at its July, 1999 meeting. A small group of KSBE members made substantial revisions to the writing team's fifth draft in late July, and the board approved this revised document at its August meeting (Staver, 2000).

Several years prior to the development of science standards, in 1989, the KSBE embraced a continuous improvement process, including the adoption of strategic directions that launched the development of an outcomes-based framework for quality instruction called Quality Performance Accreditation (QPA) (KSBE, 1993). The board's motivation for this action stemmed from its dissatisfaction with past accreditation criteria that focused on, for example, the number of books in libraries and the square footage of buildings. QPA focuses on the continuous improvement of academic achievement and evaluates the quality of a school on the basis of students' continuous improvement and performance at specific points in their K-12 education. Continuous improvement as a process is as important in QPA as is assessed achievement at a point in time.

QPA was brought on line in stages in the early 1990s. It requires all independent school districts in the state—currently there are 301—to assess students' progress in reading and mathematics and to select a third subject (e.g., science or social studies) as an elective assessment. Each subject, required or elected, must be assessed via three separate sources of data. State level science standards and assessments existed prior to the development of state standards based on the *NSES*; however, when standards were developed and adopted by the KSBE for all school subjects (reading, mathematics, social studies, and science), the KSBE also commissioned the development of new state level assessments based on its new standards. Long before the advent of NCLB, Kansas school districts were required to meet achievement standards in required and elected subjects and to submit and carry out school improvement plans for those subjects when school performance was below acceptable levels.

In sum, clearly, national science standards have influenced state policy in Kansas via development of state science standards and science assessments. To the extent that the process and products of these developments in Kansas are representative of other states, I hypothesize that national science standards have similarly influenced other state science policies.

On Classroom Practice

A second question is: In what ways does the influence of national standards reach all the way to local school districts and into individual classrooms of K-12 teachers of science? Again, I will use Kansas as an example. I have been a faculty member and director of a Kansas State University center whose mission is improving K-16 science, mathematics, and technology education since 1988, and I interact regularly with K-12 teachers in several venues. Included are local, state, regional, and national science teacher professional society meetings, visits to local and nearby schools in the Kansas state professional development school partnership to observe and supervise prestudent teaching field experiences and student teachers, to work on externally funded science education projects, and as a member of a new 26-person committee that is now revising the current state science standards. Topics such as science standards and science assessment rarely fail to come up, and the phrase most often used by teachers is "teaching to" either the standards, the assessments, or to both. Moreover, many teachers report feeling pressured by building and district administrators to do so (e.g., science standards writing team members, personal communication, June 21, July 26, August 30, 2004).

Regarding professional development of K-12 teachers of science, even a cursory perusal of the conference programs of the Kansas Association of Teachers of Science (KATS), other state science teachers associations, the area and national convention programs of the NSTA, as well as conference programs of the American Association of Physics Teachers, the American Chemical Society, the National Association of Biology Teachers, and the National Association of Geology Teachers since 1996 reveals a steep growth and now a steady presence of presentations by K-12 teachers, science educators, and scientists on an extensive variety of topics related to national and/or state standards.

In sum, wherever teachers of science get together, the topic of aligning science content and instruction with state standards and assessments is sure to be discussed at length. My observations, conversations with colleagues, and other experiences tell me that school districts and teachers are hard at work aligning their programs and practices with state science standards.

On Science Curriculum

A third question is: What have been the reactions of publishers of K-12 science curriculum materials to national science standards? Publishers of science curricula and other instructional materials regularly exhibit their products at professional society meetings of K-12 teachers of science, and they assure all who stop to talk or examine the materials that their products meet standards (e.g., Gerald Wheeler, personal communication, February 24, 2005). Beginning with an appropriate measure of scientific skepticism, perhaps topped with a sprinkle of cynicism, I assert that believing publishers' claims without triangulating through research that their materials meet standards is akin to a customer who is considering the purchase of a used car. Let the buyer beware!

There are, however, several excellent—meaning standards-based or standards-aligned—science curriculum materials currently on the market. I will briefly share two examples, largely because I have interacted with and participated in curriculum and/or professional development projects with two organizations for over 20 years. The two organizations are the BSCS, and the Lawrence Hall of Science (LHS) at the University of California at Berkeley. BSCS has developed science curricula since the days of Sputnik. The LHS emerged as a major player in science curriculum development with its post Sputnik K-6 program *Science Curriculum Improvement Study (SCIS)* (Karplus, Thier, Knott, Lawson, & Montgomery, 1970) under the direction of Robert Karplus, a University of California, Berkeley physicist turned science educator. In today's era of standards and NCLB, BSCS and LHS represent exemplars in developing science curricula that are consistent with national standards.

Two K-6 science curricula, one from BSCS and one from LHS, serve as evidence. Several years prior to the *NSES*, BSCS developed *Science for Life and Living* (BSCS, 1992), now known as *BSCS Science Teaching Relevant Activities for Concepts and Skills (TRACS)* (BSCS, 1999); also prior to the *NSES*, LHS developed the *Full Option Science System (FOSS)* (Regents, 1993). After the publication of the *NSES* in 1996, BSCS and LHS worked to align their curricula with national standards. Employing a biological metaphor, I point out that these two K-6 science curricula are somewhat different phenotypic expressions of the same genotype. This genotype includes a research base of children's thinking and development, an intent to include all students, an emphasis on constructing understanding, a guided inquiry instructional model, use of collaborative learning, assessment of understanding and abilities, and encouragement of equity.

An example of somewhat divergent expressions of this genotype is the use of different forms of the venerable Learning Cycle (e.g., Lawson, Abraham, & Renner, 1989), a well known, effective guided inquiry instructional model, which was born in *SCIS* (Karplus, et al., 1970). *FOSS* uses the Learning Cycle in its original three-stage format (explore, invent, apply), whereas BSCS modified the Learning Cycle into a 5-sequence model known as the 5-E Instructional Model (the Es are engage, explore, explain, elaborate, evaluate). The original Learning Cycle and the 5-E Instructional Model are grounded in children's thinking (e.g., Bransford,

Brown, & Cocking, 2000) and Piagetian developmental concepts (e.g., Bybee & Sund, 1982). Implemented as hands-on, minds-on instruction, the Learning Cycle allows all learners to participate and motivates them to understand through active, collaborative learning experiences. Both curricula employ a collaborative learning system that is incorporated into the hands-on, guided inquiry instruction. Both curricula have taken seriously the admonition of national standards that *less is more* by including fewer, more central concepts in more depth. *FOSS* has no student text, although its current edition now includes a small trade book called *FOSS Science Stories* with each kit-based unit. *BSCS TRACS* has had a student text since its original edition. Both have paid serious attention to developmental issues in deciding the grade level at which specific concepts should be introduced. For example, prior to standards, K-6 science textbook series from large publishing houses such as Merrill (Hackett, Moyer, & Adams, 1989); Silver, Burdett, and Ginn (Mallinson, Mallinson, Froschauer, Harris, Lewis, & Valentino, 1991); and Scott Foresman and Company (Cohen, Cooney, Hawthorne, McCormack, Pasachoff, Rines, & Slesnick, 1991) introduced simple machines in third grade. *FOSS* introduces only levers and pulleys and does so at the fifth-sixth grade due to developmental considerations. Current editions of science text series from large publishers have moved simple machines to a higher grade.

Bringing this section to closure, I assert that, in my opinion, the principal influence of national standards has been to increase the focus on aligning science curriculum, instruction, and assessment with state standards, which of course were developed on the shoulders of national science standards. This focus is inclusive, in that it includes teachers, principals, superintendents, state departments of education, higher education, publishers, parents, and communities. Essentially, the focus now includes all who hold a stake in K-12 education.

INFLUENCES ON STANDARDS

No Child Left Behind—The Influence of Law

The phrase *no child left behind* immediately calls to mind two questions: What children are currently being left behind? Why? According to Thomas and Bainbridge (2002), public educational policy presently leaves children behind because: (1) Approximately 10.5 million learners live in poverty and do not have access to health insurance; (2) the majority of the nation's schools do not receive adequate funding, particularly those schools in relatively poor states and school districts, and less qualified teachers often teach in these schools; (3) according to research, disadvantaged children struggle to develop cognitively, to acquire an adequate vocabulary, and to learn sounds necessary to learn to read; (4) child-care centers that choke creativity and development are attended by millions of children; (5) the federal government substantially under funds accommodations targeted for learners with disabilities, yet federal law requires such special accommodations; and (6) students in affluent schools score higher than students in poor schools according to all recent state-level standardized tests.

The Elementary and Secondary Education Act (ESEA), first enacted in 1965, is the primary federal law that influences K-12 public education across the nation. On January 8, 2002, President Bush signed into law a reauthorization and major revision of the ESEA. Titled the No Child Left Behind (NCLB) Act of 2001 (NCLB, 2002), the reauthorized law delineates sweeping federal reforms to improve K-12 education across the length and breadth of the U.S. Prior to NCLB, the federal government served K-12 education primarily as a source of funding—currently about 9 cents of each dollar that schools receive from all sources. With the advent of NCLB, the federal government has taken a much stronger role in shaping how teachers teach and what the nation's students learn (Bloomfield & Cooper, 2003).

The NCLB law is based on four principles: (1) accountability for results in terms of students' achievement; (2) more emphasis on instruction whose effectiveness is based on scientific research; (3) availability of options for parents in the education of their children; and (4) more emphasis on local control and flexibility. I will restrict my discussion of the law's influence on state science standards, programs, and practices in school science in Kansas primarily to principle 1 due to limitations of space and scope. Again, individual readers must judge the extent to which developments in Kansas are relevant to other states.

Accountability for students' achievement and assessment in science under NCLB is not required until 2007; however, Kansas has already begun to address the requirements of NCLB for school science. Presently, Kansas schools are accountable for making adequate yearly progress, known as AYP, toward the goal of proficient status in reading and mathematics (Kansas State Department of Education, 2005a). AYP has not yet been operationally defined for science, but AYP is defined conceptually in Kansas as the growth rate of the percentage of students in individual schools, districts, and the state toward the state's defined status of proficient. KSDE (2005b) defines five levels of achievement, which are on its Website (http://online.ksde.org/rcard/definitions.aspx?org_D%&rpt_type=3) as follows:

- *Exemplary:* Students who perform at the exemplary level on the Kansas state assessments consistently demonstrate high performance. These students have a well-developed ability to apply knowledge and skills in all situations. Their work is superior.

- *Advanced:* Students who perform at the advanced level on the Kansas state assessments frequently demonstrate high performance. These students effectively demonstrate the ability to apply knowledge and skills in most situations. They have a command of difficult, rigorous and challenging material.
- *Proficient:* Students who perform at the proficient level on the Kansas state assessments demonstrate a mastery of core skills. These students exhibit competence in applying knowledge and skills in most problem situations. They show evidence of solid performance.
- *Basic:* Students who perform at the basic level on the Kansas state assessments show partial mastery of fundamental skills. These students have a basic knowledge of content, but struggle in applying knowledge and skills in problem situations.
- *Unsatisfactory:* Students who perform at the unsatisfactory level on the Kansas state assessments demonstrate a lack of core knowledge, skills and concepts. Their command of the content is very limited and their ability to apply knowledge or demonstrate understanding is minimal.

A brief snapshot of AYP in mathematics may provide a hint of what may come to pass in science. Determination of AYP for mathematics includes results on the state's mathematics assessment, participation rates on the state's mathematics assessment, attendance rates for elementary and middle schools, and graduation rates for high schools. To meet AYP in mathematics a school must fulfill four criteria: (1) all of its students and all appropriate disaggregated subgroups must meet or exceed the annual measurable target in mathematics; (2) 95% or more of its students must take the state mathematics assessment; (3) at least a 75% graduation rate for a high school; and (4) at least a 90% attendance rate for an elementary or middle school. The AYP criteria for mathematics from 2001-2002 through 2013-2014 are presented in Table 7.1. The growth rate target for 2003-2004 AYP in mathematics is 53.5% at or above proficient for grades K-8, and 38.0% at or above proficient for grades 9-12. The minimum disaggregated group size is 30 students, except for students with disabilities, which is 40.

If a school does not meet AYP under the target percentage criterion, there exist two alternative avenues for meeting AYP. One is via a 99% confidence interval statistical analysis of the school's mean percentage score and the AYP target; the other is via safe harbor. Establishment of safe har-

Table 7.1. Adequate Yearly Progress Percentage Targets of Proficient or Above for Mathematics

	Grade Levels	
School Year	*K-8*	*9-12*
2001-02	46.8	29.1
2002-03	46.8	29.1
2003-04	53.5	38.0
2004-05	60.1	46.8
2005-06	60.1	46.8
2006-07	66.8	55.7
2007-08	73.4	64.6
2008-09	77.8	70.5
2009-10	82.3	76.4
2010-11	86.7	82.3
2011-12	91.1	88.2
2012-13	95.6	94.1
2013-14	100.0	100.0

bor involves two conditions: (1) at least 95% participation rate on the state mathematics assessment; and (2) at least a 90% attendance rate (elementary and middle school) or a (75% graduation rate (high school) or demonstration of improvement from the previous year. A school is considered in safe harbor if the percentage of students below proficient decreases at least 10% from the previous year. If the decrease is less than 10%, a 75% confidence interval is applied. If safe harbor is established, the school is considered to have met AYP.

All of this information is contained in various report cards for individual schools, districts, the state as a whole, and in a variety of reports available on the KSDE (2005c) Website. The state requires report cards for each school; the federal government requires district and state reports. Each report card is published every fall; it must present at least 2 years of data as a whole and disaggregated in several categories: (1) socioeconomic status; (2) disability status; (3) gender; (4) race/ethnicity; (5) English proficiency; and (6) migrant status. Below are three tables of state science data for 2001 and 2003, which is the most recent year for which data are available. Science is assessed at grades 4, 7, and 10. Table 7.2 contains the aggregated state data for all students at all levels of achievement on the state science assessment. Tables 7-3 and 7-4 present the data

Table 7.2. Percentages of All Students at All Levels of Achievement across All Assessments

		Levels of Achievement				
Grade	*Year*	*Unsatisfactory*	*Basic*	*Proficient*	*Advanced*	*Exemplary*
4	2001	11.5	25.6	24.4	26.1	12.5
4	2003	8.7	23.3	24.6	28.3	15.1
7	2001	16.3	21.4	26.1	23.7	12.5
7	2003	14.4	21.0	26.0	24.7	14.0
10	2001	25.4	23.4	21.8	17.3	12.1
10	2003	23.4	22.6	22.0	18.4	13.6

Note: In 2003, 0.2% not tested at grade 4; 0.3% at grade 7; 1.1% at grade 10.

for two disaggregated groups, economically disadvantaged students and limited English proficient students, respectively.

Examination of the percentages in Table 7.2 indicates that, as a whole group, Kansas' students are making progress in terms of their achievement. At all assessed grades (4, 7, and 10), the percentage values in the unsatisfactory and basic categories decrease from 2001 to 2003. The proficient level remains nearly constant from 2001 to 2003, with 0.2% increases at grades 4 and 10 and a 0.1% decrease at grade 7. At the advanced and exemplary levels, percentages increase from 2001 to 2003 at all grades assessed. A reasonable interpretation of this pattern of change in the percentages is a general movement upward in achievement at all grades assessed, with movement into proficient from below (unsatisfactory and basic) being roughly equal to movement out of proficient and into the higher levels (advanced and exemplary) achievement.

Table 7.3 contains achievement information for the disaggregated group of Kansas students who are classified as economically disadvantaged. When compared with the pattern of values in Table 7.2, the pattern of values in Table 7.3 shows higher percentages in the unsatisfactory and basic categories, lower percentages at the proficient level, with one exception (grade 4 in 2003), and lower percentages in the upper levels of achievement. The well documented national achievement gap between economically disadvantaged students and the group as a whole is readily apparent. The general pattern of change in percentage values observed in Table 7.2 is repeated in Table 7.3, with two exceptions. First, at grade 10, the percentage of economically disadvantaged students at the basic level of achievement remains unchanged (25.6%) from 2001 to 2003. Second, the proficient level shows net gains for all assessed grades. Moreover, the decreases in the percentage values for the unsatisfactory and basic lev-

Table 7.3. Percentages of Economically Disadvantages Students at All Levels of Achievement across All Assessments

		Levels of Achievement				
Grade	*Year*	*Unsatisfactory*	*Basic*	*Proficient*	*Advanced*	*Exemplary*
4	2001	20.9	35.0	22.6	16.4	5.2
4	2003	15.9	33.3	24.9	19.1	6.7
7	2001	31.2	28.4	22.6	13.6	4.2
7	2003	27.3	27.9	23.0	15.5	6.3
10	2001	43.5	25.6	16.4	9.4	5.1
10	2003	40.7	25.6	17.5	10.4	5.7

Table 7.4. Percentages of Limited English Proficient Students at All Levels of Achievement across All Assessments

		Levels of Achievement				
Grade	*Year*	*Unsatisfactory*	*Basic*	*Proficient*	*Advanced*	*Exemplary*
4	2001	26.9	41.8	19.6	8.7	2.9
4	2003	25.8	41.7	20.6	10.0	1.9
7	2001	64.9	25.4	7.7	1.3	0.7
7	2003	52.9	29.1	12.2	4.2	1.7
10	2001	66.7	23.8	4.8	4.8	0.0
10	2003	66.3	20.0	9.4	2.8	1.4

els are substantially larger than the corresponding values for all Kansas students. Finally, values at the upper levels of achievement (advanced and exemplary) show increases in all instances. Like the group as a whole, economically disadvantaged students exhibit a general movement upward in terms of achievement, but movement into proficient from below (unsatisfactory and basic) is larger than movement out of proficient and into the higher levels (advanced and exemplary) of achievement.

Table 7.4 contains achievement information for the disaggregated group of Kansas students who are classified as limited English proficient (LEP). The pattern of percentage values in Table 7.4 compared to Table 7.2 reveals a more severe gap in achievement for LEP students compared to the group as a whole. The pattern of change between 2001 and 2003 values, however, is similar in general to the growth patterns indicated in Tables 7-2 and 7-3.

All of this provides a clear snapshot of how NCLB has influenced standards as policy as well as programs and practices, including science and beyond, in Kansas. Despite the controversy surrounding the law, policy makers at the state level have implemented NCLB because it is now the law, and education in Kansas will abide by it. Regarding the nature of the implementation, NCLB has become the broad umbrella under which the existing QPA framework now operates. Included, of course, are all of the issues discussed earlier herein, such as aligning instruction and assessment with standards. All of these are now being brought into alignment with NCLB.

Controversy Over NCLB—The Influence of Belief Systems and Values

The increased federal role in K-12 public education through NCLB has brought controversy as strong supporters and equally strong critics debate the pros and cons of the law as well as the motivations of its advocates. Supporters of NCLB cite the need to be able to assess student progress, to hold schools accountable for student progress, and to provide parents with choices when the progress of their schools fails to meet acceptable levels. Critics acknowledge supporters' views but they express strong concerns about the motivations behind and implementation of NCLB.

Freeman (2005) points out the existence of a poorly hidden agenda behind NCLB, an agenda that is the latest expression of a 20-year lobbying effort to privatize public education. According to Freeman, the motivation of this lobbying effort is substantially ideological:

> they dislike anything that smacks of government control, the more so if the service is effective, for such examples repudiate the theological superiority of all things private. Some of its motivation is directed toward right-wing social engineering: they want to control the curriculum that future generations of American students must absorb. And much of it is simply economic: these "prophets of profit" want to get their hands on the $500+ billion that is spent every year in the U.S. on public K-12 education.

This is not, per se, bad. We do, after all, live in at least a quasi-capitalist society where the pursuit of profit is not a social evil. But it's the bashers' hypocrisy that rankles. They do not declare any of these motives openly. Rather, they talk of such vaguely incongruous motives as "empowering minorities" and "streamlining" education. These, of course, are the same corporate zealots who brought the "magic of the market" to a formerly vibrant public health system. They are the pious do-gooders (remember Enron?) who bestowed energy privatization on California, the better to reap the "'efficiencies' of competition. They are the same bleeding-heart altruists who profess wanting to 'save' social security by turning it over to the tender mercies of the financial services industry" (pp. 1-2).

Freeman's (2005) thesis is embodied in the title of his article: "Is Public Education Working? How Would We Know?" He argues that the venerable Scholastic Aptitude Test (SAT) alone possesses sufficient stature as an assessment instrument due to its extensive history of use, a nation wide scope, and comprehensive character. The SAT possesses five specific qualities that would permit decision makers to conduct a sound analysis of his thesis. First, SAT data can be used to examine trends across several generations because it has been used for over 40 years. Second, the SAT can function as a cumulative assessment of K-12 education because it is administered in a student's junior or senior year of high school. Third, SAT data are not susceptible to variations in states' assessments and accountabilities because the SAT has been and can be administered across the country. Fourth, the concept of grade inflation is not part of SAT data. Fifth, the focus of the SAT is broad intellectual development, which includes cultural knowledge, logic, communication, mathematical computation, and specific academic content.

Freeman then offers some results from the SAT that shed light on his thesis. First, the 2004 SAT overall scores set 30-year highs, including 28-year highs in English, and 36-year highs in mathematics. These record scores include not only Caucasians but also Native Americans, Latinos, African Americans, and Asian Americans. A record 48% of 2.9 million seniors across the United States took the SAT, and a record 36% of these were minorities. All of these records were achieved with a larger, more diverse cadre of students compared to 30 years ago, when only elite students, about 15% of the general student population took the test.

Several societal changes make these results even more impressive in Freeman's eyes. First, most mothers are now members of the workforce. Second, the last decade has witnessed the largest infusion of immigrants into U.S. public education in the nation's history. Most were nonEnglish speakers on their arrival and possessed poor educational backgrounds relative to the United States. Third, public schools are now responsible for many more services than in the past, and many of these services are not related to academic achievement. Fourth, the nation's students are now socialized by television more than ever before. Only 9% of their time is spent in class. Fifth, the advent of equal opportunity has seen millions of the nation's top teachers leave teaching, especially females who now have open doors to careers in business, medicine, law, and other fields. Freeman concludes by saying,

> the question of whether public schools can deliver should no longer be open for debate. The only question is whether we have the courage to now properly fund public education so that it can take our children and our society to even higher levels of achievement.... Public education is not only the most important democratizing institution in American today. It is the foundation of our economic future as well. (p. 4)

Robert Sternberg (2004), noted professor of psychology and education at Yale University, views NCLB as well intended because it responds to schools' need for accountability and calls for rigorous, scientifically research-based practice in education. Sternberg laments, however, that NCLB is failing to achieve its intended goals because "it flies in the face of much of what we know about the science of education" (p. 56). He goes on to identify and briefly discuss 12 reasons why NCLB will continue to fail: (1) No clear standards of accountability for the NCLB accountability standards, which are arbitrary and perhaps punitive; (2) penalizing schools for student characteristics related to achievement over which schools have no control, such as socioeconomic background; (3) an inflexible accountability system with only one measurement for all students that penalizes schools with students who possess diverse learning needs; (4-5) extremely high stakes encourage schools to cheat and to urge low achieving students to drop out; (6) NCLB assessments emphasize low level knowledge such as memorization rather than how to use knowledge in complex ways; (7) basic reading, writing, and arithmetic are receiving increased emphasis to the extent that higher level learning and applications are being omitted from the curriculum; (8) good science must be determined within scientific communities rather than by politicians; (9) conventional multiple-choice and short-answer assessments are deemed sufficient to measure all knowledge; (10) schools are turning curricula into preparation centers for tests; (11) NCLB is a requirement that schools cannot afford but must enact; and (12) although NCLB was enacted into law with bipartisan support, that support no longer exists, nor do many of the schools across the country support NCLB. Sternberg closes his criticism as follows:

> No Child Left Behind is an act used to produce the nation's educational report card. But, it, itself, receives a failing grade. Schools are being straightjacketed in attaining what is best for our children, and straitjackets cannot produce the kind of flourishing education system our children need and deserve. (p. 42)

Controversy Over Evolution—The Influence of Belief Systems and Values

Examining the career and work of Elliott Eisner, Uhrmacher and Matthews (2005) cite Eisner's view that the source of controversies in education as school reform occurs is competition between sets of values and goals. Arguing that values and goals are founded in ideologies, Eisner asserted that, "Ideologies are belief systems that provide the value premises from which decisions about practical educational matters are made" (Jackson, 1992, p. 302). In a democracy such as the United States, there exists extensive diversity among citizens as well as leaders in terms of their values and underlying belief systems.

Two values that most Americans share as a culture are faith and reason. Sometimes faith and reason collide, and the subject of teaching evolution in public schools is a prime example of how these values and their underlying belief systems have influenced state science standards. Moreover, arguments over evolution seem to lie at the core of Sternberg's (2004) eighth reason why NCLB is failing.

Controversy in public schools concerning the teaching of evolution moved to center stage on a national level in 1925 with the trial of John Scopes, a biology teacher in Dayton, Tennessee. Although widespread public attention has waxed and waned in the 80 years since the Scopes trial, the controversy itself has continued steadily, merely submerging and surfacing in the consciousness of the public. Whereas widespread public controversy over evolution did not surface during development of the *NSES*, the development of state science standards has sometimes been caught up in divergent, conflicting beliefs among citizens and decision makers about what constitutes good science. Nowhere has such controversy been more open to public viewing, and nowhere has it been a clearer snapshot of school reform in a democracy in the era of standards as in Kansas. I noted earlier herein the nature and extent of my involvement in the development of state science standards for the KSBE in 1998-1999. Returning to that endeavor, I now focus on the controversy that erupted as well as its sources in discordant values and belief systems.

Christian fundamentalists and other religious conservatives have long served on the KSBE. The religious beliefs of the expert committees of science teachers, university science educators, and scientists that have developed science standards for the board have not been consistent with fundamentalist Christian doctrine. When Christian fundamentalists and religious conservatives have represented a majority, controversy has erupted. Perhaps the only commonalities between the two groups—Christian fundamentalist and conservative board members and members of science standards development committees—were their goals and the importance of their work. Each group had the same goal in mind, to produce the best possible science standards and through these standards to serve all children in all 301 Kansas school districts. Both groups also considered the development and adoption of science standards to be of prime importance. Controversy developed because the two groups worked toward a common goal from different belief systems and values.

The writing team's view was that evolutionary theory in particular and the science standards developed by the team in general reflected well established, mainstream science. Early on, the writing team embraced the vision of the *NSES* (NRC, 1996) and adopted the *NSES* content framework, including the unifying concepts and processes. Evolution is a central concept of the content standards in the *NSES*, and evolution is one-half of the backbone—equilibrium is the other half—of one of five unifying concepts and processes in the *NSES*. The writing committee built on the foundation of the *NSES* because its members felt that the 5-year development process was fundamentally sound, and that the process had produced an excellent visionary document, one that could guide reform in Kansas.

The *NSES* development process included extensive peer and client review in the form of critiques and reviews by 22 science education and scientific organizations as well as wide spread state and local participation by over 18,000 scientists, science educators, teachers, school administrators, and parents (NRC, 1997).

I stated earlier herein that the writing team delivered its initial draft of standards to the KSBE in January, 1999. Throughout the winter months, several KSBE members expressed their concern with the writing committee's characterization of evolutionary theory as good science in the writing committee's early drafts. During March and April, 1999, staff and members of the Creation Science Association for Mid-America (CSAMA) developed an alternative draft (CSAMA, 1999a) for a member of the KSBE, who then presented it to the full board at its May, 1999 meeting (Ackerman & Williams, 1999). The CSAMA alternative draft (CSAMA, 1999a) was used as a negotiating tool in board members' discussions among themselves and with the writing team throughout the spring and summer months of 1999. Unable to garner a six-vote majority for the CSAMA alternative draft, supportive board members asked the CSAMA authors to prepare a revision of their alternative draft, and the CSAMA authors complied (CSAMA, 1999b). After examining the CSAMA revision, at least five board members indicated in informal conversations with their board colleagues that they would not support the revised CSAMA draft in an official board vote. Board members also attempted to convince the writing team to remove evolution. Failing in this quest, these board members then edited the writing team's work in late July, 1999 to produce a document that could get six votes. On August 11, 1999, the KSBE approved this edited version of the writing committee's work by a 6-4 vote.

The board's action led to immediate attention in the local, regional, and national print and electronic media. Kansas Governor Bill Graves, a moderate Republican, immediately issued a news release stating, "This is a terrible, tragic, embarrassing solution to a problem that didn't exist" (Myers, 1999). On August 14, 1999, I appeared on the "Today Show" with a board member who voted to approve the edited standards to discuss the issues with host Matt Lauer.

I assumed then, and still do, that the board member consulted the CSAMA to develop an alternative draft of the writing team's science standards because his beliefs shared much in common with the beliefs of the organization. A reading of the CSAMA Website (http://www.csama.org) on December 23, 2004 provides much insight about the organization. Under the category "What is CSA?":

> CSA is a non-denominational, independent, non-profit, educational and research corporation whose members are concerned about the widespread false teaching called "evolution." The widespread acceptance of this false notion of origins has resulted in physical harm to millions in this century alone, in lawlessness in our society, in the deprivation of a proper relationship with their Creator for countless people.

Regarding CSA's goals:

> To educate people regarding the vast amount of scientific evidence that supports Biblical Creation as the true account of origins, and that the General Theory of Evolution is not only a false notion of history, it is an extremely dangerous one, the fruits of which have destroyed entire nations including the wanton slaughter of at least 100 million people in this century.
>
> To inspire faith in unbelievers and encourage the faith of believers, in the Bible as the Word of God, and therefore the only trustworthy source of information regarding the meaning, purpose, destiny and conduct of human lives. To show that Biblical Creation, because it is true, is the only "scientific" explanation of origins, and therefore is the only account of origins that can possibly be useful to science.

Regarding CSA's beliefs:

> The Bible is the divinely inspired written Word of God. It is the supreme authority in all matters of faith and conduct.
>
> The final guide to the interpretation of Scripture is Scripture. Human "scientific pursuits" have shed precious little light on the meaning, intent, or content of the Revelation of God. Whether those pursuits were in the so-called "historical sciences" or in the "testable, repeatable, and falsifiable sciences," in virtually all cases of claimed "improved knowledge" over revelation, the "improvement" was either already available to the believing and discerning reader of Scripture, or the "improvement" was illusory. The account of origins presented in Genesis is a simple, but factual, presentation of actual events, and, therefore, it provides a reliable framework for scien-

> tific research both in the historical sciences and in the operational sciences. The fact of Creation and the fact of its Creator are an integral part of the Gospel of Jesus Christ.
>
> The scientific implications of Creation are important, but they are SECONDARY to the proclamation of the Gospel of Jesus Christ.

Clearly for CSAMA as an organization, faith and reason clash over evolution and CSAMA takes the position that faith in Holy Scripture trumps reason in the form of science.

Although neither version of the CSAMA's alternative standards was adopted by the board, the board members' edits of the writing committee's fifth draft shed a great deal of light on their values and beliefs. The text of the board-approved standards contained numerous modifications and additions; the origin of much of the new text is the CSAMA standards. Further, numerous pieces of text in the writing team's standards were removed in producing the board-approved standards. Results of a content analysis show that board editors made approximately 300 changes. Moreover, the content analysis shows that board editors targeted three science content themes with their changes: evolutionary theory; ancient earth and universe concepts; and the nature of science (Staver, 2000). Although the national, regional, and local print and electronic media emphasized the removal of the core of evolutionary theory in their coverage of the Kansas board's action, I viewed the portrayal of the nature of science in the board-approved standards as the most pervasive change. This description represents an understanding of the nature of science that approving board members and the CSAMA writers view as consistent with their fundamentalist Christian views.

What of the beliefs of the 27 members of the expert writing committee? Although no formal census of team members' religious beliefs was ever taken, my informal conversations over the time that all of us worked together led me to conclude that a great deal of diversity existed. One team member's beliefs held much in common with beliefs expressed by religious right board members and the CSAMA. This member also expressed concerns about evolution in the committee's science standards but was the only one to do so. A majority of writing committee members held religious beliefs, but they did not view evolution as in conflict with their faith. Finally, some members were agnostic or atheistic about God.

I am very familiar, however, with my own religious beliefs. And, I take time herein to share my beliefs because they are consistent in general with the majority of writing team members who themselves believe in God and simultaneously see no conflict between evolution in particular or science in general and their religion. To begin, I was baptized in a Presbyterian tradition, and I grew up and was confirmed in a Lutheran tradition. As an adult, I have been a member of a Presbyterian church as well as a member of its session (church governing body), and I have also been and am now a member of a United Methodist church. Thus, I describe myself as a Christian within a mainstream Protestant tradition. The official doctrines of each variation of these traditions—Lutheran, Presbyterian, and United Methodist—although quite different in some respects, share a common position with respect to evolution, in that none are in conflict with science in general or with evolution in particular.

Perhaps the most succinct manner in which I can illustrate this no-conflict position is to present a short sequence of questions that I often use to make this point when I am giving presentations to science teachers about creationists' arguments that science, its methods, and its underlying philosophy, are all atheistic and cannot be separated. I assert that scientific methods and the variety of philosophical positions that underlie these methods can be separated, and I make my point by analogy. Depending on more than 15 years of Gallup Poll data showing that 9 of 10 Americans are believers, I ask: "Please, hold up your hand if you consulted your personal religious views (e.g., prayed) in deciding where to sit in the room?" Only once has a teacher held up his or her hand, and anyone who does so is excluded from the next question. I then point out that those who did not use their personal religious views in deciding where to sit made their decisions using an informal version of the same methodological naturalism that scientists use to conduct their research. Then I ask: "If you did not hold up you hand before, please raise your hand if you believe that you rejected your personal religious views because you did not use them in deciding where to sit?" No one has ever held up his or her hand in response to my second question. I concur with their response, and by analogy, believing scientists (and there are many) who do not consult their personal religious views in doing science do not become atheists because they do so.

I often support my point with a quote from Pope John Paul II (1984). This quote is from a speech given by John Paul II to the Pontifical Academy of Sciences on October 3, 1981:

> Cosmogony itself speaks to us of the origins of the universe and its makeup, not in order to provide us with a scientific treatise, but in order to state the correct relationship of man with God and with the universe. Sacred Scripture wishes simply to declare that the world was created by God, and in order to teach this truth, it expresses itself in terms of the cosmology in use at the time of the writer. The sacred book likewise wishes to tell men that the world was not created as the seat of the gods, as was taught by other cosmogonies and cosmologies, but was rather created for the service of man and the glory of God. Any other teaching about the origin and makeup of

the universe is alien to the intentions of the Bible, which does not wish to teach how heaven was made but how one goes to heaven. (p. 97)

The words of John Paul II reveal a central commonality and difference in Christian theology that seems to lie at the core of the controversy. Both fundamentalist and mainstream Christian doctrines view Holy Scripture as the ultimate written source for understanding our relationship as humans with God, our Creator. Fundamentalist Christian doctrine, however, also sees the Bible as a factual account of origins and, therefore, as a science book. Mainline Christian doctrine does not share this view. It comes, therefore, as no surprise that fundamentalist Christian doctrine and science are in conflict, because both share the same goal, explaining reality, but they rely on different ways of knowing. The interface of science and religion need not be a zone for clashes in the minds of many scholars (e.g., Gould, 1999; Polkinghorne, 1998), and I have relied on some of this thinking to construct my own position. Most helpful in this personal quest was a small conference on the emerging concept of intelligent design that I attended in the summer of 2000. Diogenes Allen, a professor of philosophy of religion at Princeton Theological Seminary, gave the banquet speech, during which he pointed out:

God is indeed the ultimate source of order, but there must be in nature organizing principles and properties as proximal sources of order as well. For God to effect God's will, nature must have many more properties than the mechanical philosophy granted. In order to act in accordance with God's will there must be in the creation the capacity to carry out what God wills by the exercise of God given powers. So, in addition to the ultimate source of order, there must be proximal sources of order as well. (Diogenes Allen, 2000, June 22-24, Banquet Speech)

These words helped me understand science as a powerful and useful proximal source of order, one that is consistent with God's creation and will. That science and God's will and creation are consistent remains a matter of faith, but I see no reason why religious scholars and scientists should not pursue further understanding of the interface of science and religion. Establishing new scientific and theological knowledge at this interface, however, must be allowed to take place in and abide by the accepted procedures of the respective scholarly communities. Sadly, current attempts to insert Intelligent Design Theory, a well known 200 year old religious concept, into school science as an alternate scientific theory to evolution have yet to include the necessary scientific scrutiny of the concept. In Sternberg's (2004) words, "science has always proceed best when it is left totally independent of the political process, and when competing schools of thought are left to slug it out on the scientific battle field free of political influence or interference" (p. 56).

In the 3 months following the board's action on August 11, 1999, I spent perhaps 25% of a typical work day responding to inquiries from members of the print and electronic media. One point that I repeatedly made to reporters was that democracy got us into this and democracy would get us out, too. Whereas this assertion was based on my faith in the presence of a heretofore silent, moderate-to-conservative, and open minded Kansas citizenry as well as the capacity of the democratic process itself to redirect seemingly extreme actions, my faith was confirmed as we moved through the winter months of 2000 and to the summer primaries and fall general elections.

The KSBE is an elected body of 10 members. Each member is elected by voters in one of 10 geographical districts throughout the state. A member's term is 4 years, and five members, one-half of the board, stand for election every 2 years. Kansas voters expressed their embarrassment and disagreement with three of the five members in the 2000 elections. Four of the five members whose terms required them to stand for election in 2000 were part of the six majority votes on August 11, 1999. One moved out of the state, and a new, moderate Republican was elected to replace him. Two others, including the board president, were defeated in their August, 2000 Republican primary elections, and the winners were then elected in the general election in November, 2000. The fourth person, who had submitted the CSAMA alternative drafts, was reelected. The fifth person, who had voted with the four-member minority on August 11, 1999, was reelected.

The results of the Republican primary elections, held in early August under the hot Kansas sun, clearly suggested a shift toward moderation for the board that would take its seat in January 2001. Seeing this, I contacted my cochair immediately following the August primaries, and we reconvened the writing committee, worked through the late summer and fall months, and produced a sixth draft of our standards. This draft was presented to the new KSBE at its January, 2001 meeting. The board studied the draft for a month, then approved it at its February, 2001 meeting. Coincidently, this occurred on February 14, Valentine's Day, and CNN cameras were present to share the discussion and decision with the world.

Time marches on, however, and as I write this chapter in November and December of 2004 and early January of 2005, I am distressed to say that democracy has gotten Kansas back into this again. The 2002 elections left the board split evenly (5-5) between the religious right and moderates. The results of the 2004 primary and general elections gave the religious right a six-member majority on the KSBE beginning in January, 2005. Moreover, a new, 26-member team of teachers of science, university science educators, and scientists has been revising the state science standards since summer. This new writing team presented its first draft of revisions to the current board in November, 2004. Following a 20-minute presentation by the writing team's cochairs (I am a member but not a cochair of this team), religious right members of the board questioned the cochairs for almost 2 hours. I predict that Kansas will have returned to center stage of national public attention by the time this is published.

In sum, controversy over evolution in Kansas clearly illustrates how differing sets of values and beliefs as well as democracy in action have influenced and continue to influence state policy in the form of the state's science standards. When able to do so, KSBE members who hold fundamentalist Christian beliefs and values enact policy that is consistent with their religious beliefs but is at odds with the beliefs and values of their own committees that were selected to develop science standards for the board because of their represent expert knowledge in modern science and science education.

CONCLUDING THOUGHT

The process of reform in a democracy will never be perfectly coherent because of the diversity of values and belief systems of those who are involved. But, diversity is an essential strength in a democracy just as it is an essential strength in biology. For democracy to thrive, all citizens must participate. All the while, we must remember John Dewey's (1916) admonition almost a century ago that the concept of communities rests on the things that community members have in common. Moreover, we communicate in order to share and hold onto common values and beliefs. Prime among these commonalities are language, history, values, and knowledge. Together these help form a common culture (Hlebowitsh, 2001). We must also continue to communicate with and respect those with whom we do not share common values and beliefs.

REFERENCES

American Association for the Advancement of Science Project. (1993). *Benchmarks for science literacy.* New York: Oxford University Press.

Ackerman, P., & Williams, B. (1999). *Kansas tornado: The 1999 science curriculum standards battle.* El Cajon, CA: Institute for Creation Research.

Allen, D. (Speaker). (2000). *Banquet speech* (Cassette Recording No. TS-51-10). Springdale, Arkansas: Orthodox Christian Cassettes.

Atkin, J. M., & Black, P. (1997). Policy perils of international comparisions: The TIMSS case. *Phi Delta Kappan, 79*(1), 22-28.

Biological Sciences Curriculum Study. (1992). *Science for life and living*. Dubuque, IA: Kendall/Hunt.

Biological Sciences Curriculum Study. (1999). *BSCS science t.r.a.c.s.* Dubuque, IA: Kendall/Hunt.

Bloomfield, D. C., & Cooper, B. S. (2003). NCLB: A new role for the federal government: An overview of the most sweeping federal education law since 1965. *T H E Journal, 30*(10), Supp. 6-9.

Bransford, J. D., Brown, A. L., & Cocking, R. R. (Eds.). (2000). *How people learn: Brain, mind, experience, and school.* Washington, DC: National Academy Press.

Bybee, R. W., & Sund, R. B. (1982). *Piaget for educators* (2nd ed.). Columbus, OH: Merrill.

Bybee, R. W. (1997). *Achieving scientific literacy: From purposes to practices.* Portsmouth, NH: Heinemann.

Center on National Educational Policy. (1996). *Do we still need public schools?* Bloomington, IN: Phi Delta Kappa.

Cohen, M. R., Cooney, T. M., Hawthorne, C. M., McCormack, A. S., Pasachoff, J. M., Rines, K. L., & Slesnick, I. L. (1991). *Discover science.* Glenview, IL: Scott, Foresman & Company.

Creation Science Association for Mid-America. (1999a). *Kansas science education standards working trial draft 4a*. Cleveland, MO: Author.

Creation Science Association for Mid-America. (1999b). *Kansas science education standards working draft CDC 8a*. Cleveland, MO: Author.

Dewey, J. (1916). *Democracy and education.* New York: Free Press.

Freeman, R. (2005, January 3). *Is public education working? How would we know?* [On-line], pp. 1-4. Available: http://www.commondreams.org/views05/0103-22.htm

Gould, S. J. (1999). *Rocks of ages: Science and religion in the fullness of life.* New York: Ballantine.

Hackett, J. K., Moyer, R. H., & Adams, K. K. (1989). *Merrill science.* Columbus, OH: Merrill.

Hlebowitsh, P. S. (2001). *Foundations of American education: Purpose and promise* (2nd ed.). Belmont, CA: Wadsworth.

Jackson, P. (Ed.). (1992). Curriculum ideologies. In *Handbook of research on curriculum* (pp. 302-326). New York: Macmillan.

Kansas State Board of Education. (1993). *Quality performance accreditation document—Revision 1993*. Retrieved January 3, 2005, from http://www.ksde.org/outcomes/qpa.html

Kansas State Department of Education. (2005a). *Report card 2003—2004.* Retrieved January 3, 2005, from http://online.ksde.org/rcard

Kansas State Department of Education. (2005b). *Definitions*. http://Kansas State Department of Education. (2005c). Retrieved January 3, 2005, from http://ww.ksbe.state.ks.usonline.ksde.org/rcard/definitions.aspx?org_D%&rpt_type=3

Karplus, R., Thier, H. D., Knott, R., Lawson, C. A., & Montgomery, M. (1970). *Science curriculum improvement study.* Chicago, IL: Rand McNally.

Klopfer, L. P., & Champagne, A. B. (1990). Ghosts of crisis past. *Science Education, 74*(2), 133-154.

Lawson, A. E., Abraham, M. R., & Renner, J. W. (1989). *A theory of instruction: Using the learning cycle to teach science concepts and thinking skills.* Monograph #1 of the National Association for Research in Science Teaching. Cincinnatii, OH: National Association for Research in Science Teaching.

Lerner, L. S. (2000). *Good science, bad science: Teaching evolution in the United States.* Washington, DC: Thomas B. Fordham Foundation.

Mallinson, G. G., Mallinson, J. B., Froschauer, L., Harris, J. A., Lewis, M. C., & Valentino, C. (1991). *Science horizons.* Morristown, NJ: Silver, Burdett, & Ginn.

Merriam-Webster. (1970). *Webster's seventh new collegiate dictionary.* Springfield, MA: G & C Merriam.

Myers, R. (1999, August 14). Evolution de-emphasized. *The Topeka Capital-Journal,* p. 1A.

National Council of Teachers of Mathematics. (1989). *Curriculum an evaluation standards for school mathematics.* Reston, VA: Author.

National Research Council. (1996). *National science education standards.* Washington, DC: National Academy Press.

National Research Council. (1997). *Introducing the national science education standards.* Washington, DC: National Academy of Sciences.

No Child Left Behind. (2002). *No child left behind act of 2001.* Pub. L. No. 107-110, 115 Stat. 425.

Paul, J., II. (1984). Roman Catholic Church. In M. Matsumura (Ed.), *Voices for evolution* (p. 97). Berkeley, CA: National Center for Science Education.

Polkinghorne, J. (1998). *Belief in God in an age of science.* New Haven, CT: Yale University Press.

Regents. (1993). *Full option science system.* Chicago, IL: Encyclopaedia Britannica Educational Corporation.

Rutherford, F. J., & Ahlgren, A. (1990). *Science for all Americans.* New York: Oxford University Press.

Staver, J. R. (2000, May). *Evolutionary theory: A clash between public understanding of science and standards-based science education.* Paper presented at the annual meeting of the National Association for Research in Science Teaching (NARST), New Orleans, LA.

Sternberg, R. J. (2004, October 27). Good intentions, bad results: A dozen reasons why the no child left behind act is failing our schools. *Education Week,* pp. 56, 42.

Thomas, M. D., & Bainbridge, W. L. (2002). No child left behind: Facts and fallacies. *Phi Delta Kappan, 83*(10), 781-782.

Uhrmacher, P. B., & Matthews, J. (2005). *Intricate palette: Working the ideas of Elliot Eisner.* Upper Saddle River, NJ: Pearson Merrill Prentice Hall.

PART III

IMPACT OF SCIENCE STANDARDS ON TEACHING

Part three examines the impact of science standards on learning and teaching behaviors in K-12 classrooms. Chapter eight discusses a pedagogical shift in viewing teaching and learning science brought about by the science standards. Connie Gable explores the impact of the reform efforts on K-12 science inquiry from a systems perspective. Chapter nine describes the impact of science standards on the development of students' reasoning and argumentation skills with peers during classroom inquiry activities. In this chapter, Cynthia Sunal investigates how students develop the skills necessary to ask questions, gather and analyze evidence, and argue their interpretations of the evidence with their peers. Chapter ten discusses impact and process of curriculum alignment and assessment to the science standards. Luli Stern and Jo Ellen Roseman describe a procedure and research-based criteria for judging the alignment of curriculum materials and their assessments to specific ideas in the national science standards and for evaluating the quality of instructional support tied to those ideas.

CHAPTER 8

IMPACT OF THE REFORM EFFORTS ON K-12 SCIENCE INQUIRY

A Paradigm Shift

Connie Gabel

This chapter explores the impact of the reform efforts on K-12 science inquiry from a systems perspective. Research indicates that the epistemological and pedagogical paradigm shift that is advocated in the *National Science Education Standards* is not being met. The problem stems from the complexity of science inquiry and a lack of knowledge regarding the implementation of science inquiry in the classroom. A science inquiry model is proposed. A number of key elements are incorporated into the science inquiry model. These key elements include using an inquiry continuum, scaffolding with fading, linking science inquiry skills, using a schema to anchor learning, utilizing questions appropriately, relating prior knowledge, applying brain-based research, teaching for depth of understanding, using a group process, communicating to others, constructing knowledge, and integrating knowledge. Research and discussion of this model as well as implications are provided.

The Impact of State and National Standards on K-12 Science Teaching, 215–256

State and national education standards and the ensuing accountability of educators in the public schools have been among the most hotly debated topics in the history of United States education. Media sources have brought constant reminders, updates, and accountability scorecards to our attention on this subject. The origin of the current reform efforts can be traced to the National Science Board's report entitled *Today's Problems, Tomorrow's Crises* (National Science Foundation, 1982) in which a crisis was declared in K-12 mathematics, science, and technology education. Secretary of Education, Terrell Bell, responded to the report by appointing an 18-member National Commission on Excellence in Education to examine the crisis. The commission's report, *A Nation at Risk: The Imperative for Educational Reform* (1983) sent shock waves throughout the nation with its dismal findings. The report stated: "Our Nation is at risk. Our once unchallenged preeminence in commerce, industry, science, and technological innovation is being overtaken by competitors throughout the world" (p. 5).

A major response to the identified risks occurred in 1989 at an education summit of the nation's governors, assembled by President George H. Bush, where a plan was discussed for curriculum reform and standards-based education. For the first time, national education goals were set and science was included in those goals: "By the year 2000, U.S. students will be first in the world in science and mathematics achievement" (Malcom, 1993, p. 4). The realization that solutions must be comprehensive, not piecemeal prevailed. As a result of the summit, federal funds were made available for the development of the *National Science Education Standards*, with the *National Research Council* (NRC), the research arm of the National Academy of Sciences, asked to coordinate the development of the science standards (Collins, 1997).

Collins (1997), who served as director of the NRC development team, noted that there are four distinctive features—concerning the science standards. The first emphasizes that all students can learn and understand science. A grounding in science inquiry is a second feature that includes an expanded definition of science incorporating facts, concepts, laws, theories of science subject matter, and the human aspects of science. The third feature focuses on linking standards for science content, teaching, and assessment. Professional development, program, and system standards to support the vision of science education are included as the final feature (Collins, 1997).

Since the publication of the national science standards in 1996, science inquiry has received the least attention from science educators and has slid under the radar screen of public discussion regarding science education in the United States. However, an analysis of the impact of state and national science standards on science teaching would not be complete without a discussion of science inquiry as a valuable component of instruction, which is the focus of this chapter. A major change such as this takes time and often is met with obstacles and resistance. Because the standards promote a new direction from the way that science inquiry has been taught in the past, a new way of thinking about science inquiry in the K-12 classroom is needed. This chapter will first explore the implementation of science inquiry as outlined in the *National Science Education Standards* (NRC, 1996), through the lens of systems change theory. The second part, illustrates a successful method for teaching science inquiry in the classroom that embraces the changes in epistemology and pedagogy contained in the national science standards.

THE LENS OF SYSTEMS CHANGE THEORY

What happens when institutions, such as school systems, try to put a change in place? Rogers (1995), an expert in systems change theory, indicates that innovations may take a long period of time, often years, to become widely adopted. A problem facing many organizations is how to speed up the diffusion rate of the innovation. To become adopted the innovation must be communicated through channels over time among the members of the educational community. Change agents, such as university science educators, can influence a school's informal leaders in adopting and spreading new ideas.

Rogers (1995) notes that whether or not the new idea is adopted depends on a number of factors. First is the relative advantage of the innovation. Second is the compatibility of the innovation, and third is its complexity. Trialability and observability are other factors that influence adoption of an innovation such as the standards for science inquiry. Time is also an element in the adoption process. Rogers also states:

> The *innovation-decision process* is the mental process through which an individual (or other decision-making unit) passes from first knowledge of an innovation to forming an attitude toward the innovation, to a decision to adopt or reject, to implementation of the new idea, and to confirmation of this decision. (p. 36)

In terms of the science inquiry standards, a school system may have adopted an innovation because of requirements imposed by state mandates and/or because of state assessment programs. Therefore, adoption at the district level in states with standards legislation and state assessment programs, such as the *Colorado Student Assessment Program* (CSAP), where the state assessments include science does not appear to be a problem for adoption. However, the problems may appear during implementation when there are misinterpretations about the meaning of science inquiry standards as they apply to curriculum and instruction.

Issues with implementing science inquiry are illustrated in a research study of nine representative school districts in Michigan, conducted by Spillane and Callahan (2000), where adoption was not a problem. The state science standards in Michigan emphasize the construction of science knowledge through inquiry, as well as replacing cookbook laboratory methodology with more authentic science inquiry approaches. Spillane and Callahan found that district policy makers were aware of the state efforts to reform science education. Accordingly, they followed the policy initiatives, and implemented the reforms (as they understood them) in their districts' policies and programs. Even though the district policy makers indicated they responded to the state mandates, only one third of the districts' efforts to develop or revise district policies focused on substantive alignment required by the state standards. Two-thirds of the districts only aligned their curriculum with those obligatory topics in state standards and did not address inquiry. Spillane and Callahan (2000) concluded that limited district level implementation of science inquiry developed from what district policy makers understood about the science standards. The theme *hands-on* predominated the district policy makers' description of their understanding of the reforms contained in national and state standards, but the constructivist methodology called for in the state standards in support of science inquiry was rarely mentioned. Instead district policy makers referred to implementing more laboratories in response to the standards movement rather than a fundamental change in the districts' approach to science inquiry. The majority of districts interpreted the science standards as the need to change instructional strategies to accommodate the different learning styles of students and to interest and motivate students, rather than changes in epistemology and pedagogy that promote science inquiry. Overall, Spillane and Callahan (2000) found that one-third of the districts are in tune with some of the fundamentals of the science standards, but only at a surface level. Therefore, when examined through the systems change lens, the problem is not one of resistance to adoption by these Michigan district policy makers. However, there is a large problem with the adoption. The majority of the districts are not implementing the key changes specified in the science standards because they are misunderstanding the necessary fundamental epistemological and pedagogical changes set forth in the standards.

Change theorists, Watzlawick, Weakland, and Fisch (1974), refer to first-order change and second-order change. A first-order change is one that occurs within a system in which the system itself remains unchanged and a second-order change is one that changes the system. Many of the school districts examined by Spillane and Callahan (2000), therefore, are only undergoing a first-order change. According to Watzlawick et al. (1974):

> The history of science shows very clearly that in the course of time, scientific theories tend to become more and more complex as scientists try to accommodate more and more exceptions and inconsistencies into the overall premises of a theory. It may then take a genius to throw this patchwork out and find a new, elegant set of premises to account for the phenomena under study. But this kind of simplification is then precisely a second-order change. (p. 41)

Preparing both teachers and students for conducting science inquiry is a very complex process. Consequently, many school districts are responding by implementing only a first-order change by adding more hands-on activities and laboratories, rather than the second-order change that was intended in the standards.

Using the problem formulation/problem resolution method of looking at science inquiry modifications through the systems change lens provides additional insight. Problem formulation/problem resolution involves four steps (Watzlawick et al., 1974):

1. a clear definition of the problem in concrete terms
2. an investigation of the solutions attempted so far
3. a clear definition of the concrete change to be achieved
4. the formulation and implementation of a plan to produce this change (p. 110).

An examination of the problem and events leading up to the implementation of the *National Science Education Standards* was provided in the introduction. The second step of the problem formulation/problem resolution methodology that provides a historical perspective of incorporating science inquiry into the curriculum is discussed next for the purpose of identifying solutions that have already been attempted.

HISTORICAL PERSPECTIVE

Science inquiry is certainly not a new phenomenon in K-12 science education. Reviewing past efforts can help put current curriculum reform effort in perspective. Three distinct periods of teaching science inquiry can be identified (Gabel, 2001b, 2004). First, during the nineteenth century, science became part of the K-12 curriculum under the persuasion of scientists who believed that science provided students with the opportunity to make observations and draw conclusions about the natural world using the inductive process (DeBoer, 2000). Intellectual development would be supported when students carried out independent inquiries and investigations in science laboratories (DeBoer, 2000). Therefore, in the late 1800s, laboratories were part of the science curriculum in the United States (Linn, 1997) and students were encouraged to inquire by observing physical objects in nature such as insects, rocks, and leaves (Mintzes & Wandersee, 1998). During this same time period, Armstrong outlined courses of instruction for developing inquiry, observation, and reasoning skills, and in 1903, his book, *The Teaching of Scientific Method and Other Papers on Education* was published (Armstrong, 1903). In the 1890s, The Committee of Ten recommended 4 years of science in high school and advocated the replacing of textbook science with experiences involving natural phenomena (Mintzes & Wandersee, 1998). In the early 1900s, the teaching of the scientific method was also encouraged by science teacher John Dewey (1916) who penned an article in the first edition of the *General Science Quarterly* expressing his belief that methods of science involving problem solving through reflective thinking should be the outcome of science teaching. In the 1930s, process skills became a focus in the curriculum with influence from the Progressive Education Association and the establishment of the Committee on the Function of Science in General Education of the Commission on Secondary School Curriculum (Bybee, 1997). Thus, this first distinct period of incorporating science inquiry into the curriculum occurred in the late 1800s and early 1900s and continued to influence science teaching until the middle of the twentieth century (Gabel, 2001b, 2004).

The second period of the science inquiry instruction is marked by the launch of Sputnik in 1957. The launch of Sputnik provided the impetus to generate public support for math and science teaching as well as increased federal funding (Bybee, 1997; Parker, 1993). The ensuing reform efforts of the 1960s placed a strong focus on inquiry (Parker, 1993; Roth & Bowen, 1993) as there was a perceived need to train students to become scientists (DeBoer, 2000). One common thread among developers of the science curriculum of the 1960s was that they were united in their belief that science inquiry is a major function of the laboratory (Tamir & Lunetta, 1981). Never before in history was there such a large, organized intellectual effort to change science education. The National Science Foundation provided support for curriculum development that would provide the students with many opportunities for inquiry (Roth & Roychoudhury, 1993). The curricular materials that were developed during the 1960s, such as the Biological Sciences Curriculum Study, Chemical Education Material Study, Earth Science Curriculum Project, and Physical Science Study Committee, attempted to create inquiry laboratory-based problems that could be investigated by students of all ability levels. During this time period, scientists developed abstract models of the natural world. Inquiry was the preferred pedagogy because it mirrored the way that scientists did their work and teaching focused on preparing students to be scientists and to understand the work of scientists (DeBoer, 2000). Curriculum reform efforts of the 1960s were also heavily influenced by psychologists such as Joseph Schwab and Jerome Bruner (Mintzes & Wandersee, 1998).

The curricular materials emphasized critical thinking skills and the nature of science, but they gave an erroneous impression of the science inquiry contained in the curriculum. The inquiries were highly structured (Tamir & Lunetta, 1981) and at the lowest level of inquiry (Kyle, 1980) such as having students write down their observations for a burning candle or follow a procedure for chemical reactions. Therefore, the curriculum of the 1960s rarely asked the students to formulate a question to be investigated, to formulate a hypothesis to be tested, to predict experimental results, to design their own experiments, to apply the experimental technique, or to formulate a new question based on the results. However, through the metaanalysis of many research studies, Shymansky, Hedges, and Woodworth (1990) determined that the reformed curricula were more effective in improving student performance on cognitive measures and improving attitudes about science. The new curriculum was definitely better than that which it replaced. DeBoer (1991) noted: "The impact was impressive. Never before had a single curriculum initiative had such a wide-spread effect on science teaching in this country" (p. 167). However, inquiry was not extensively implemented, especially in chemistry and physics (DeBoer, 1991), and the inquiry that was achieved did not continue to exist in the classroom (Bybee, 1997).

The third phase of the science inquiry evolution is the one that began in the 1980s and continues to be part of K-12 science education today. The expectations set forth in the *National Science Education Standards* (NRC, 1996) bring dramatic changes in science education to the table, especially with the emphasis on inquiry as a method for understanding the world and achieving knowledge.

Furthermore, according to the *Standards,* inquiry is a process in which "students describe objects and events, ask questions, construct explanations, test those explanations against current scientific knowledge, and communicate their ideas to others" (NRC, 1996, p. 2). Students "identify their assumptions, use critical and logical thinking, and consider alternative explanations" (NRC, 1996, p. 2). Through the use of inquiry, "students actively develop their understanding of science by combining scientific knowledge with reasoning and thinking skills" (NRC, 1996, p. 2).

This broad characterization of inquiry is definitely a paradigm shift from the cookbook-style laboratories that were incorporated into the 1960s curriculum revisions. Inquiry is prominently featured in the *Standards* with unique tactics as noted in the preface of *Inquiry and the National Science Education Standards:*

> The term "inquiry" is used in two different ways in the *Standards*. First, it refers to the *abilities* students should develop to be able to design and conduct scientific investigations and to the *understandings* they should gain about the nature of scientific inquiry. Second, it refers to the teaching and learning strategies that enable scientific concepts to bemastered through investigations. In this way, the *Standards* draw connections between learning science, learning to do science, and learning about science. (NRC, 2000, p. 15)

Consequently, major adjustments in both epistemology and pedagogy were needed to move from the cookbook-style laboratories used in the past to the expansive science inquiry approach outlined in the *National Science Education Standards*.

A definition of science inquiry as stated in the *National Science Education Standards* provides a starting point.

> Scientific inquiry refers to the diverse ways in which scientists study the natural world and propose explanations based on the evidence derived from their work. Inquiry also refers to the activities of students in which they develop knowledge and understanding of scientific ideas, as well as an understanding of how scientists study the natural world. (NRC, 1996, p. 23)

Therefore, when students are engaged in inquiry, they make observations, pose questions, and examine books and other sources of information to see what is already known. Additionally, they plan investigations and use tools to gather, analyze, and interpret data. Based on experimental evidence, they propose answers, explanations, and predictions and communicate the results to others (NRC, 1996).

According to the *Standards*, "Inquiry requires identification of assumptions, use of critical and logical thinking, and consideration of alternative explanations" (NRC, 1996, p. 23). The *Standards* also stress the importance of conducting complete inquiries by stating, "Students will engage in selected aspects of inquiry as they learn the scientific way of knowing the natural world, but they also should develop the capacity to conduct complete inquiries" (NRC, 1996, p. 23).

Even though inquiry is delineated in the *Standards*, Spillane and Callahan (2000) found that there is a major problem with its clarity in terms of the epistemological and pedagogical paradigm shift that is needed to transform this concrete definition of science inquiry into classroom action. Thus, as we arrive at Watzlawick's Step 3 (a clear definition of the concrete change to be achieved), the process bogs down moving from the passive science inquiry on paper to the active version—carried out by students and teachers in the classroom.

Likewise, the *National Science Education Standards* (NRC, 1996) provide the formulation of a plan (Step 4, Watzlawick et al., 1974) with many details regarding the teaching of science by grade levels, professional development of teachers, science education program standards, and science education system standards. However, as noted earlier, there is a problem with the implementation of the standards in the classroom. Because the standards have not been implemented as intended, perhaps there has been little change in student learning. The next section examines from available assessment data the impact of implementing science inquiry on student learning,

ASSESSING CHANGE IN STUDENT LEARNING

In order to evaluate the impact of a shift from a cookbook to an inquiry approach, Ruiz-Primo, Shavelson, Hamilton, and Klein (2002) evaluated results obtained by a school district in a National Science Foundation supported state systemic initiatives (SSI) in California. The researchers were concerned that "Evaluation often contributes to the rise and fall of reform prematurely, by not finding effects when, although small, they are there and the reform that produced them should have been nurtured" (p. 370), so they explored the sensitivity of a multilevel, multifaceted approach for detecting the outcomes of the California SSI inquiry science program. For a population sample of 481 fifth graders they determined that inquiry instruction had an impact on student performance, and the magnitude of the treatment effects decreased as the distance of the assessment's direct connection to the classroom increased. Immediate (close) assessments were students' science notebooks. The close assessments were directed at the curriculum content and activities of the classrooms tested. Proximal assessments contain the knowledge and skills of the curriculum, but content differed. The distal assessment was devised by the *California Systemic Initiative Assessment Collaborative* and was more general. Gains from pretest to posttest on the close assessment were approximately 7.2% (fall) and 14% (spring), and on the proximal assessment, 2.7% (fall) and 5% (spring) [calculated using data from Ruiz-Primo et al., 2002].

Full option science system units were used to implement the reforms in science inquiry. The researchers (Ruiz-Primo et al., 2002) found evidence that implementation of the inquiry units varied across classrooms. Tasks required of students by the teachers of the inquiry units were generally at a low level with teachers asking students to copy definitions or to record results, but not interpret the results or draw conclusions. The science notebooks revealed low student performances across the units taught. The school district has received funding from the National Science Foundation since 1990 to institute the inquiry reforms. The researchers conclude that the classrooms should receive closer monitoring based on students' opportunities to learn (Ruiz-Primo et al., 2002). Therefore, once again implementation is a major problem even with reform-based science kits containing a teacher guide and a video to help prepare the teacher to use inquiry methods.

Colorado's state assessment and accountability data provide a pictorial representation of statewide efforts to implement the standards and the resulting impact on student learning. Colorado has been working to put state science standards in place since 1992. A number of excellent and experienced science educators were part of the state standards writing process and even though they acknowledge the interdependence and interconnectedness of the state science standards, they nevertheless, chose science inquiry as Standard 1 of the science standards. In the *Colorado Model Content Standards for Science* (Colorado Department of Education, CDE, 1995) Standard 1 reads thusly, "Students understand the processes of scientific investigation and design, conduct, communicate about, and evaluate such investigations."

Since 1997, the CDE has been testing students using the CSAP. Eighth grade students were first tested in science in 2000. Table 8.1 indicates the CSAP results (CDE, 2000, 2001, 2002, 2003b).

Students in Colorado need to achieve a proficient or advanced level to be considered as passing the state science standards. Therefore, approximately one-half of the eighth graders in the state passed the state science

Table 8.1. State Summary of Colorado Eighth Grade Science Scores

Year	*# Students Tested*	*% Unsatisfactory*	*% Partially Proficient*	*% Proficient or Advanced*
2003	56,498	21	27	49
2002	55,421	19	27	50
2001	54,642	18	29	49
2000	53,878	20	31	45

Note: Percents may not total 100% due to rounding by CDE

standards on the 2003 CSAP. In comparison, in 1996, on the *National Assessment of Educational Progress* (NAEP) Grade 8 Science test, 47% of the Colorado students were proficient in science (CDE, 2004). Thus, these results show that CSAP is getting similar scores as those from the NAEP testing. Therefore, the CSAP scores indicate a first-order change rather than a second-order change that would be associated with more dramatic differences. A more in-depth examination of CSAP results may reveal more insights.

The CSAP assessments are aligned with the science content standards. The eighth grade students are tested on subcontent areas on the Colorado science inquiry standard, Standard 1 (CDE, 1995). One subcontent area is experimental design and investigations, which states "Students understand and apply scientific questions, hypotheses, variables, and experimental design." A second subcontent area includes results and data analysis. This subcontent area expects that "Students organize, analyze, interpret, and predict from scientific data to communicate the results of investigations."

On the 2002 CSAP testing (CDE, 2002), the science scores were highest for Colorado Standard 1, Scientific Investigations. An analysis of the subcontent areas of Standard 1 indicates that students did the best on results and data analysis. The mean percentage score for Colorado Science Standard 1 is 69.1%. The mean percentage score for the entire science assessment in 2002 is 61.4%. Thus, in the third year of testing, Colorado students perform best on science inquiry indicating more progress in this area when compared to average science scores.

There is much debate about how students are assessed. Lawrenz, Huffman, and Welch (2000) recently examined the cost of student assessments in science using various formats. They included the time spent preparing the items, developing a scoring consistency, and scoring the assessments. A base of 1,000 students was used to evaluate each of the assessments. Their findings indicate that multiple-choice items cost the least and a having students conduct a full inquiry type of investigation, the most. An open-ended item costs eighty times as much as a multiple-choice item. Using a laboratory content station costs 300 times as much as one multiple choice. A full inquiry type of investigation costs 500 times as much as the multiple-choice item. Perhaps these results explain why multiple choice testing has dominated large scale assessments of student learning in science. Multiple-choice tests are also readily available. Likewise, they are easier to administer and score.

Colorado has a mixture of the formats on the CSAP that cost the least, ranging from multiple choice to constructed response. On the 2002 eighth grade science CSAP, a total of 63 items were multiple choice and 21 were constructed responses. Colorado Science Standard 1, Scientific

Investigations had twelve multiple-choice items and nine constructed responses for a total of 27 points/98 total points. Therefore, about 28% of Colorado's CSAP science questions were directly related to Standard 1. On the constructed responses, the number of points varied from one to four points with most of the items counting as one or two points. Because of the interdependence and interconnectedness of the Colorado science standards, answers to other questions may have been learned through scientific investigations but are concealed from direct analysis. The Cronbach Alpha Reliability Coefficient for the 2002 Science CSAP is 0.93 and for Standard 1, it is 0.81 (CDE, 2002).

Colorado also requires the 11th grade students to take the ACT. The results for science reasoning for 2001, 2002, and 2003 are as follows: 18.8, 19.3, and 19.2. Consistent with the Science CSAP results, the ACT scores are slightly higher for science reasoning in 2002 than in 2003. The ACT 2002 science reasoning score is 0.5 point higher than the 2001 science reasoning score.

PROFESSIONAL DEVELOPMENT OF SCIENCE TEACHERS

Universities can promote a paradigm shift in inquiry by addressing the way that science has been done in the K-12 classroom. For example, at the University of Colorado, Denver, Michael Marlow has used endangered lake fish (ELF) to provide teachers and students with opportunities to learn science through authentic science inquiries and to support the teachers with implementing inquiry standards (Marlow, Nass, Stevens, Clark, Gabel, & McWilliams, 1999). He worked through a science inquiry network that links classrooms in Michigan and Colorado so the teachers and students could communicate with each other about the Lake Victoria endangered cichlids. Content case studies were used to stimulate interest about the inquiry and also to create a knowledge base to support students in their inquiries. The case studies used real-life problems with multiple variables that helped them develop critical thinking skills (Gabel, 1999a).

Marlow and Stevens (1999) analyzed interviews of Colorado ELF science teachers for the purpose of assessing their views concerning the impact of the authentic science inquiries on themselves and their students. First, the teachers understood the paradigm shift that was being asked of them regarding the teaching of science inquiry because their definitions of science inquiry were similar. They defined authentic science inquiry as happening when students engaged in real-life, open-ended problems paralleling the work that scientists do in their laboratories. Teachers were asked to identify the important components of an authentic science inquiry project. The teachers responded with the following characteristics: open-ended, built on previous knowledge, containing choices for students, related to student lives, problem-based, and resource rich. However, observations by university staff found that the teachers had difficulty putting their views into practice. Teachers could implement the activities designed by the university scientists and educators, but they had difficulty creating their own inquiries and even asked for help in this area.

Additional science inquiries with the ELF were designed by a university scientist/science educator and reviewed by ELF teachers (Gabel, Hassler, & Muller, 1999). The scientist/science educator provided direct in-class support and modeling for selected ELF teachers' middle school classrooms with the result being that students learned to design their own experiments, which they presented at a student science research day at the university (Gabel, 2000a). The verbal presentations by the students at the research day indicated that the students had developed a solid understanding of how science is done (Marlow & Stevens, 1999). A change in perception of the teachers became apparent in post Research Day interviews and questionnaires with most of the teachers recognizing the importance of open-ended inquiry in both acquisition and understanding of science knowledge (Marlow & Stevens, 1999). Teachers noted that a change in their views occurred because of increased student interest and success during the inquiry experience (Marlow & Stevens, 1999).

Another example of a university's involvement with professional development to help teachers implement the science standards has occurred in Louisiana. Radford (1998) observed that most of the elementary and middle school teachers in Louisiana were not teaching the reform-based science that is called for in the *National Science Education Standards* (NRC, 1996), the *Benchmarks for Science Literacy* (American Association for the Advancement of Science, AAAS, 1993), and the National Science Teachers Association's scope, sequence, and coordination (Aldridge, 1996; Pearsall, 1992). He believed that professional development of teachers was needed to implement these changes. Therefore, a multiyear research project (Radford, 1998) in Louisiana was funded by the Louisiana systemic initiatives program to develop a teacher training model to ensure that the recommended educational reforms were implemented in the classroom. The project was designed to improve teachers' science content knowledge, science process skills, and confidence in doing science inquiry.

The Project Laboratory Investigations and Field Experiences (LIFE) were developed to provide professional development for the identified population of middle school life science teachers (Radford, 1998). LIFE's goals focused on changing the way the teachers taught by improving their science content knowledge, science process skills, and attitudes toward science teaching. University science and science education faculty in collaboration designed LIFE with an exemplary middle school teacher. LIFE introduced middle school life science teachers to science reform strategies and equipped them with the knowledge, process skills, and confidence to change their way of teaching. LIFE embraced the following belief: "If we want students to understand the nature of scientific inquiry, teachers must have the experience of working as scientists" (Radford, 1998, p. 74).

With this philosophy in mind, the university staff immersed teachers in the program by modeling classroom instruction that incorporated both the process of doing science and the appropriate pedagogy. Furthermore, the instruction was organized using a constructivist philosophy. The teachers were also instructed in methods to minimize misconceptions. LIFE was composed of four components (Radford, 1998):

1. a 3-week summer course
2. an independent science research project
3. academic year follow-up workshops
4. a leadership institute during the second summer

In the 3-week summer course, the instructors described a situation, for example production of heat when two chemicals are mixed that naturally generated a question that needed to be answered, and then teachers worked in small inquiry groups. Each teacher was then allowed to order $300 of supplies to do these same inquiries in their classrooms the following year.

Radford (1998) assessed the effectiveness of LIFE including the impact on student learning. Over 2,100 students and 90 project teachers in grades 4-10 were part of the research study. Data were collected over a 3-year period using a variety of techniques including surveys, objective tests, and classroom observations. Findings indicated that LIFE had a positive impact on the LIFE teachers' classrooms in Louisiana. When compared to students of nonproject teachers. Students of LIFE teachers improved significantly on an objective test measuring science inquiry process skills ($F = 6.5$, year 1; 51.1, year 2; 87.8 in year 3) when compared to students of nonproject teachers. The LIFE teachers' students had superior performance on science attitude as well when compared to the students in the control group. The science attitude survey consisted of Likert-type items.

Additionally, data were collected by outside evaluators through observation, individual and group interviews, and open-ended questionnaires. Students of Life teachers indicated that they did more experiments than previous years, they learned more science, and science was more fun. These students also noted the importance of working in collaborative groups and discussing science ideas. The LIFE teachers indicated that they were more actively engaged in learning with the students, and they guided students in their science inquiries. Teachers also expressed how much LIFE assisted them by improving their science content and helping them to gain confidence in teaching students using inquiry-based science. The teachers also noted that they understood the scientific method for the first time. Therefore, the teachers' understanding of the importance of students designing and conducting independent scientific inquiries in LIFE is similar to the findings from teacher interviews at the University of Colorado after those teachers had university assistance in implementing inquiries in their classrooms (Marlow & Stevens, 1999).

In addition to professional development of in-service teachers, universities are changing their preservice teacher programs. For example, the Arizona Collaborative for Excellence in the Preparation of Teachers (ACEPT) Program (Adamson et al., 2003) is a systemic reform effort that uses month long summer workshops to teach university and community college faculty about instructional reforms, particularly those of Project 2061 (AAAS, 1989, 1998). The goal of the collaborative was for faculty participants to modify their courses so learning is active, learner-centered, and inquiry-oriented, and consistent with the nature of scientific inquiry (AAAS, 1989). The purpose of this systemic reform is to improve science and mathematics instruction at Arizona State University and the surrounding community colleges, particularly in courses in which preservice teachers are enrolled. The ACEPT program has also supported the creation of a new middle school science teacher preparation program along with a campus center for science, mathematics, and technology education. Adamson et al. (2003) examined the effect on preservice biology teachers enrolled in the ACEPT program after they accepted teaching positions. They concluded that biology teachers, with 1-11 years of experience, who had taken one or more ACEPT course during their teacher preparation program had significantly higher scores on a reformed instructional measure. Years of teaching experience was controlled for in the analysis of covariance by using it as a covariate. Students of the ACEPT program teachers also had significantly higher achievement on science reasoning, nature of science, and biology concepts when compared to students of nonACEPT influenced teachers who also taught in the Phoenix metropolitan area.

MAKING THE PARADIGM SHIFT

As the previous research suggests, making the paradigm shift is not easy. Spillane and Callahan (2000) learned that Michigan districts believe they are implementing the science standards, but the paradigm shift has not occurred in the majority of the districts. Colorado has been working on standards since 1992, but the test results (CDE, 2000, 2001, 2002, 2003b) do not signify a statewide second order change (Watzlawick et al., 1974). Marlow tried to obtain a paradigm shift by putting ELF in classrooms so that science teachers would have a means for doing science inquiries with their students, but the teachers generally were not able to do the inquiries without assistance from the university (Marlow & Stephens, 1999). Those who have tried to use computers for science inquiries have also encountered intervening factors in achieving the paradigm shift advocated in the national science standards (Hoffman, Wu, Krajcik, & Soloway, 2003; Keselman, 2003).

Why is Teaching Science Inquiry so Difficult?

The vision for teaching science inquiry to students now spans 3 centuries. As a nation, we advanced from an agrarian society through the industrial era to the informational age, but teaching science inquiry to students has remained a mystery that has been virtually unresponsive to many efforts during these national transitions. Our current knowledge regarding how to promote science inquiry is very limited (Fradd & Lee, 1999), and the opportunity for students to obtain these intellectual skills has been rare (Jungwirth & Dreyfus, 1990; Roth & Bowen, 1994; Welch, Klopfer, Aikenhead, & Robinson, 1981). Open inquiries, where students design their own experiments are rarely found in laboratory manuals, textbooks, and supplementary materials for classrooms (Pizzini, Shepardson, & Abell, 1991). Research data are not encouraging (Jungwirth & Dreyfus, 1990; Roth & Bowen, 1994) primarily because researchers have found that science inquiry is resistant to analysis, and implementing science inquiry is a complex problem (Germann, Aram, & Burke, 1996) with few programs succeeding in encouraging science inquiry learning with student-designed experiments (Pizzini, Shepardson, & Abell, 1991). Thus, we are reminded of the factors that Rogers (1995) presents in regards to change. One of these is complexity. Teaching science inquiry is a complex process that requires students to perform multiple tasks well to conduct a meaningful scientific inquiry. This is in agreement with prior research in science education, which indicates limited resources and opportunities for inquiry and points to problems with implementation due to the complexity of science inquiry.

Implementing standards-based science inquiry in K-12 classrooms is a major problem that needs to be overcome. A science inquiry model is needed. (Gabel, 2001b, 2004; Gabel, Buckley, Hassler, Muller, & Marlow, 2000; Gabel, Hassler, & Muller, 1999; Gabel, Marlow, Stevens, & Clark,

1997) analyzed the complexity of teaching science inquiry and uncovered some of the key elements for implementing science inquiry in the classroom as a strategy for teaching students how to design their own experiments (Gabel, 2001b). These key elements include using an inquiry continuum, scaffolding with fading, linking science inquiry skills, using a schema to anchor learning, utilizing questions appropriately, relating prior knowledge, applying brain-based research, teaching for depth of understanding, using a group process, communicating to others, constructing knowledge, and integrating knowledge (Gabel, 1999b, 2000a, 2000c, 2001a, 2001b, 2002a, 2002b, 2004; Gabel, Buckley, et al., 2000a; Gabel, Hassler, et al., 1999; Gabel, Marlow, et al., 1997; Gabel, Marlow, Stevens, Clark, & McWilliams, 1999; Marlow, Nass, Stevens, Clark, Gabel, & McWilliams, 1999).

In a research study, Gabel (2001b) compared middle school students who used the science inquiry model to a control group. The quasi-experimental study contained three independent variables and three dependent variables. The main independent variable was teaching method. Gender and ethnicity were the other independent variables. The three dependent variables, ability to design an experiment, science inquiry process skills, and attitudes toward science, were measured using instruments designed for that purpose. Students were assigned to the teaching method based on voluntary participation by their teachers. The control group teachers were in the same buildings and same grade level as the experimental group teacher. The control group teachers taught science inquiry in compliance with the state standards on science inquiry as they understood them.

Three hundred students (number of male and female students was approximately equal) from two urban schools districts in a major metropolitan area, participated in the study. The majority of the students was from low socioeconomic areas and came from ethnic minority groups (78% of the population in the study). Students were administered both a pre- and posttest for each of the three dependent variables. The pretest scores were used as a covariant in the analyses of covariance.

Only the primary research question: Do students in the experimental group perform significantly better than the control group on ability to design an experiment, yielded a statistically significant F ratio ($F = 132.5$, $df = 1, 270$; $p < .001$), with the experimental group performing significantly better than the control group. Attitudes toward science were measured using a survey with Likert-type responses, and science inquiry process skills were measured with a multiple-choice instrument. The results for these three research questions are given in Table 8.2 (Gabel, 2001b).

Table 8.2. Summary of Results—Dependent Variable: Ability to Design an Experiment

Source	*Df*	*Mean Square*	*F*	*Significance*
Teaching Method	1	41,343.7	132.5	<0.001
Ethnicity	1	96.0	0.3	0.6
Gender	1	26.5	0.08	0.8
Teaching Method × Ethnicity	1	530.5	1.7	0.2
Teaching Method × Gender	1	345.6	1.1	0.3
Ethnicity × Gender	1	10.8	0.04	0.9
Teaching Method × Ethnicity × Gender	1	285.1	0.9	0.3

A dependent t-test was used to answer the following research question: Do students in the experimental group exhibit better science inquiry process skills after treatment than before treatment? There was a significant gain between the pretest and posttest scores. The mean on the pretest on the science inquiry process skills assessment was 59.6% and the mean for the posttest was 66.4%. This is a difference of 6.8 percentage points. Dependent t-test results indicate that there is a significant difference (t = 3.02, df 131, p = .003) between the pretest and posttest scores on science inquiry process skills for the experimental group. The science inquiry process skills assessment was a multiple choice instrument using unfamiliar material. Qualitative data also suggested that these students performed better in a laboratory setting (Gabel, 2001b).

Are the science inquiry skills linked? Factor analysis of the science inquiry skills assessment reveals the extraction of one major factor using the principal component analysis method. This factor has an eigenvalue of 5.03. The scree plot also indicates the extraction of one major factor. These results provide support that the science inquiry process skills are linked. Also, qualitative data from student work, videotapes, teacher interviews, classroom observations, and the ability to design an experiment assessment indicate that the students in the experimental group linked science inquiry process skills. The students in the control group exhibited a more fragmented approach on the ability to design an experiment assessment (Gabel, 2003b).

The science inquiry model schema (Figures 8.5a and 8.5b) was analyzed to determine if the students retained it. Sixty-four percent of the students in the experimental group retained the complete schema and an additional 9% missed only one or two steps. This contrasts with 6% of the students in the control group who recalled the scientific steps used by their teachers and 43% who listed no steps (Gabel, 2001b).

Teacher Directed → → → → → → → → → → → → → → → → → Student Directed

Teacher Demonstration   Structured Inquiry   Guided Inquiry   Open Inquiry   Discovery

Figure 8.1. Science inquiry continuum.

Science Inquiry Continuum

The science inquiry model promotes science inquiry as a continuum. As students gain experience with science inquiry, they become more independent of the teacher in conducting their science inquiries (Gabel, 2001b). The science inquiry continuum illustrated in Figure 8.1 (Gabel, 2001b, 2002a) portrays science inquiry ranging from a teacher-directed approach to a student-directed approach. On the left end of the continuum teachers demonstrated, directing the inquiry. Next is structured inquiry and guided inquiry. The teacher plays a large role in directing the students in the structured and guided inquiries. In an open inquiry, the student directs the inquiry

Recognizing the continuum's various levels teachers incorporate specific levels of science inquiry into their classrooms. Perhaps with a better understanding of science inquiry, teachers can move their students through various levels of science inquiries without feeling so overwhelmed. The science inquiry model moves the students from a structured, teacher-directed inquiry to a student-directed inquiry. Teachers who implemented the science inquiry model need to be supported, especially with open inquiries (Marlow & Stephens, 1999). With support from a science specialist, middle school teachers perform guided inquiries with their students, but open inquiries are a challenge for them; therefore, these teachers find open inquiries too difficult (personal communication, December, 2003). Radford (1998) found that having teachers acting like scientists along with learning the appropriate pedagogical skills had a significant impact on students' science inquiry process skills in those teachers' classrooms.

Scaffolding with Fading

The science inquiry model incorporates scaffolding with fading. Scaffolding is a process that enables a student to solve a problem or carry out a task that the student could not do unassisted. By identifying what information or skills a student lacks, a teacher can bridge the gap with scaffolding to allow the student to achieve new learning. Scaffolding is like providing a bridge across a river so that the student can move from one side to the other side. Figure 8.2 bridging the gap with scaffolding is a pictorial representation depicting the teacher bridging the knowledge gap so that a student attempting a science inquiry with insufficient knowl-

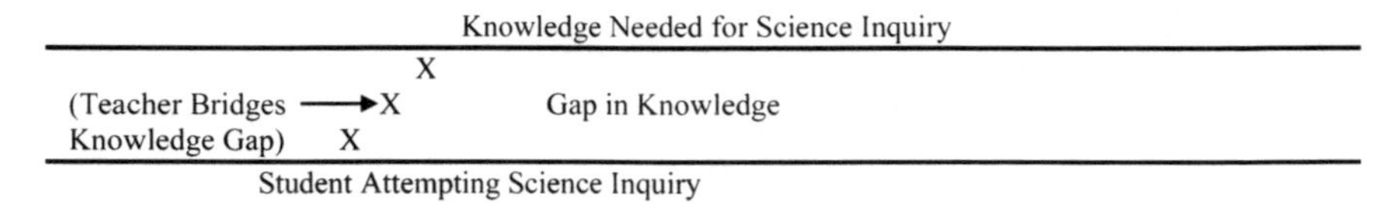

Figure 8.2. Bridging the gap with scaffolding.

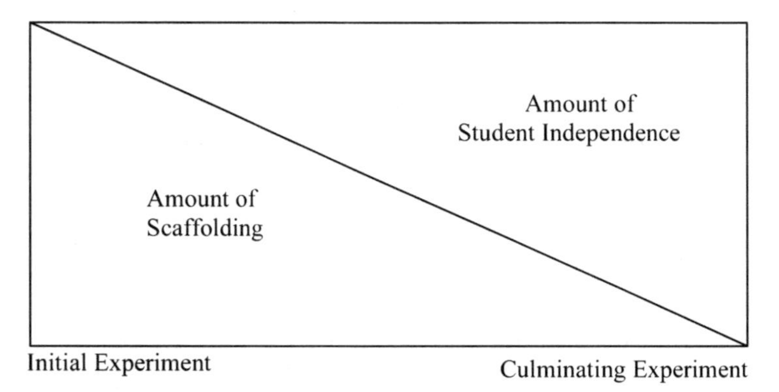

Figure 8.3 Conceptual models for a scaffolded inquiry set with fading.

edge is assisted in performing the inquiry. However, fading is also important. As a student's competency in the inquiry process increases, the teacher needs to provide less scaffolding.

Thus, the student moves from a structured, teacher-directed approach to the more open, student-directed mode in an inquiry set. Figure 8.3 conceptual model for a scaffolded inquiry set with fading (Gabel, 2001a, 2001b) provides a pictorial representation of scaffolding with fading. In the initial experiment, the amount of scaffolding is high, and in the culminating experiment, the amount of scaffolding is low. Thus, as the student moves from the initial experiment to a culminating experiment in an inquiry set, the amount of scaffolding a teacher provides should decrease, or fade and student independence should increase.

In a set of four experiments, a progression in which students have maximum help at Level 1 in the beginning when they need it most, and little, if any, assistance at Level 4 is shown in Figure 8.4, model for implementation of a scaffolded inquiry set with fading (Gabel, 2001b).

For example, four experiments used the concept territorial behavior of ELF. In the first experiment, the teacher guided the students step by step in performing an experiment. In the second experiment, the students

Level 1 – Maximum scaffolding.
Student mimics teacher.
Teacher provides links to prior knowledge.
All of steps provided for student with complete details.

Level 2 – Amount of scaffolding decreases.
Teacher helps student to recall steps from previous experiment.
Teacher provides other links to prior knowledge as appropriate.
One or more of big steps is missing.
Student refers to earlier work to help recall missing steps.

Level 3- Minimal scaffolding.
Only topic given.
Student supplies remaining steps.
Student refers to earlier work as needed.
Teacher monitors challenge level, and provides scaffolding if needed to prevent high level of frustration.
Teacher provides links to prior knowledge, if necessary.

Level 4 – Scaffolding only if needed.
Student designs experiment.
Teacher becomes facilitator.
Teacher monitors experimental designs for safety.
Teacher may serve as sounding board to guide students in making decisions.

Figure 8.4. Models for implementation of scaffolded inquiry set with fading.

were asked to make a change to the fishes' environment and use the method they had been taught in the first experiment to answer a second experimental question. In the third experiment on territorial behavior, the students were directed to make a different change to the fishes' environment and to conduct an experiment similar to the first two. Guidance was removed for all steps except to guide the students in deciding the change to the fishes' environment. In the fourth experiment, the students were asked to design their own experiment regarding territorial behavior of the fish. They were instructed that they could only make one change from a previous experiment.

The science inquiry model assumes that when students are first learning new concepts, they need to be taught in a structured sequence rich in details about how to perform a multitasked assignment in order to be successful. Once students have successfully performed this multitasked assignment, they can move to the next level and do part of it on their own following the same structure that has been provided by the teacher. In other words, scaffolding means providing students with the necessary support so that they can move to the next step. Thus, a scaffold is the requisite knowledge and skill to move to the next step in the learning process.

In the Science inquiry model study (Gabel, 2001b), attitudes toward science was used as a measure to assess student differences. The experimental group's mean on the attitudes toward science assessment was slightly more positive than the control group, but the difference was not statistically significant. Analysis by repeated measures yielded similar results. The implications of this finding are that when students are gradually given responsibility for designing their own experiments, they did not become frustrated, and overall their experiences were positive.

The scaffolding with fading was successful with sixth grade students, and investigations with higher grade level students signify that four experiments in an inquiry set is the most successful for development of a similar open inquiry by students (Gabel, 2004).

Linking Science Inquiry Skills

Typically, teachers have taught inquiry skills separately. It was thought that by teaching the skills separately, starting from the simplest skills such as observing and measuring, that the science inquiry skills could then be linked (Roth & Roychoudhury, 1993). It was further believed that when students were taught the science inquiry skills separately, they could learn to do more complex tasks such as designing an experiment. However, teaching the skills separately and expecting students to be able to use them to conduct an open inquiry has not worked. Empirical research (Gabel, 2001b, 2003b, 2004) indicates that the most successful students link one step to the next to form a complete science inquiry. Thus, these students approach their experimental designs in a holistic manner and link the science inquiry. Therefore, separately teaching students to observe, to measure, to make a data table, an so forth, is not as successful as giving students an overview of the complete experiment and having them construct the experimental steps to answer a scientific question.

Schema to Anchor Learning

A schema is incorporated in the science inquiry model to guide students. Schema theory implies that humans use frameworks for organizing information in memory, and then new information is connected to existing cognitive structures for better understanding (Anderson, Spiro, & Anderson, 1978; Gallini, 1989; Glaser, 1984, 1991). In designing scientific inquiries, Gabel (2000a, 2001b; Gabel, Hassler, et al., 1999) found

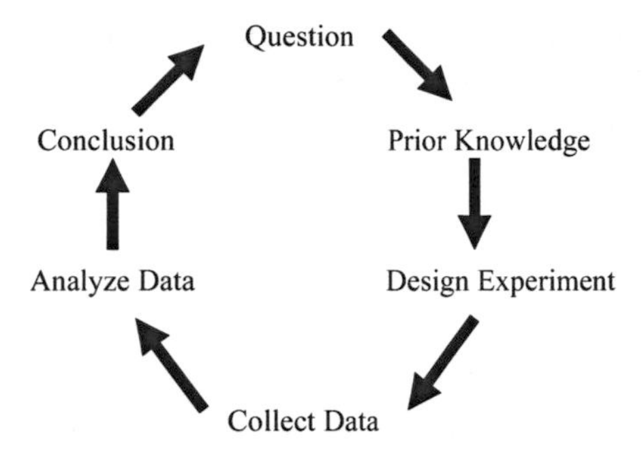

Figure 8.5a. Science inquiry model schema for middle school students.

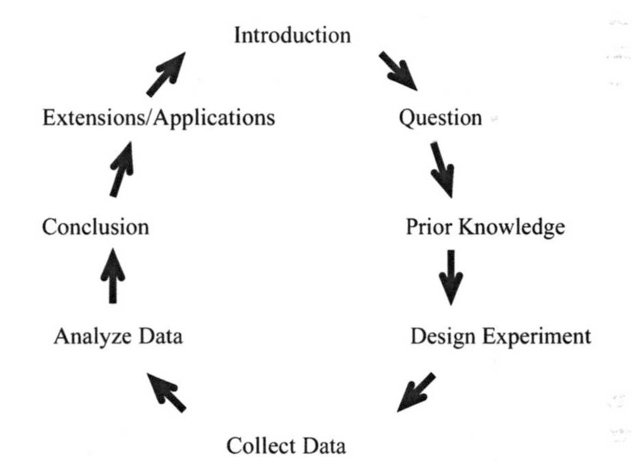

Figure 8.5b. Science inquiry model schema for high school students.

that a schema plays an important role. Students used a schema as shown in Figure 8.5a and 8.5b (Gabel, 2000a, 2001b; Gabel, Hassler, et al., 1999). The schema was posted in each classroom, and students became very accustomed to it. Incorporating a schema into the process gave students a framework for organizing their information. Research results (Gabel, 2001b) indicate that the students stored the schema into their long-term memory because even when they were not asked to complete an analysis and a conclusion (since they were lacking data), the students still wanted to write these steps so that the process would be complete. They even invented imaginary data so that they would have a complete schema.

To help students incorporate the schema into long-term memory, they practiced writing and verbalizing the various steps in the schema as part of the inquiry. The teachers also helped students to recall the schema through class dialogue. Therefore, the students were very familiar with the schema. When a procedure is repeated frequently, the brain stores the information for easy access but in a science lab, the experience only becomes effective if the work is repeated often enough for it to become a procedure (Holloway, 2000). By having the schema stored in long-term memory, it appears that the students could then move on to more difficult tasks. Since they already had a framework for designing scientific inquiries, they could then devote their mental energies to more complex aspects of designing the scientific inquiries.

The science inquiry model is also successful at the high school level (Gabel, 2004). With high school students, an introduction is added to the schema. This helps them to link new information to the investigation that they are about to do. This is also a method for linking lecture notes and the textbook content so that students can see how the content learned in lecture and the textbook fits in with the experiment they are doing. Likewise, adding extensions/applications at the end helps high school students to see how the experiment or scientific concept can be applied in other situations. Thus, using a schema helps students to anchor information in a cognitive framework and appears to free mental energies for more complex tasks.

Guided Questioning

The use of guided questions by the teacher is a technique that helps students to construct knowledge and to reflect on what they have learned. A teacher helping students understand that real-life scientific problems are complex with many interconnections utilized the following dialogue. Students were learning content for science inquiries by using case studies (Gabel, 1999a).

T: What's the big problem?
S1: Sewer water and chemicals causing problems with the Nile perch and other fish and the lake.
S2: They all relate: sewer water, smoking the Nile perch, over catching, and the Nile perch eating the cichlids.
T: Let's say that if you fix one of the problems, which one would you start to fix?
S2: That's a hard one.
T: So, you could probably start with any one of them?
S2: The British should give the Africans nets to catch the Nile perch for the restaurants.
S1: How could the British have done the research?
T: The point that you're trying to make is the British didn't have the means to do the research. But, there have been ways to do

the research in the last 10 years. How can we fix one of the problems?

S3: Africans could cut up the Nile perch and dry the strips.

Other students entered the discussion, and the teacher continued with the guided questions until the students realized that all of the problems were interconnected.

Techniques such as questioning depend upon teacher facilitation (Gabel, 1999a). Teachers who have more training in such techniques help students to construct meaningful learning by guiding the discussion as Teacher A did in the classroom dialogue above. However, Teacher B had minimal training in such techniques and in a similar discussion in his classroom, he gave the students too much information. The teacher also gave the students the conclusion rather than guiding them to the conclusion through questions and discussion such as that used by Teacher A. With the science inquiry model, teachers were taught to guide the discussion by using questions similar to the way they were used by Teacher A in the classroom dialogue above. For example, to help the student link the previous experiment to the present as part of their prior knowledge, the teacher might ask the students, "What did you learn from last week's experiment?" rather than the teacher telling the students, "Last week we learned that the cichlids like to stay in the bottom of the aquarium and you can record that for your prior knowledge."

If a student asked how to make a data table after the first experiment, the teacher would reply, "How did you make the one last week?" The student would respond by telling or showing the teacher the data table. Then the teacher might say, "Would one similar to last week's work for this experiment?" If the student said yes and a change was needed, the teacher would respond with, "Would you need to change anything from last week's lesson?" This process takes longer, but students use higher order thinking skills following this method (Gabel, 1999a).

Striving for an acceptable answer motivates learning, and authentic questioning may be a source of energy for an investigation. According to Harpaz and Lefstein (2000):

> Questioning involves an ability to transcend given information, an understanding of knowledge, and a mental willingness to undermine and rebuild existing knowledge structures and to set up the conceptual framework in which to answer the question. Learning and teaching must focus on questioning, rather than on producing correct answers. (p. 55)

The basic characteristics of questioning (Harpaz & Lefstein, 2000) are: (1) creating an atmosphere that both enables and encourages creativity, (2) facilitating the acquisition of knowledge in a way that will lead to understanding, (3) undermining the learners' cognitive constructs to motivate learning, and (4) binding the knowledge to questioning to show how knowledge is conceptually and motivationally determined by the questions.

Prior Knowledge

Another key element of the science inquiry model is prior knowledge. Linking new information to previously stored information helps students to see that they already have information about the new topic. This lends meaning to the information, makes it more relevant, and likewise makes it more interesting. If the information is interesting to the student, it is more easily learned. When students began a new experiment each week, they linked to previously stored information (Gabel, 2001b). This helped them to see that they already had some knowledge about the new experiment and caused them to be less apprehensive (Westwater & Wolfe, 2000).

In the science inquiry model schema shown in Figure 8.5a and 8.5b, prior knowledge is shown as the second step. Actually, prior knowledge can appear at all parts of the schema, but in the model provided, students were specifically asked to focus on prior knowledge as the second step. At the beginning of the set of inquiries, students had a difficult time including prior knowledge, but after a couple of experiments, students easily grasped that they now had prior knowledge that related to the experiment at hand. After completing several inquiries, students remarked that they had a lot more prior knowledge now than they had at the beginning. Therefore, students understood that they had learned many science inquiry process skills.

The teacher focused on having students learn to make a data table and a line graph from the beginning. The teacher spent much class time ensuring that students were doing the data table and the line graph correctly. These then became prior knowledge and faded as part of the scaffolding process.

Brain-Based Research

The importance of prior knowledge is also confirmed by brain research. When the brain encounters new information, the first thing that the brain does is search for a familiar pattern or feature. The brain then hunts for established neural networks to find a fit for the new information (Westwater & Wolfe, 2000). It is easier for the brain to make sense of the new information if in the retrieval process; stored information that is similar to the new information is found (Westwater & Wolfe, 2000). Connecting new knowledge is highly dependent upon prior knowledge (Jensen, 2000a). If the input is familiar, then the existing connections are reinforced, and through repetition the neural connections are strengthened to form lasting memories. Rest and emotions are also important in strengthening the neural connections (Jensen, 2000a). Brain research also indicates that the strongest neural networks are formed from actual, concrete experiences (Westwater & Wolfe, 2000). In this research on science inquiries the use of ELF ensured that the students had concrete experiences.

The science inquiry model applied other brain-based research. For example, routine and ritual are also necessary for optimal learning (Jensen, 2000a). The same science inquiry model schema (Figure 8.5a and 8.5b) was repeated each week. Frequent repetition of a procedure triggers the brain to store this information for easy access (Holloway, 2000), but experiences in the laboratory only become effective if the work is repeated frequently enough to become a procedure (Holloway, 2000). Learning involves groups or networks of neurons that develop over time through the process of making connections, developing the right connections, and strengthening the connections (Jensen, 2000a). When the brain recognizes incoming information as similar to previously stored information, this leads to increased understanding and retention (Westwater & Wolfe, 2000). In essence, the brain learns from practice (Meyer & Rose, 2000). Thus, the way in which the experiments were structured using the science inquiry model schema (Figure 8.5a and 8.5b) and the conceptual model for a scaffolded inquiry set with fading (Figure 8.3) led to more optimal conditions for learning.

The students' experiments discussed earlier that used the science inquiry model occurred at one-week intervals. This time frame is supported by current brain research. The brain needs time for processing, and the human brain is limited in the amount of explicit information it can learn. Otherwise it becomes overloaded, and there is no new learning (Jensen, 2000a, 2000b). Also, the amount of details in the inquiries was restricted to prevent cognitive overload. The brain can learn short bursts of information, and then it needs time for information processing so that memory formation can take place (Jensen, 2000b). Once the neurons make the connections then the brain surrounds and insulates the nerve cells with myelin thus allowing the conduction to go much faster (D'Arcangelo, 2000). Therefore, the brain needs time for myelination, or strengthening of its existing neural pathways (Jensen, 2000b).

Observations indicate that the partial repetition each time also led to greater retention and mastery of key concepts. This result is supported by neuroscience research findings that repetition strengthens neural connections in the brain to form lasting memories (Jensen, 2000a). Since the brain had to keep retrieving stored information that was similar to the new information, this also increased understanding and retention (Westwater & Wolfe, 2000).

The learner needs to be emotionally involved for effective learning to occur, even if the learner views the situation as slightly stressful (D'Arcangelo, 2000). If there is not a challenge, the brain finds it difficult to engage (Walsh, 2000). The teachers' use of questions rather than giving direct answers challenged the students and helped them to stay emotionally involved. Questioning is also a motivator for learning (Harpaz & Lefstein, 2000).

Depth of Understanding

The science inquiry model focuses on depth of understanding, which takes time and must involve more than surface knowledge. Meaningful understanding is important in science inquiry. In studying science inquiry in the classroom with diverse language backgrounds, Fradd and Lee (1999) note that the learning of science is dependent on the student's ability to comprehend and communicate concepts and understandings. In research on inquiry learning and the standards, Layman, Ochoa, and Heikkinen (1996) state that students need to know both content and process. In trying to understand how to move students toward inquiry, Short and Armstrong (1993) assert that students need to know what they are learning, how they are learning, and why they are learning. In their research on open inquiry with eighth graders, Roth and Bowen (1993) emphasize the importance of the contextual situation and they found that student learning was meaningful when it was situated in the context of the student's experiences.

Constructing a depth of understanding is important to the new science inquiry paradigm advocated in curriculum reforms. Case studies (Gabel, 1999a) help students to understand real-life situations that are complex and multivariable. Case studies, therefore, can facilitate students in achieving deeper and more meaningful understanding of the contextual setting for the science inquiry. Jonassen (1994) also stresses purposeful knowledge construction occurs best in learning environments that provide for a multiple representation of reality and those which avoid oversimplification of the instruction through a representation of the real world's natural complexity. According to Jonassen (1994), case-based real-world learning environments support the collaborative construction of knowledge.

Decreasing misconceptions and linking to prior knowledge facilitate students in acquiring a depth of understanding. It is believed that guided inquiries also help students to achieve deeper meaning through development of background knowledge of the problem, possible hypotheses, relevant variables, and experimental design (Arons, 1993; Germann, 1991). This in turn is supported by brain research indicating that the strongest neural networks are formed from actual, concrete experiences (Westwater

& Wolfe, 2000). Integration of other areas of the curriculum reinforces the neural networks and provides additional linkage. These methods, therefore, increase depth of understanding.

When researching the use of case studies with the ELF (Gabel, 1999a), Bloom's taxonomy (1956) was used as an indicator of critical thinking to gauge depth of understanding. Classroom observations that indicated analysis, synthesis, and evaluation were considered higher order thinking skills in contrast to defining words, stating facts, or explaining materials discussed. In the first and second case studies about the ELF, the students primarily stated facts that came from the case studies. In the first case study, only 28% of the student discussion statements could be classified at the higher levels of thinking. By the end of the second case study, students began to use more statements that involved analysis, synthesis, and evaluation. Consequently, in the third case study, students voiced these types of statements much more frequently, and 74% of the statements involved higher order thinking skills. The discussion indicated that students had achieved much more depth of understanding regarding the plight of the ELF by reading and discussing the case studies about the fish. Case studies, therefore, can create depth of understanding about the content.

Depth of understanding about science process skills takes time. With the science inquiry model, scaffolding with fading was used to help students increase depth of understanding in a sequence of inquiries at one-week intervals. Thus, as the students moved from the first experiment to the fourth in the inquiry set, their depth of understanding of science process skills increased as evidenced by observations indicating that the students could do these skills by themselves in their small groups.

Group Dynamics

Group dynamics are an important element of the science inquiry model. Group dynamics play a major role in achieving an inquiry paradigm shift. The key characteristics important to the group process involved structuring, scaffolding, authenticity, cooperative learning, high expectations, divergent thinking, a high stakes public performance, peer expectations, anticipation, a safe environment, and novelty.

The students were beginning scientists doing authentic, real world experiments using methods like those of scientists to discover answers to questions unknown to both teachers and students. Within the structured small groups, the students were given certain tasks to complete. The tasks were such that the students had to help each other, and this led to scaffolding by the more capable peer. Before the students did their experiments, they anticipated what would happen and how the fish would react. The fish preferred the bottom of the aquarium. Therefore, many groups tried to think of different experiments to encourage the fish to go to the top. One small group used a red ball to entice the fish to the top of the water and they were very excited when the fish swam to the top of the aquarium to check out the red ball.

Divergent thinking was encouraged, and the students felt safe in voicing a wide range of thoughts that yielded many rich ideas. Also, the novelty of working with the ELF made the inquiries very exciting for the students and the teachers. However, expectations were high, and students were challenged with a high stakes public performance at the university where they shared their research with their peers. In order to complete the display boards to present, cooperation within the groups was necessary.

When asked what made the small groups so successful, students identified these additional characteristics (Gabel, 2001b; Gabel, Buckley, et al., 2000):

- Students were allowed to talk with each other.
- Each took responsibility for his/her part of the work.
- They had to have patience with one another.
- It was ok to copy from each other.
- They reflected on what they learned when they explained their thought processes to their group members.

The student identified characteristics offer additional insights for understanding the group dynamics. Routinely students took responsibility and learned to be patient with both each other and the process. Key factors that emerged for the students were talking with each other and sharing written information. By sharing, they had a chance to reflect on both the process and the experiments.

With the science inquiry model, much work was completed as a whole group as the students learned new skills. They sat in their small groups and helped each other to complete the process of a step, but the focus of the discussion alternated between the whole class and the small groups. As responsibility shifted from the teacher to the student, more of the process was completed in the small groups. At the beginning of the inquiry set, the small groups focused mostly on data collection and later moved to the higher-level skills involved in designing an experiment. There were too many tasks for a student to complete alone. Therefore, the students had to work as a team to complete their scientific investigations. Consequently, this was an aspect of the design of the process that contributed to the successful group dynamics that were observed.

Communication

The *National Science Education Standards* (NRC, 1996) emphasize communication of scientific inquiry. For grades 5-8, this section of the standards is as follows: "Communicate scientific procedures and explanations. With practice, students should become competent at communicating experimental methods, following instructions, describing observations, summarizing the results of other groups, and telling other students about investigations and explanations" (p. 148).

The students wrote the various parts of the inquiry as contained in the science inquiry model schema in Figure 8.5a and 8.5b. In keeping with the conceptual model for a scaffolded inquiry with fading in Figure 8.3, the students followed the model provided by the teacher for the majority of the first experiment in the set. The teacher provided less and less information as students moved from experiment one to experiment four in an inquiry set. Thus, the students composed their own prior knowledge, design of experiment, data analysis, and conclusions as they became more competent. They were always permitted to look at their prior work, as any scientist would do.

Recognizing the importance of communicating with other students as included in the *National Science Education Standards*, the science inquiry model students were also linked to each other through an Internet conference (Gabel, 2001b; Gabel, Marlow, et al., 1997; Gabel, Marlow, et al., 1999). Frequently the students communicated with each other about their fish and their experiments. This part of the process gave the students another avenue to reflect on their experiments and to communicate their findings with their peers who were also engaged in authentic scientific research like themselves.

Additionally, students communicated their results at a research day at the university. For this research day, each small group explained a science inquiry through a display board and verbally. In order to create the display board, students within the small groups had to communicate with each other regarding which experiment that were going to highlight. Each display board was required to contain all of the parts of the science inquiry model schema (Figure 8.5a and 8.5b) for one experiment. The time frame was arranged so that students had to divide the duties of completing this display board. Many of the students used word processing for the final copy and some even taught themselves how to make a graph on the computer. Some of the students were not satisfied with their display boards. They asked the teachers to allow them to work during lunch and after school, and they even asked one teacher to come in on Sunday so they could redo their display board. Others convinced nonscience teachers that their display boards contained artwork, language arts, and math. Students argued that they should be allowed to work during that class time. Needless to say, the students were proud of their scientific research and wanted to display it in the best light. Students like to do science inquiry as scientists do it, and they will often put in much extra work to communicate their results to others.

Construction of Knowledge

An important element of the science inquiry model is the construction of knowledge. With the teacher's assistance, students constructed their own knowledge during the fish inquiries. As part of this process, the students learned to construct experimental designs, data tables, and graphs. They also learned how to construct questions, record their prior knowledge, analyze data, and formulate conclusions. Consequently, on the posttest, the students using the science inquiry model were better able to construct experimental designs with data tables, and graphs than the control group students. This difference is similar to that found in research studies (Glaser, 1984, 1991; Schiano, Cooper, Glaser, & Zhang, 1989) between novices and experts where the more expert problem solvers worked forward in a constructive fashion in contrast to the less advanced problem solvers. According to Glaser (1984, 1991), the knowledge available to the problem solver, and the way in which the knowledge is organized makes a difference in thinking skills.

In writing about constructivism, Brooks and Brooks (1993) note that "Designing, thinking, changing, and evaluating—most particularly in response to a felt need—create interest and energy.... A constructivist framework challenges teachers to create environments in which they and their students are encouraged to think and explore" (p. 30).

The key then is for the teacher to create an atmosphere that is conducive to constructing understandings. By posing a problem, the teacher can create student interest. The first problem posed to students in the ELF inquiries regarded whether the fish spent more time in one part of the aquarium than another. Additionally, it was noted that other students believed this to be true, but we needed some evidence. How could we gather scientific evidence regarding this question? Thus, the students began to construct knowledge as they figured out how to divide the aquarium into sections, make observations, record the data, and make conclusions based on the data. In the classrooms, they also constructed a frame of reference for their work. Students were taught to label the front section of the aquarium in terms of quads; thus, they could communicate understandable descriptions with each other in small groups, as a class, as a research group of classrooms through the Internet, or at a research day. Since these students had all constructed the same frame of reference, they could meaningfully communicate in terms of quads. Constructing this frame of reference allowed students to observe the fish during nonexperimental days and talk with each other and their teachers. In turn, this led to increased student interest in the topic and the construction of more knowledge.

Integration of Knowledge

Science inquiry provides many opportunities to integrate knowledge and anchor learning. For example, in the fish inquiries, there were many opportunities to integrate mathematics (Gabel, Buckley, et al., 2000). Students needed to measure the aquarium in the first experiment and divide it into sections. One group wanted to walk toward the aquarium at an angle in another experiment. This gave the teacher a great opportunity to introduce some geometrical concepts on angles. Evaporation led to mathematical discussions on area and volume and an additional science inquiry to test the hypotheses.

Students integrated language arts in written work in their science journals and also in communicating their results to others. In one classroom, the teacher also integrated language arts by asking the students to write five paragraph essays on the cichlid experiments; and one student who had struggled all year with writing, easily wrote five paragraphs (Gabel, 2001c, 2002c; Gabel, Hassler, et al., 1999). When elementary schools focus primarily on reading and writing at the expense of science, they lose the opportunities to richly integrate science into these academic areas. The end result is that when students reach middle school they have lost the opportunity to learn much science at the elementary level. Consequently, these students have to start near the bottom of the ladder in learning science skills when they reach middle school.

Constructing the display boards involved the integration of knowledge from many fields. Students used art to enhance their boards, for example, through drawings of fish and the use of color. Of course, language arts, such as grammar, spelling, punctuation, and constructing sentences as well as paragraphs were integrated. Technology was incorporated as they used word processing and learned to make graphs. Data analysis involved the use of mathematics and activities such as finding a map of Africa integrated social studies skills. Some students' research involved using music and others integrated their knowledge from other units in science and applied it in an experiment (i.e., using pumice because it would float on water). Thus, the integration of knowledge adds to the richness of the students' experiences and their depth of knowledge. Additionally, integration strengthens neural networks in the brain as it reinforces and connects learning from another subject area or experience.

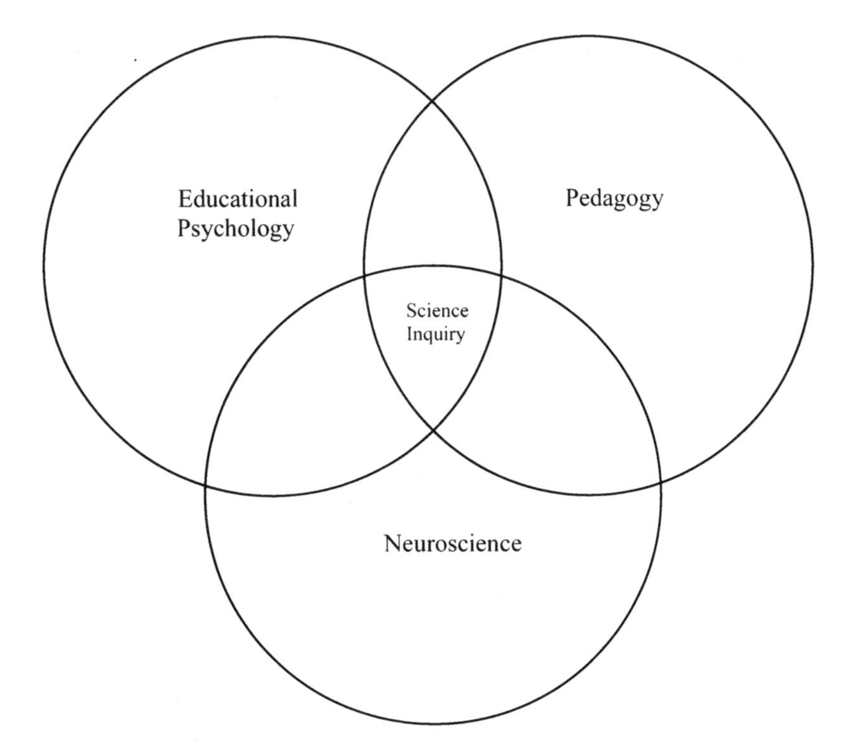

Figure 8.6. Complexity of science inquiry.

Successful Science Inquiry

Successfully teaching science inquiry is a complex process that involves integration of subject matter as well as integration of findings from the fields of educational psychology, neuroscience, and pedagogy. Figure 8.6 depicts the integration of the fields of educational psychology, pedagogy, and neuroscience as components contributing to the complexity of science inquiry (Gabel, 2002a). However, because science inquiry also includes both content and process, this dichotomy adds to the complexity when teaching science inquiry to students. Additionally, the various fields of science contribute their own intricacies to teaching science inquiry.

The beginnings of a successful path to teaching science inquiry may look like chaos, and teachers may want to throw up their hands and declare, "This is not worth it." During such times, the relative advantage and compatibility of the innovation (Rogers, 1995) in the classroom are being weighed. However, as students become more competent, the chaos subsides and students may start asking for materials or may want to stay after school to do an inquiry. They may even request to redo a science display board because they have not done their best. Capra (1996) reminds us about the importance of chaos being transformed into order within the complexity of science.

> In the new science of complexity, which takes its inspiration from the web of life, we learn that non-equilibrium is a source of order. The turbulent flows of water and air, while appearing chaotic, are really highly organized, exhibiting complex patterns of vortices dividing and subdividing again and again at smaller and smaller scales.... Throughout the living world chaos is transformed into order. (p. 190)

Thus, so it is with science inquiry. Over time, chaos is transformed into order amidst the ever-present complexity. The *National Science Education Standards* (NRC, 1996) refer to the complexity of teaching science and to the importance of science as an "active process." The *Standards* emphasize that "Hands-on activities are not enough—students also must have 'minds-on' experiences" (p. 20). The overall focus of the science inquiry model is on depth of understanding as students construct knowledge on how to design their own science inquiries. This model provides minds-on experiences for both teachers and students.

Conclusion and Implications

In conclusion, the pedagogical barrier to teaching K-12 science inquiry has been broken, but there is much work that remains. The diffusion process could take a good part of this century if substantial efforts are not put forth to increase the diffusion rate. A paradigm shift, as outlined in the *National Science Education Standards* and in state science standards such as Colorado and Michigan, is not pervasive in our nation's classrooms according to current research on science inquiry. Primary reasons for this absence of a paradigm shift appear to be lack of understanding regarding what is being asked, what the paradigm shift looks like in the classroom, and how to implement this shift. Although there still is resistance to the change, other teachers really do not know how to teach science inquiry. Perhaps this is because they have not had the opportunity to practice inquiry as scientists do. However, schools are adopting the science standards, as they perceive them. Therefore, many classrooms are simply instituting a first-order change by including more science activities, rather than the second-order change that would occur when the way science is taught has been transformed. Consequently, current in-service science teachers as well as the preservice teachers need to be taught how to teach science inquiry. Also, there is the concern that many teachers are still not very comfortable using computers to delve into the science inquiry realm (Gabel, 2000b).

As indicated in the *National Science Education Stand*ards (NRC, 1996), "Becoming an effective science teacher is a continuous process that stretches from preservice experiences in undergraduate years to the end of a professional career" (p. 55) and "teachers will need ongoing opportunities to build their understanding and ability" (p. 56).

Looking back to the first attempts at including science inquiry in the curriculum, we have made much progress. As a nation, we have the *National Science Education Standards* as well as many state science standards and Project 2061 benchmarks to guide us. This is the first time in our nation's history that we have such an extensive written plan in place. A number of science educators know what science inquiry in the classroom should look like, and schools are adopting science standards, as they understand them. However, large changes such as this take time to diffuse to individual classrooms. As state board of education Chairman Clair Orr noted in response to Colorado's release of the first CSAP science results:

> These results are consistent with our past experience of first time results for new subject matters at previously untested grade levels. Nonetheless, we are again reminded that this time of higher standards and higher expectations is a steep mountain that we must all climb together.

Where do we go from here to continue our climb on that steep mountain? What is needed is a team of those who understand how to teach science inquiry in the classroom to act as change agents to help with the paradigm shift that is required. Some teachers are comfortable with making the inquiry paradigm shift and can act as change agents for others in their school. However, the vast majority of teachers are having difficulty making this paradigm shift (Marlow & Stevens, 1999; Radford, 1998; Spillane & Callahan, 2000), and do want help from an expert in inquiry implementation (Gabel, 1999b; Marlow & Stevens, 1999). Therefore, these teachers need scaffolding to show them how to implement science inquiry in their classrooms in collaboration with experts. Recall that Rogers (1995) mentions observability as one of the factors important for change. Teachers do best when they have someone in their classrooms helping them to teach science inquiry (Gabel, 1999b). Research is indicating that this lack of in-class support is a major obstacle to overcome. Master science inquiry instructors who would provide in-class support for teachers could be trained at regional centers.

Teachers also are citing a lack of curriculum and science inquiry examples and an inability to figure out inquiries by themselves (Gabel, 1999b; Marlow & Stevens, 1999), and studies by the AAAS (Kesidou & Roseman, 2002; Stern & Ahlgren, 2002) are also indicating deficiencies in curriculum. Therefore, we also need some of the nation's top science educators to develop curriculum. If we are not willing to commit to these essentials, we will have to be content with a first-order change in most classrooms with more of the same type of science activities being taught in the vast majority of our nation's schools rather than the inquiry paradigm shift in epistemology and pedagogy that is advocated in the standards.

In summary, implications include classroom implementation of the science inquiry model, development of additional curriculum, professional development of science teachers, and systematic research for assessment purposes. Implementing the science inquiry model in middle school and high school classrooms would teach students how to design scientific inquiries. This implementation process would require curriculum to fit the needs of students at each grade level or science course. Professional development of science teachers would be essential for them to learn to use the science inquiry model. Research would be necessary on an ongoing basis to assess whether or not the teachers were achieving the paradigm shift in epistemology and pedagogy that is indicated in the *Standards* (NRC, 1996). Research is also needed to learn how to best extend the science inquiry model to the elementary grades.

In light of the significant progress that has been made in putting a roadmap, in the form of standards and benchmarks, in place nationally and at the state level, to not do what is necessary to implement the changes at the classroom level would be unthinkable. From a historical perspective, we have made significant progress in implementing science inquiry, but the many obstacles given earlier, such as the complexity of teaching science inquiry, have slowed the diffusion rate of the change recommended. Thus, we need to look to those who have been successful in implementing science inquiry to provide significant leadership to move us forward in achieving the paradigm shift in science inquiry.

REFERENCES

Adamson, S. L., Banks, D., Burtch, M., Cox, F., III, Judson, E., Turley, J. B., Benford, R., & Lawson, A. E. (2003). Reformed undergraduate instruction and its subsequent impact on secondary school teaching practice and student achievement. *Journal of Research in Science Teaching, 40*(10), 939-957.

Aldridge, B. G. (Ed.) (1996). *Scope, sequence, and coordination: A framework for high school science education.* Arlington, VA: National Science Teachers Association.

American Association for the Advancement of Science. (1989). *Science for all Americans.* Washington, DC: Author.

American Association for the Advancement of Science. (1993). *Benchmarks for science literacy.* New York: Oxford University Press.

American Association for the Advancement of Science. (1998). *Blueprints for reform.* New York: Oxford University Press.

Anderson, R. C., Spiro, R. J., & Anderson, M. C. (1978). Schemata as scaffolding for the representation of information in connected discourse. *American Educational Research Journal, 15*(3), 433-440.

Armstrong, H. E. (1903). *The teaching of scientific method and other papers on education.* London: Macmillan.

Arons, A. B. (1993). Guiding insight and inquiry in the introductory physics laboratory. *The Physics Teacher, 31*(5), 278-282.

Baird, W. E., & Borich, G. D. (1989). Authors' response to "confirmative factor analysis for validity consideration." *Science Education, 73*(6), 657.

Bloom, B. (1956). *Taxonomy of educational objectives.* New York: McKay.

Brooks, J. G., & Brooks, M. G. (1993). *In search of understanding: The case for constructivist classrooms.* Alexandria, VA: Association for Supervision and Curriculum Development.

Bybee, R. W. (1997). *Achieving scientific literacy.* Portsmouth, NH: Heinemann.

Capra, F. (1996). *The web of life.* New York: Doubleday.

Collins, A. (1997). National science education standards: Looking backward and forward. *The Elementary School Journal, 97*(4), 299-313.

Colorado Department of Education. (1995). *Colorado model content standards for science.* Denver, CO: Author.

Colorado Department of Education. (2000). *Colorado student assessment program results.* Denver, CO: Author.

Colorado Department of Education. (2001). *Colorado student assessment program results.* Denver, CO: Author.

Colorado Department of Education. (2002). Colorado student assessment program results. Denver, CO: Author.

Colorado Department of Education. (2003b). *Colorado student assessment program results.* Denver, CO: Author.

Colorado Department of Education. (2004). Student assessment results. Denver, CO: Author.

D'Arcangelo, M. (2000). How does the brain develop? A conversation with Steven Petersen. *Educational Leadership, 58*(3), 68-71.

DeBoer, G. E. (1991). *A history of ideas in science education: Implications for practice.* New York: Teachers College Press.

DeBoer, G. E. (2000). Scientific literacy: Another look at its historical and contemporary meanings and its relationship to science education reform. *Journal of Research in Science Teaching, 37*(6), 582-601.

Dewey, J. (1916). Method in science teaching. *General Science Quarterly, 1*(1), 3-9.

Fradd, S. H., & Lee, O. (1999). Teachers' roles in promoting science inquiry with students from diverse language backgrounds. *Educational Researcher, 28*(6), 14-20.

Gabel, C. (1999a, March). *Using case studies to teach science.* Paper presented at the National Association for Research in Science Teaching, Boston, MA.

Gabel, C. (1999b, April). *The use of scaffolding to teach students to design inquiry-based experiments.* Paper presented at the American Association for the Advancement of Science, Annual SWARM Division Meeting, Santa Fe, NM.

Gabel, C. (2000a, April). *Connecting communities of learners in implementing an instructional method for teaching inquiry-based science.* Paper presented at the National Association for Research in Science Teaching, New Orleans, LA.

Gabel, C. (2000b, April). *An analysis of the use of technology in the science classroom.* Paper presented at the American Association for the Advancement of Science, annual SWARM division meeting, Las Cruces, NM.

Gabel, C. (2000c, September). *Key elements for teaching science inquiry.* Paper presented at the Colorado Science Convention, Denver, CO.

Gabel, C. (2001a, March). *Analyzing the effectiveness of a scaffolded approach for teaching science inquiry to middle school students.* Paper presented at the National Association for Research in Science Teaching, St. Louis, MO.

Gabel, C. (2001b). *Effectiveness of a scaffolded approach for teaching students to design scientific inquiries.* Unpublished doctoral dissertation. University of Colorado, Denver.

Gabel, C. (2001c, October). *Exciting kids with real-world learning.* Paper presented at the Colorado Science Convention, Denver, CO.

Gabel, C. (2002a, April). *Science Inquiry: A vision for practice in the twenty-first century.* Paper presented at the National Association for Research in Science Teaching, New Orleans, LA.

Gabel, C. (2002b, November). *Brain-based science teaching.* Paper presented at the Colorado Science Convention, Denver, CO.

Gabel, C. (2002c, November). *Teaching science with an integrated approach.* Paper presented at the Colorado Science Convention, Denver, CO.

Gabel, C. (2003b, March). *Linking science inquiry skills in a holistic approach.* Paper presented at the National Association for Research in Science Teaching, Philadelphia.

Gabel, C. (2004, April). *Structuring science inquiry for optimal student learning.* Paper presented at the American Association for the Advancement of Science, annual SWARM division meeting, Denver, CO.

Gabel, C., Buckley, P., Hassler, J., Muller, L., & Marlow, M. (2000, April). *Linking science inquiry, process, and content.* Paper presented at the National Science Teachers Association, Orlando, Fl.

Gabel, C., Hassler, J., & Muller, L. (1999, March). *Building frameworks in science experiments.* Paper presented at the National Science Teachers Association Convention, Boston, MA.

Gabel, C., Marlow, M., Stevens, E., & Clark, J. (1997, November). *How does open-ended science inquiry impact the classroom?* Paper presented at the National Science Teachers Association Convention, Denver, CO.

Gabel, C., Marlow, M., Stevens, E., Clark, J., & McWilliams, S. (1999, March). *Linking hot science topics.* Paper presented at the National Science Teachers Association Convention, Boston, MA.

Gallini, J. K. (1989). Schema-based strategies and implications for instructional design in strategy training. In C. B. McCormick, G. E. Miller, & M. Pressley (Eds.), *Cognitive strategy research: From basic research to educational applications* (pp. 239-268). New York: Springer-Verlag.

Germann, P. J. (1991). Developing science process skills through directed inquiry. *The American Biology Teacher, 53*(4), 243-247.

Germann, P. J., Aram R., & Burke, G. (1996). Identifying patterns and relationships among the responses of seventh-grade students to the science process skill of designing experiments. *Journal of Research in Science Teaching, 33*(1), 79-99.

Glaser, R. (1984). The role of knowledge. *American Psychologist, 39*(2), 93-104.

Glaser, R. (1991). The maturing of the relationship between the science of learning and cognition and educational practice. *Learning and Instruction, 1*(2), 129-144.

Harpaz, Y., & Lefstein, A. (2000). Communities of thinking. *Educational Leadership, 58*(3), 54-57.

Hoffman, J. L., Wu, H., Krajcik, J. S., & Soloway, E. (2003). The nature of middle school learners' science content understandings with the use of on-line resources. *Journal of Research in Science Teaching, 40*(3), 323-346.

Holloway, J. H. (2000). How does the brain learn science? *Educational Leadership, 58*(3), 85-86.

Jensen, E. (2000a). *Brain-based learning.* San Diego, CA: The Brain Store.

Jensen, E. (2000b). Moving with the brain in mind. *Educational Leadership, 58*(3), 34-37.

Jonassen, D. H. (1994). Thinking technology: Toward a constructivist design model. *Educational Technology, 34*(4), 34-37.

Jungwirth, E., & Dreyfus, A. (1990). Diagnosing the attainment of basic enquiry skills: The 100-year old quest for critical thinking. *Journal of Biological Education, 24*(1), 42-49.

Keselman, A. (2003). Supporting inquiry learning by promoting normative understanding of multivariable causality. *Journal of Research in Science Teaching, 40*(9), 898-921.

Kesidou, S., & Roseman, J. E. (2002). How well do middle school science programs measure up? Findings from Project 2061's curriculum review. *Journal of Research in Science Teaching, 39*(6), 522-549.

Kyle, W. C., Jr. (1980). The distinction between inquiry and scientific inquiry and why high school students should be cognizant of the distinction. *Journal of Research in Science Teaching, 17*(2), 123-130.

Lawrenz, F., Huffman, D., & Welch, W. (2000). Policy considerations based on a cost analysis of alternative test formats in large-scale science assessments. *Journal of Research in Science Teaching, 37*(6), 615-626.

Layman, J. W., Ochoa, G., & Heikkinen, H. (1996). *Inquiry and learning: Realizing science standards in the classroom.* New York: College Entrance Examination Board.

Linn, M. (1997), The role of the laboratory in science learning. *The Elementary School Journal, 97*(4), 401-417.

Malcom, S. (1993, November). *Promises to keep: Creating high standards for American students* (National Education Goals Panel No. 94-01). Washington, DC: U.S. Government Printing Office.

Marlow, M. P., Nass, J., Stevens, E., Clark, J., Gabel, C., & McWilliams, S. (1999, March). *Learning science through authentic inquiries.* Paper presented at the National Association for Research in Science Teaching, Boston, MA.

Marlow, M. P., & Stevens, E. (1999, March). *Science teachers' attitudes about inquiry-based science.* Paper presented at the National Association for Research in Science Teaching, Boston, MA.

Meyer, A., & Rose, D. H. (2000). Universal design for individual differences. *Educational Leadership, 58*(3), 39-43.

Mintzes, J. J., & Wandersee, J. H. (1998). Reform and innovation in science teaching: A human constructivist view. In J. J. Mintzes, J. H. Wandersee, & J. D. Novak (Eds.), *Teaching science for understanding: A human constructivist view* (pp. 29-58). San Diego, CA: Academic Press.

National Commission on Excellence in Education. (1983). *A nation at risk: The imperative for educational reform.* Washington, DC: U.S. Government Printing Office.

National Research Council. (1996). *National science education standards.* Washington, DC: National Academy Press.

National Research Council. (2000). *Inquiry and the national science education standards: A guide for teaching and learning.* Washington, DC: National Academy Press.

National Science Foundation. (1982). *Today's problems, tomorrow's crises: A report of the National Science Board Commission on precollege education in mathematics, science, and technology.* Washington, DC: U.S. Government Printing Office.

Parker, F. (1993). *Turning points: Books and reports that reflected and shaped U.S. education, 1749-1990s.* Cullowhee, NC: Education and Psychology. (ERIC Document Reproduction Service No. ED 369 695)

Pearsall, M. K. (Ed.) (1992). *Scope, sequence, and coordination of secondary school science. Vol. II: Relevant research.* Washington, DC: National Science Teachers Association.

Pizzini, E. L., Shepardson, D. P., & Abell, S. K. (1991). The inquiry level of junior high activities: Implications to science teaching. *Journal of Research in Science Teaching, 28*(2), 111-121.

Radford, D. L. (1998). Transferring theory into practice: A model for professional development for science education reform. *Journal of Research in Science Teaching, 35*(1), 73-88.

Rogers, E. M. (1995). *Diffusion of innovations.* New York: The Free Press.

Roth, W. M. (1989). Confirmatory factor analysis for validity consideration: A critique. *Science Education, 73*(6), 649-655.

Roth, W. M., & Bowen, G. M. (1993). An investigation of problem framing and solving in a grade 8 open-inquiry science program. *The Journal of the Learning Sciences, 3*(2), 165-204.

Roth, W. M., & Bowen, G. M. (1994). Mathematization of experience in a grade 8 open-inquiry environment: An introduction to the representational practices of science. *Journal of Research in Science Teaching, 31*(3), 293-318.

Roth, W. M., & Roychoudhury, A. (1993). The development of science process skills in authentic contexts. *Journal of Research in Science Teaching, 30*(2), 127-152.

Ruiz-Primo, M. A., Shavelson, R. J., Hamilton, L., & Klein, S. (2002). On the evaluation of systemic science education reform: Searching for instructional sensitivity. *Journal of Research in Science Teaching, 39*(5), 369-393.

Schiano, D. J., Cooper. L. A., Glaser, R., & Zhang, H. C. (1989). Highs are to lows as experts are to novices: Individual differences in the representation and solution of standardized figural analogies. *Human Performance, 2*(4), 225-248.

Short, K. G., & Armstrong, J. (1993). Moving toward inquiry: Integrating literature into the science curriculum. *The New Advocate, 6*(3), 183-199.

Shymansky, J. A., Hedges, L. V., & Woodworth, G. (1990). A reassessment of the effects of inquiry-based science curricula of the 60's on student performance. *Journal of Research in Science Teaching, 27*(2), 127-144.

Spillane, J. P., & Callahan, K. A. (2000). Implementing state standards for science education: What district policy makers make of the hoopla. *Journal of Research in Science Teaching, 37*(5), 401-425.

Stern, L., & Ahlgren, A. (2002). Analysis of students' assessments in middle school curriculum materials: Aiming precisely at benchmarks and standards. *Journal of Research in Science Teaching, 39*(9), 889-910.

Tamir, P., & Lunetta, V. N. (1981). Inquiry-related tasks in high school science laboratory handbooks. *Science Education, 65*(5), 477-484.

Walsh, P. (2000). A hands-on approach to understanding the brain. *Educational Leadership, 58*(3), 76-78.

Watzlawick, P., Weakland, J., & Fisch, R. (1974). *Change: Principles of problem formation and problem resolution.* New York: W. W. Norton & Co.

Welch, W. W., Klopfer, L. E., Aikenhead, G. S., & Robinson, J. T. (1981). The role of inquiry in science education: Analysis and recommendations. *Science Education, 65*(1), 33-50.

Westwater, A., & Wolfe, P. (2000). The brain-compatible curriculum. *Educational Leadership, 58*(3), 49-52.

CHAPTER 9

ARGUMENTATION AND THE SCIENCE STANDARDS

The Intersection of Scientific and Historical Reasoning and Inquiry

Cynthia Szymanski Sunal

National science standards emphasize students' involvement in inquiry using reasoning skills and argumentation with peers of the results of ongoing investigations. How do students develop the skills necessary to ask questions, gather and analyze evidence, and argue their interpretations of the evidence with their peers? The skills needed by scientifically literate adults are broad and interrelated with those needed for literacy in other disciplines. Skills used in historical inquiry may enhance and connect to scientific inquiry. This chapter presents a model describing the intersections of scientific and historical inquiry. It considers the skills historians have had to develop in order construct and argue an evidentiary case when the evidence cannot be replicated. Applications of these skills in science are discussed. Finally, the chapter examines the use of skills discussed in the model by students as they build a case to argue their response to a problem. This study considered the ability of small groups of students aged 13-15 years to structure and argue a scientific evidentiary case in which the limitations of the evidence were recognized. The findings of this study indicate that these stu-

The Impact of State and National Standards on K-12 Science Teaching, 257–299

dent groups had developed some of the characteristics of argumentation and group process skills used in building and arguing an evidentiary case. The standards' emphasis on inquiry, reasoning and argumentation is appropriate for these students. However, these students and their student groups need additional guidance and lesson scaffolding not ordinarily found in traditional science classrooms if they are to continue to develop their use of evidence and recognition of its limitations in argumentation and fully meet the national science standards.

INTRODUCTION

Three primary emphases in teaching science are described in the *National Science Education Standards* (*NSES*) (National Research Council, NRC, 1996); (1) all students must have an understanding of science, (2) all students must have grounding in science inquiry and (3) science content, teaching, and assessment must follow guidelines that result in understanding and inquiry by students. While much attention has been given to the first and third *NSES* emphases, the nature of science and science inquiry has received less attention.

> Implementing the standards will require major changes in much of this country's science education. The standards rest on the premise that science is an active process. Learning science is something that students do not something done to them. Hands-on activities, while essential, are not enough. (NRC, p. 2)
>
> The standards call for more than science as process in which students learn such skills as observing, inferring, and experimenting. Inquiry is central to science learning. When engaging in inquiry students describe objects and events, ask questions, construct explanations, test those explanations against current scientific knowledge, and communicate their ideas to others. They identify their assumptions, use critical and logical thinking, and consider alternative explanations. In this way, students actively develop their understanding of science by combining scientific knowledge with reasoning and thinking skills. (NRC, p. 2)

INQUIRY IN THE NATIONAL SCIENCE STANDARDS: RATIONALE

The *NSES* (NRC, 1996) use the idea of inquiry in three primary ways, *scientific inquiry, inquiry learning,* and *inquiry teaching*. Scientific inquiry is described in *NSES* as the processes scientists use in investigation and description of the natural world based on evidence. Inquiry learning is described as involving an active student learning process that is "something that students do, not something that is done to them" (NRC, 1996,

p. 2). Inquiry learning by students is connected to scientific inquiry in that "Inquiry ... refers to the activities of students in which they develop knowledge and understanding of scientific ideas, as well as an understanding of how scientists study the natural world" (NRC, p. 23). Students "must actively participate in scientific investigations, and they must actually use the cognitive and manipulative skills associated with the formulation of scientific investigations" (NRC, p. 173). Inquiry teaching involving planning for hands-on activities, while essential, is not enough. Students must have minds-on "experiences as well.... Inquiry is central to science learning" (NRC, p. 2).

Science inquiry engages students in meaningful science learning by helping them to identify the question of concern, devise ways and means to carry out their investigation, gather and interpret data, generate ideas and communicate their findings for further criticism (NRC, 1996). Criticism is central to the nature of science and to the meaningful understanding of science. It requires students to be open to and accept comments from peers for the revision of their science ideas or to critique others' science ideas. Engagement in science as inquiry requires that students refine their abilities to pose questions, predict, make observations, measure, interpret data, make inferences and use creative and critical thinking to develop their understanding of science concepts. Science inquiry involves not only the understanding of scientific knowledge, but how that knowledge was produced and the overall inquiry skills needed to accomplish this production of knowledge.

A primary factor contributing to the importance of inquiry incorporating argumentation rests on the constructivist idea that meaningful learning can be viewed as a process of conceptual change. The process begins when a student's existing ideas are seen as inadequate to allow the student to understand a newly experienced event or idea. This has been interpreted as occurring when a student must replace a specific knowledge schema (Posner, Strike, Hewson, & Gertzog, 1982). Alternatively, the student reorganizes by adding, modifying, or deleting concepts and generalizations and the links between them to reorganize the knowledge schema in a continual conceptual growth process (Shymansky et al., 1997). A student judges the worthiness of personal prior knowledge by the extent to which a new idea is understandable, plausible, and fruitful (Hewson & Thorley, 1989).

School science is intended to assist a student in this conceptual change process. Research has demonstrated that conceptual change, especially with key science ideas that are goals of the national science standards, is difficult and takes time. Thus, conceptual change rarely occurs in traditional school science lessons and textbooks (Alvermann & Hynd, 1989; Beeth & Hewson, 1999; Pearsall, Skipper, & Mintzes, 1997). Factors related to conceptual change with key science ideas are; intrinsic motivation (Pintrich & Schunk, 1996), epistemological beliefs about where knowledge comes from (Driver, Leach, Scott, & Wood-Robinson, 1994) and metacognitive awareness of the ideas involved (Beeth & Hewson, 1999). Researchers, as well as the national science standards, have proposed that conceptual change requires the purposeful incorporation of the factors creating cognitive conflict and argumentation in an inquiry teaching/learning strategy.

Argumentation of evidence is a part of scientific inquiry that is present throughout the national science standards (American Association for the Advancement of Science, AAAS, 1993; NRC, 1996). Inquiry does not occur without argumentation. When students or scientists are involved in inquiry, the resulting evidence is argued with peers since data must be interpreted in terms of existing or possible scientific hypotheses and in the social context of the times. Yet, students and teachers of science are not well prepared to consider evidence and use it to build a case arguing for a particular response to a question or problem (Driver, Newton, & Osborne, 2000). Application of a framework for characterizing features of students' discourse on epistemological reasoning about inquiry in science, found most elementary through high school students used inquiry at or below the least sophisticated level, *phenomenon-based* reasoning, simply making the phenomena happen so that the results can be observed (Driver, Leach, Miller, & Scott, 1996). Few students used *relation-based* reasoning about inquiry where an investigation seeks a cause or predicts an outcome by correlation or a linear causal sequence. Argumentation is limited here to supporting a claim based on description of variables, and the quality and accuracy of data. Few students use *model-based* reasoning to justify their inquiry. Such reasoning tests or develops a model or theory or compares theories. In such reasoning, argumentation includes querying about data, methods of analysis, assumptions about the model being investigated, and alternative explanations.

Evidence also indicates that alternative ideas about inquiry, actions that do not characterize the pursuits of scientists, dominate the beliefs and actions of students and teachers, and the fundamental message of textbooks and the media (Windschitl, 2004). In conducting science inquiry projects and in classroom teaching, Windschitl noted, "Almost entirely absent from the participants' (pre-service science teachers with majors in science) journals and interviews were references to the epistemological bases of inquiry—talk of arguments tying data to claims, alternative explanations, the development of theories" (p. 503). Examination of upper elementary, middle school and high school science textbooks found that few, if any, activities provided students with even basic inquiry opportunities such as developing their own questions, selecting their own variables, or consideration of controlling variables (Chinn & Malhotra, 2002; Germann, Haskins, & Ails, 1996). Researchers have concluded that a key factor in inquiry experiences for students and teachers is "helping them conceptualize science as a way of knowing rather than as a canon of content. They should learn how to argue claims in science and engage in such arguments based on their own inquiry" (Windschitl, p. 508).

Rosalind Driver and Paul Newton (1997) noted that "Science teaching has paid little attention to argument and this has led to important shortcomings" (p. 1). In particular, they stated that the lack of attention to argumentation in science has led to a false impression of "science as the unproblematic collation of facts about the world" (p. 1). When many people encounter controversies in science these are puzzling since science is viewed as a set of facts. This context for science instruction leads to students' lack of "ability to argue scientifically through the kinds of socioscientific issues that they are increasingly having to face in their everyday lives" (p. 1) (see also Norris & Phillips, 1994).

Associated research has examined the effects of cognitive intervention in formal cognition skills. In the United Kingdom, Shayer and Adey (1993) examined cognitive intervention in science during middle school with students aged 11 and 12. The intervention focused on variables and multiple variables, proportionality, compensation, probability, correlation, classification, equilibrium, and formal models. Nontraditional, cooperative group oriented, inquiry-based training methods were used. By the end of the intervention period, the experimental group had achieved a significantly greater gain in levels of cognitive processing skills than had the control group. Shayer and Adey (1993) reported long-term effects 4 years after the initiation of their intervention in terms of increased performance by experimental students on the relatively objective norms of the science, mathematics, and English measures administered as part of the United Kingdom's national examination system, the General Certificate of Secondary Education. While the intervention reported large and permanent effects on students' achievement of cognitive processing skills, the repertoire of cognitive skills needed to use evidentiary data in building a case for its interpretation and in arguing that case were not part of Shayer and Adey's research effort.

REASONING AND ARGUMENTATION IN SCIENCE

Argumentation is a web of skills used to organize evidence into a framework supporting a hypothesis, viewpoint, or stance. As a question is asked, evidence is gathered, analyzed and evaluated to build a case arguing for a particular resolution of the question. The evidence is subjected to analysis using a set of criteria that may or may not have been predetermined. Often, the criteria are established before the evidence is examined. But, the evidence may first be examined without criteria in order to try to find patterns or otherwise establish criteria by which to analyze and evaluate it. When criteria are established, they take a form that enables others to examine the criteria, evidence, analysis, and evaluation conducted using the criteria. Students' development of the ability to structure, argue, and restructure a case built on evidence is important to their development as inquirers.

> An important stage of inquiry and of student learning is the oral and written discourse that focuses the attention of students on how they know what they know and how their knowledge connects to larger ideas, other domains, and the world beyond the classroom. (NRC, 1996, p. 36)

To facilitate such development it is necessary to examine how students interpret what the available evidence means (Osborne, Erduran, Simon, & Monk, 2001). Identification and consideration of limitations of the evidence is a key component to its interpretation.

Inquiry, as described in *NSES*, "requires identification of assumptions, use of critical and logical thinking, and consideration of alternative explanations" (NRC, p. 23). *Benchmarks for Science Literacy* indicates,

> But as important as it is that students come to understand the nature of logic, it is even more important that they learn how to use logic and evidence in making valid, persuasive arguments and in judging the arguments of others. This will only happen if students have a lot of practice in formulating arguments, presenting them to classmates, responding to their criticisms, and critiquing the arguments of others. (p. 231)

Science can be seen as making progress through the replacement of hypotheses by newer ones with greater empirical content; that is, they account for a larger number of observations. This progress occurs as current hypotheses are recognized as deficient through experimentation. The key criterion a hypothesis must meet is that it has testable consequences that could lead, in principle, to its falsification (Popper, 1934). The limitations of the evidence are considered carefully during the formation and testing of the hypothesis. While Popper focuses on the refutation aspect of scientific investigation, Kuhn (1962) considers how hypotheses frame the experimental design and indicate expected observations thereby influencing data collection and the possibility of bias. Both Popper and Kuhn consider hypotheses and their testing as central to scientific investigation. Kuhn's work, however, also leads to consideration of hypothesis generation. A hypothesis is generated based on evidence.

Initially, pieces of evidence may suggest a trend or events. Further work is carried out to identify evidence supporting the possibility of the trend or events. Such work can be used to construct a hypothesis. Limitations of the evidence are important in hypothesis generation because their consideration structures the hypothesis being generated.

INQUIRY AND REASONING IN THE DISCIPLINES

To be able to use inquiry and conceptualize inquiry experiences, a clearer understanding of the inquiry is needed. All disciplines are based on investigation; questions are asked, evidence is gathered and analyzed, and its interpretation is argued with peers. Differences between disciplines are based in the questions asked and the kind of evidence used. History, for example, is a discipline using evidence from the past that cannot be closely replicated while physics uses evidence from current phenomena, objects, or events that can be replicated. Although the evidence used varies by discipline, all share the use of evidence in arguing the questions investigated by the discipline.

Scholars in fields ranging from history to physics have called for greater attention in the curriculum to the use of evidence and its argumentation if students are to become inquirers (Osborne et al., 2001; Yerrick, 2000). A lack of attention to using evidence to build a case for a response to a question can lead to students' unquestioning acceptance of material delivered through direct instruction in a lecture or through a book (Driver et al., 1996; Kuhn, 1993). It is easier to accept or reject someone else's viewpoint or stance outright than to consider the evidence for and against it, the limitations of that evidence, and how that evidence with its limitations was interpreted. Indeed, many students seem to have little understanding that a case is built slowly, using and arguing evidence (Kuhn, 1993). They do not understand that the process of building a case for a specific response to a question is the primary process of the scientific disciplines. The concept of a case used throughout this chapter is that of a narrative framework centered on an event or events for which interactions are analyzed and conclusions are reached. The analysis incorporated within a case involves an understanding of context, chronology and the interpretation of evidence.

Commonalities exist between disciplines in the processes used with investigation. For example, deciding whether the evidence is too scant to support an interpretation is a process shared among disciplines. The use of many cognitive processing skills, however, varies between disciplines. For example, the physicist uses measurement skills to a level beyond that typically used by historians. The historian, on the other hand, often is better able to order events in large scales of time than is the physicist. Both disciplines use cognitive processing skills, but at different levels of application. Each discipline must teach students its primary investigatory skills and how to apply them to the discipline.

Strengths exist in each discipline's application of its primary investigatory skills that may be useful in the development of investigations in other disciplines. Teachers need to look for such strengths. History, for example, has developed a large repertoire of skills in the use and argumentation of evidence that differs from those typically used in science. Because historical evidence comes from the past and cannot be closely replicated, historians have had to develop a set of skills that are found to a much smaller degree among scientists. Yet, the skills of historians in using evidence and arguing the validity and reliability of their evidence are of value in the study of scientific questions. Students in science could benefit from more early training in a broader set of skills involved in using evidence to build a case and argue it with their peers. As students begin to build cases they come to better understand the need for careful investigation and data collection and to suspect quick pronouncements of conclusions until those cases have been submitted to scrutiny and argumentation with peers (Sunal, Karr, & Smith, 1998; Wineburg, 1998).

Much work has been done to better understand the development of the cognitive processing skills used by historians. This has included the study of children's development of historical understanding and interpretation of historical evidence. Researchers have examined areas such as how students develop views of historical time, how students make historical judgments, what they see as historically important and how students read history (for example: Brophy, VanSledright, & Bredin, 1993; Carretero, Jacott, Limon, Lopez-Manjon, & Leon, 1994; Epstein, 1994; Levstik, 1992; Levstik & Pappas, 1987; McKeown & Beck, 1990, 1994; Sunal & Haas, 1993; VanSledright, 1994; VanSledright & Brophy, 1992; Wineburg, 1991; Wineburg, 1998). There is a growing body of scholarship on what it means to learn and reason in history.

THE INTERSECTION OF HISTORICAL AND SCIENTIFIC REASONING

An intersection of historical thinking and scientific reasoning exists on a continuum with *pure* history on one end and *pure* science on the other. In moving along the continuum from pure history, one first moves into the social sciences which might use an historical approach in the analysis and argumentation of evidence. At the midpoint, there is an intersection of historical and scientific approaches in which both mix with neither predominating. The historical approach fades as one continues along the continuum toward pure science. The sciences for example, include a range of disciplines. One such discipline is evolutionary biology which lies somewhere along the middle of the continuum. Its evidence is mostly from the past with little opportunity for experimental manipulation of evidence except in instances such as work with ancient DNA. Recently, consideration of social aspects such as the family life of dinosaurs has begun. These social considerations relate to the understanding of finer points of evolution and are not a central focus of investigations in the field. The example of evolutionary biology, nevertheless, demonstrates the existence of a continuum rather than a dichotomy.

Intersections in Declarative Knowledge

Intersections between history and science occur in the following areas of declarative knowledge; history of science, history of ideas of science, the nature of the role culture and language play in the understanding and transmission of ideas and in the observation and interpretation of evidence, the nature of science (how science knowledge is constructed), and the use and application of science ideas in the real world (Figure 9.1). The last, use in the real world, will have many history and social science connections as well as science connections. This is the realm of science, technology, and society issues (AAAS, 1993; National Council for the Social Studies, 1994). These issues require a wide and diverse background among teachers. A strong understanding of concepts from many disciplines is needed as is an understanding of the interrelationships between those concepts. Finally, an understanding of the nature of reasoning and argumentation within the contributing disciplines is necessary. The importance of such issues is made evident by the focus accorded them in both science and social studies national standards.

History of science is an area of study by historians. It may appear in the history curriculum or in the science curriculum. In many science curricula it plays a relatively small role (AAAS, 1993). The history of ideas in science is another area of study by historians, but some scientists also study the area. It more frequently appears in science curricula than does the history of science (AAAS, 1993). The nature of the role culture and language play in the understanding and transmission of ideas and in the observation and interpretation of evidence is studied by social scientists such as psychologists and anthropologists and by some scientists and historians. In recent years it has been a part of some teaching methods textbooks used in the training of both social studies and science teachers (Sunal & Haas, 1993; Sunal & Sunal,

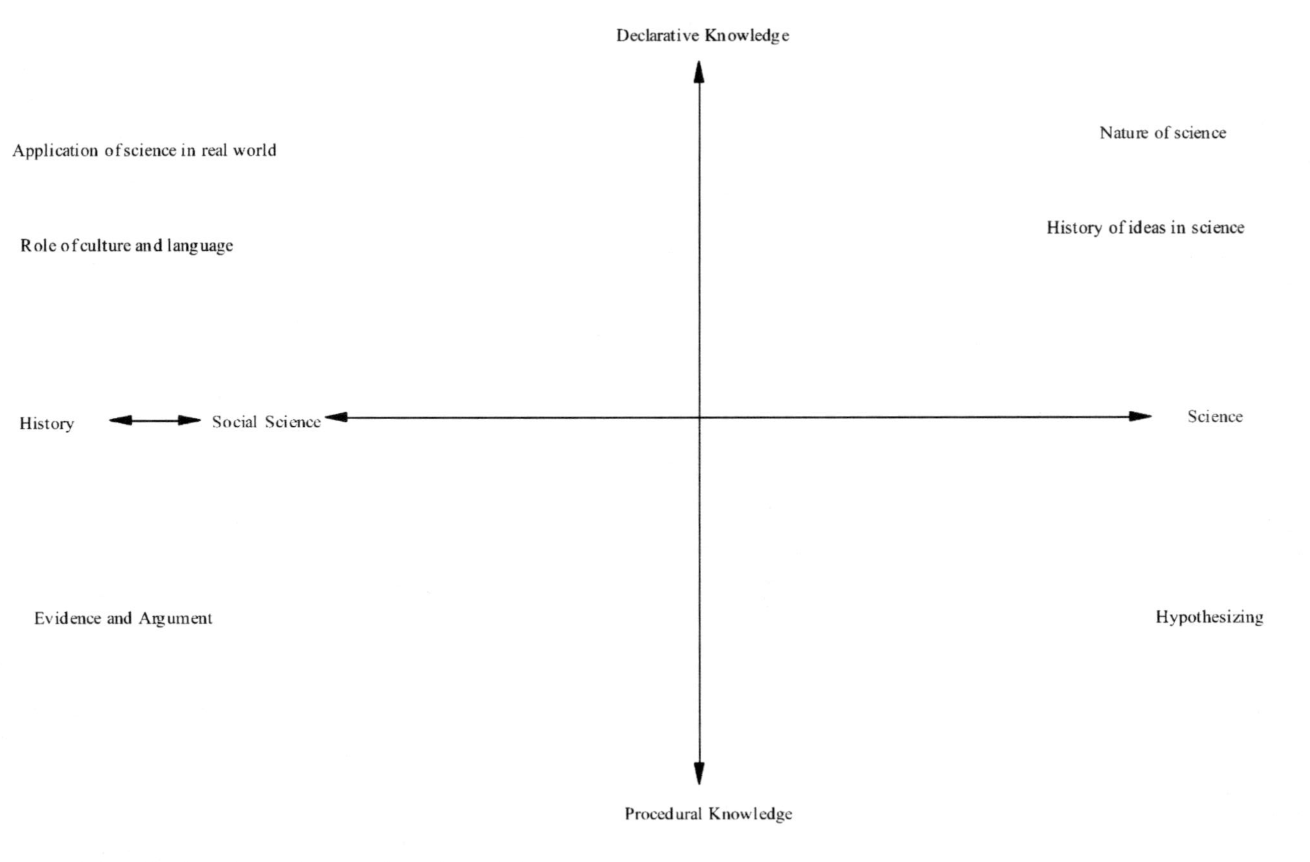

Figure 9.1. Declarative and procedural knowledge

2003). Consideration of the nature of science has been primarily the work of scientists (AAAS, 1993).

Intersections in Procedural Knowledge

The use of evidence to build and argue a case occurs in the area of procedural knowledge (Figure 9.1). The nature of the evidence used in history has led historians to extensively study the use and argumentation of evidence. Hypothesizing also is a procedural knowledge skill. It is related to the use of evidence to build and argue a case as a possible answer to the question around which an investigation is based. It has an important role in all disciplines. Historical investigations often are guided by hypotheses but frequently are used to generate an hypothesis. Scientists have extensively studied the formulation of hypotheses. In science, an investigation more often has been perceived as beginning with an hypothesis rather than leading to an hypothesis (Kuhn, 1962; Popper, 1934).

The Intersection

There are topics of study that can be classified as pure history and respond best to an historical approach. There are topics that can be classified as pure science and respond best to a scientific approach. If these are put on opposite ends of a continuum, an intersection or interface wall occurs where science and history intersect and both approaches are used and relevant. This occurs with science-related social issues such as making choices considering the health effects of foods eaten or about the probability of effects on personal health of living near a toxic waste dump. Here, science is being applied to the social world. The issue considered may be deeply embedded in the history of a region or result from the attitudes of people a century ago in North America. For example, in the past, the toxicity of materials in a waste dump may not have been well understood. Because people walked to work or used the local bus system and could not afford personal transportation, they tried to live close to work even if it meant also living close to a waste dump. Today, the toxicity of the waste dump's contents is better known and personal transportation is typical, so attitudes have changed. The nature of science becomes important as does the nature of history and the nature of the role of culture and language. The history of science and the history of ideas in science are also important when addressing science-related social issues.

THE RELEVANCE OF HISTORICAL COGNITIVE PROCESSING SKILLS IN SCIENCE

The cognitive processing skills utilized in history can help scientists address science-related social issues and theory-building in science. In research, historians use evidence to construct narratives describing and explaining historical events and developments (Booth, 1993; Fehn, 1997; Wineburg, 1991). Six clusters of ideas related to the construction of a case and reasoning in history have been identified in interviews with historians (Leinhardt, Stainton, Virji, & Ordoff, 1992). The six clusters are introduced below and followed by a discussion of the overlap of each in the teaching of science and history. Sample descriptions are given in Table 9.1. *First,* the doing of history implies a purpose or motivation. The motivating assumption underlying the construction of a case from evidence is that the case exists to help us understand both what was and also what is (Levstik & Pappas, 1987). *Second,* the construction of a case with a compelling narrative that has internal coherence is important (Leinhardt, Beck, & Stainton, 1994). *Third,* a case is exhaustive, analyzing and synthesizing all evidence that can be found and all evidence that might contradict the case (Wineburg & Fournier, 1994). *Fourth,* the case is the answer constructed in response to critical questions selected through the use of theory or as part of hypothesis development. *Fifth,* the evidence is interpreted in terms of the context of the original times, the context of the present time seeking to look at that evidence, and the implications of which evidence has survived over time. This is contextual, layered interpretation. *Sixth,* argument or debate of other competing historical interpretations occurs and includes organization of all relevant documentation.

A case also is constructed by scientists when they gather and present research to support a paradigm such as the theory of plate tectonics as originally proposed by Alfred Wegener. Fellow scientists may have differing interpretations of patterns of coastlines on either side of the Atlantic Ocean, paleontological evidence, and later magnetic rock records and earthquake location. Data is acquired and interpreted and a case supporting the new paradigm is built (Sunal & Sunal, 2003). In the forward to his book, *The Origin of Continents and Oceans,* Wegener states that,

> The book is addressed equally to geodesists, geophysicists, geologists, paleontologists.... It's purpose is not only to provide research workers in these fields with an outline of the significance and usefulness of the drift theory as it applies to their own areas, but also mainly to orient them with regard to the applications and corroborations which the theory has found in areas other than their own. (Wegener, 1966 (1929), p. viii)

Table 9.1. Six Clusters of Ideas Related to Case Argumentation

Argumentation Process	*Examples*	
	In History	*In Science*
Purpose for which evidence was collected	The plant life of today differs from the plant life of 500 years ago. This difference impacts the culture and life of people living in those regions.	The plant life of today differs from the plant life of 500 years ago in most of the world in terms of the numbers, kinds, and variety of species.
Compelling, internally coherent narrative	People cut large tracts of redwoods because farmland was more important in a subsistence economy and so were housing construction materials that were impervious to insects. Additionally, one could get "hard cash" by cutting and selling timber.	Redwoods are part of an ecosystem. If they are recognized as endangered it means that the entire ecosystem is stressed. We need to explain how the components of that ecosystem are interrelated.
Exhaustive	We include all documentation we can find that describes why people cut cypress and redwoods. These include letters, diaries, newspapers, and sales documents.	We collect field data from a diverse set of sites within the ecosystem that is being investigated.
Response to critical questions	Critical questions arise from scattered available original sources. Hypothesis development often arises from critical examination of existing primary source documents. These suggest people cut redwoods for hard cash and to try to create subsistence farmland.	Recognition of a stressed ecosystem leads to questions about causality and interrelationships. These are investigated and may be reinterpreted and then reinvestigated.
Interpreted in the context of the times	We examine primary documents trying to put ourselves into the context of those times. If the choice is between raising cash to feed our family or cutting a tree, what choice would a person living in such times make? Why?	We often investigate questions arising from the needs of our society and those for which we have the methods and equipment that enable us to investigate them. Several decades ago, large scale cutting of redwoods was not considered a problem in society because many other large tracts existed. So, the effects of cutting on the redwoods' ecosystem were not investigated.
Argumentation with competing interpretations	Others using the same primary source documents think the expansion of the railroad system was the major factor in the cutting of large tracts of redwoods. So, the argumentation of evidence supporting competing interpretations occurs.	Scientists argue the interrelationships found within the redwoods' ecosystem and the relative influence of each on the results being observed.

Purpose for Which Evidence was Collected

Each of the six clusters of ideas can be examined for their relevance to scientific inquiry. The first cluster identifies a purpose or motivation to understand both what was and also what is (Levstik & Pappas, 1987). Scientists *do* science with an underlying assumption that understanding *what is* leads to the prediction of what will be. It leads in the other direction as well, understanding *what was*. In areas of science such as the study of environmental biology or geology, what was is not necessarily what is. The plant life of today, for example, differs from the plant life of 500 years ago in most of the world in terms of the numbers, kinds, and variety of species. This difference impacts the culture and life of people living in those regions. How change has come about is part of the study of what was and its connection to what is. Knowledge of both states is needed in order to predict what will be or to determine whether such prediction is possible. So, the scientist's and the historian's motivation for the building of an evidentiary case are similar.

Students should recognize the purpose of any investigation they undertake and any case they build. Teaching from a textbook, or by lecture, provides limited opportunity for students to conceptualize the purpose or motivation of the investigation that led to the construction of a case. Instead, students are more likely to be presented with the outcomes of an investigation rather than evidence from a series of investigations (Sunal & Sunal, 2003). Their purpose becomes acquiring the knowledge represented by the outcomes. Students must be encouraged to want to identify the purpose underlying an investigation. Some efforts by schools, such as science fairs, help build students' understanding of the purpose or motivation on which investigation is built by involving students in generating their own investigation and explaining its purpose as well as its outcomes to others. To address the national standards effectively, the science curriculum needs to expand opportunities for students to seek the purpose or motivation for investigation.

Compelling Narrative for an Evidentiary Case

Second, the construction of a compelling narrative with internal coherence is important to both historians and scientists (Giere, 1991; Salter, 1997). The pieces of the case being argued must have a clearly definable, well integrated relationship. In history, the narrative is sometimes viewed as telling a story. Whether or not it reads like a story, the narrative has a basic sequence of events and includes motivations and reactions to the events. In discussing a 1960s voting rights march in the city of Birmingham, Alabama for example, a student constructed the following narrative (Sunal, 1997).

> The black people marched down Eighth Street, I think. They were pretty quiet. They wanted to be able to vote in elections but they did not want to fight with the white people. They thought if a lot of them just marched peacefully a lot of times, then white people would see that they had to let them vote. The white people would let black people vote because they would see that black people could really get together and work together and do it in a peaceful way. But, the white people got mad and yelled and threw stuff and turned water hoses on the black people marching. They weren't impressed that the blacks could have a peaceful march. The white people just didn't think black people were smart enough to be able to vote. After a lot of marches and trouble the white people finally were forced to let the black people vote because the government in Washington made them do it.

Scientists have focused less than historians on the construction of a compelling narrative. Yet, their argumentation can be conceptualized as aimed at compelling others to accept it (Sunal & Sunal, 2003).

When scientists carry out investigations they construct a narrative beginning with events that are experienced. They examine the record created by these experiences. This leads to a transformation of the facts into knowledge and value claims. The process leads to the construction of concepts, then of principles, and finally, of theory (Novak & Gowin, 1984). When the investigation is presented in textbooks, the media, or in classrooms, it typically does not take the form of a well-constructed narrative case although the investigation followed a process that could be presented as a narrative case. More emphasis on narrative could increase the impact of the argument and make interconnections between pieces of evidence more easily recognized. Communication with the public also could be enhanced. Science often is viewed as an undertaking that nonspecialists cannot comprehend. Good narrative cases may assist in the development of a higher level of science literacy among the public.

Traditional science teaching utilizes few narrative cases. Students do not perceive the whole process of investigation. The relationship between the parts and their integration is not clear (Osbourne, 1997). Osbourne has advocated the need for students, parents and teachers to have a sense of what the science curriculum is all about. He suggests that having a story to tell in answer to a big question can help us ask the right questions about the science curriculum and what it is for. He recommends the curriculum include a limited number of important narratives, for example: our place in the universe, our bodies, genetics and evolution, and materials and behavior. Such a narrative would help answer science-related questions that matter to people such as: "What is our place in the Universe? and What do we have to do to keep ourselves and the environment in a healthy state?" (Osbourne, p. 1).

By their middle/junior high school years students have encountered many narratives although their encounters mostly occur in disciplines other than science. The science curriculum can utilize this experience by presenting well constructed important narratives as Osbourne has suggested. The science curriculum could also involve students in constructing their own narrative cases describing their investigations and arguing the meaning of their findings. Such an effort is consistent with the national standards as a means of investing students in understanding how events in their investigation are related and what import these relationships have for the question addressed by the investigation.

Exhaustivity

Exhaustivity is present in the cases built by both historians and scientists. Exhaustivity involves considering all evidence supporting or contradicting the case. Chronology is a component. In cosmogony, for example, scientists seek to understand the processes and changes that created the universe we know. This includes how stars, solar systems and galaxies formed. Chronology on a huge time scale is important in cosmogony.

More is involved in constructing a case than just collecting the evidence. Wineburg and Fournier (1994) indicate that exhaustivity creates synthesis. In identifying all the evidence that might be included in the case and in analyzing the role of each and its relationship to other pieces of evidence, the case's developer must synthesize all elements into a coherent case. Bounded exhaustivity should be possible in the middle grades science curriculum.

Bounded exhaustivity sets some limits to the search for, and inclusion of, evidence (Leinhardt et al., 1994). Students have limited sources of information available to them. It may be that the teacher provides a set of sources with which they can work. Students are helped to develop an awareness of the limitations of their evidence and expected to be exhaustive within these limitations. Bounded exhaustivity occurs in many instances in science curricula when students test and discuss a set of major competing hypotheses but are not testing all possible competing hypotheses. The construction of a narrative describing scientific tests of alternative hypotheses has not been emphasized in the school science curriculum (Driver et al., 1996).

Causality is a component of the narrative. The establishment of plausible causality is the focus of experimentation (Kuhn, 1962; Popper, 1934; Sunal, 1981). Establishing internal causal links between pieces of evidence in an investigation is important to analysis. These internal links structure a narrative case with which to argue the plausibility of a hypothesis.

Concerns have been voiced about plausibility in science (Kuhn, 1993). These concerns have import for history as well. Kuhn discusses the use of pseudo evidence, defined simply as a scenario or script depicting how the phenomenon might occur (p. 324). Pseudo evidence cannot be sharply differentiated from the theory itself. Hence, responses to "What causes X?" do not differ sharply from responses to "How do you know this is the cause of X?" The individual makes an intuitively convincing case for the plausibility that is specified as leading to the outcome without providing any genuine evidence that this cause is operating during instances of the phenomenon. Kuhn found only 40% of her subjects, ages 10 through adult, were able to separate theory from evidence. She also found little change in this ability after the end of junior high school. Similar findings have been reported by Perkins (1985) and Voss and Means (1991).

In discussing historian's development of evidential exhaustivity, Leinhardt, Beck, and Stainton (1994) suggest evidence may include strong inferences about what must have been in place for a surviving piece of evidence to have been present (p. 239). Caution must be used in the inclusion of such evidence so that pseudo evidence is not used to impart plausibility to a case. The need to separate theory from evidence is basic to argumentation and the construction of a case. Kuhn's findings indicate many adults can create a case for a hypothesis that is plausible but do not recognize the need to provide evidence of the cause operating. The school curriculum has a role in helping students learn to argue and establish plausibility of an argument based on evidence (Duschl, 1990). Without such skills, students cannot investigate the questions fundamental to any discipline. Since researchers have found little change after the end of junior high school in these skills, a cognitive intervention aimed at facilitating further development of cognitive processing skills might best occur during the middle school/junior high school period.

Theory Development Leads to Critical Questions

Theory or hypothesis development is important to all disciplines. Hypothesis generation is an outcome of evidential collection and organization: it also is a guide for its collection. Hypotheses are dynamic, growing and changing within a demarcated framework. Further, the theoretical basis one operates within is the filter for interpretation and the way of imposing meaning on the data (Kuhn, 1962). Hypotheses often precede the investigation in science, guiding the collection of evidence. This process typically results in the generation of further hypotheses. To an historian, hypothesis generation most often is an outcome of evidential collection and organization although it also may serve as a guide for its collection. In both history and the sciences, hypotheses are dynamic with the expectation that they will change as new evidence is gathered or existing evidence reinterpreted (Hacking, 1983; Leinhardt, Beck, & Stainton, 1994).

Students need to develop an understanding of the role of hypothesis development in science and other disciplines. Inquiry teaching and curricula challenge students to investigate hypotheses and change them if their investigation suggests a need for change of the hypothesis.

> Teachers of science constantly make decisions, such as when to change the direction of a discussion, how to engage a particular student, when to let a student pursue a particular interest, and how to use an opportunity to model scientific skills and attitudes. (NRC, 1996, p. 33)

Middle/junior high school students who have had experience with hands-on, minds-on science in their elementary school years are ready for science curricula that involve them in identifying, investigating, and changing hypotheses. "Inquiry into authentic questions generated from student experiences is the central strategy for teaching science" (p. 31). Students with a weaker background need science curricula that provide them with experiences necessary for building basic science concepts and process skills and gradually lead them into hypothesis development and testing.

> Teachers must struggle with the tension between guiding students toward a set of predetermined goals and allowing students to set and meet their own goals.... The planned curriculum ... is modified and shaped by the interactions of students, teachers, materials, and daily life in the classroom. (p. 33)

Contextual Layered Interpretation

Contextual layered interpretation is not used as frequently in science teaching as it is by historians. In science, evidence collected in the past often is reexamined in the context of additional information available today. In studying the history of science, the history of ideas in science, the nature of science and science-related social issues, the context of the past, the context of the present time seeking to look at that evidence, and the implications of evidential survival are all important. In the formulation of cases relating to these areas, contextual layered interpretation should occur in science as well as in history.

The science curriculum typically does not address contextual layered interpretation. An example of an idea that could be part of the science curriculum and one that needs contextual layered interpretation is the health-related issue of *organic foods*. Today, the idea of pesticide-free, organically grown foods is attractive to many consumers in the United States and Europe. People are aware of damage that has been done to the health of consumers and farmers by the use of some pesticides. Rather than take the chance of ingesting chemicals that might later be found to be harmful, they indicate a preference for buying organically grown foods.

Prior to the 1950s people ate foods such as peaches, that were small, often had a worm or a hole eaten by an insect, had unsightly spots and were oddly shaped. As pesticides were able to produce a larger, better looking fruit that generally did not harbor a worm, consumers preferred to buy such fruit. Farmers saw higher yields of a better quality of fruit. A case addressing the use of pesticides in farming and public attitudes toward it, uses contextual layered interpretation when it examines the problems people in the 1940s had with the effects of pests on the food they grew and purchased, the attitudes of people in the 1950s when pesticides reduced pest damage to foods, the worry caused by the results of research studies on the effects of pesticides in the 1960s and 1970s and the favorable attitudes of consumers toward organic farming in the 1990s.

Such a case considers the context of the past. It also looks at the context of our times and biases that may effect how we interpret the events and context of past decades. Evidential survival is important. For example, do we have many photographs of fresh fruit and market records from the 1940s so that we can formulate an idea of whether most fruit really was effected by pests? Is this evidence more limited for some crops than for others? Involving students in narratives that include contextual layered interpretation enables them to perceive the complexity of scientific questions and events. Having students construct a contextual layered interpretation challenges them to address basic issues related to the role of evidence in investigations.

Argument of Competing Interpretations

Argument or debate of other interpretations is fundamental to investigation in any discipline. In science, the school curriculum traditionally has not helped students conceptualize such a dialogue nor develop the skills needed to construct it (Duschl, 1990). Interpretation follows evidence gathering and organization through which a relationship among facts is established. This is another major component of investigations in all disciplines. In history, there is an effort to establish motive among particular actors. In science much less of an effort is evident, although instances are found in history of science, history of ideas in science and in science-related social issues. School science curricula, however, generally do not address the skills involved in establishing motive (AAAS, 1993). Students need to better understand that scientists present their evidence, build an evidentiary case and argue their case as conflicting interpretations are made by other scientists.

A sample situation to help students construct understanding can involve students in considering what effects artificial intelligence systems have on society and the implications of those effects for planning and future action by society (Sunal et al., 1998). For example, artificial intelligence systems have resulted in greater standardization of products and sometimes, in cheaper costs. They have resulted in fewer jobs being available to blue-collar workers but have created a need for more highly trained workers. An implication for planning and action might be that changes need to be instituted in schooling to insure workers have the needed levels of education. Another implication might be the institution of means by which workers can continuously update their knowledge and skills. The problem places students in the position where they have to construct the schemas for themselves in order to develop a case addressing the issues associated with the problem. These schemas should include hypothesis development, construction of contextual layered interpretation and argumentation of other interpretations.

All the components of the use of evidence, its evaluation and its organization in argumentation may have faults. One's peers try to find those faults instead of endorsing even the most clearly structured, evidence-rich argument. It is only through such examination that a case is built for the response to the question under investigation. Counter-arguments are raised requiring the original investigator to reconsider the evidence, search for more evidence, and respond with a clear argument that convincingly rejects the supposed fault. The process may be lengthy and is likely to be influenced by the social context of the times in which it is occurring and the individuals who are participating in the argumentation (Driver et. al., 1996; Lee, Dickinson, & Ashby, 1994).

In daily life, cases are built for a response to many questions such as, How can I best present myself at this job interview? Is this irradiated food safe to eat? Which of these candidates deserves my vote? and Is it true that the part of the community with the lowest income people always gets the new garbage incinerator, prison and other necessary but unwanted public institutions? In daily life one can accept someone else's viewpoint or stance and use it to make decisions or one can examine the evidence and argue a case from it. Unfortunately, if one is not aware of the importance of using and arguing evidence, then the viewpoint of a perceived authority may be accepted as the answer.

Many individuals answer questions with theories for which they are unable to cite evidence (Kuhn, 1993). When asked for evidence, the theory may be restated or a single instance recalled that seems to support the theory, or there may be an insistence that it is impossible to envision any other stance. None of these responses present and argue evidence. The focus is on the plausibility of the theory not on whether any evidence can be found with which to argue its support. Such individuals cannot really understand any discipline well, nor can they rationally work through possible responses to the questions they meet in daily life. Argumentation of evidence through a well-constructed case should be a goal of science education because students experience the complexities involved in creating a coherent whole from its parts.

AN ARGUMENTATION MODEL FOR STUDENT INSTRUCTION

Argumentation involves many cognitive processing skills used to structure the case that is argued. Figure 9.2 presents a model of argumentation skills focused around the process of building a case. The model is structured around a seven-step process beginning with a question or problem, leading to hypothesis generation, then to evidence gathering, next to construction of an argument, on to further hypothesis development based on the reflection involved in argument construction, next to hypothesis

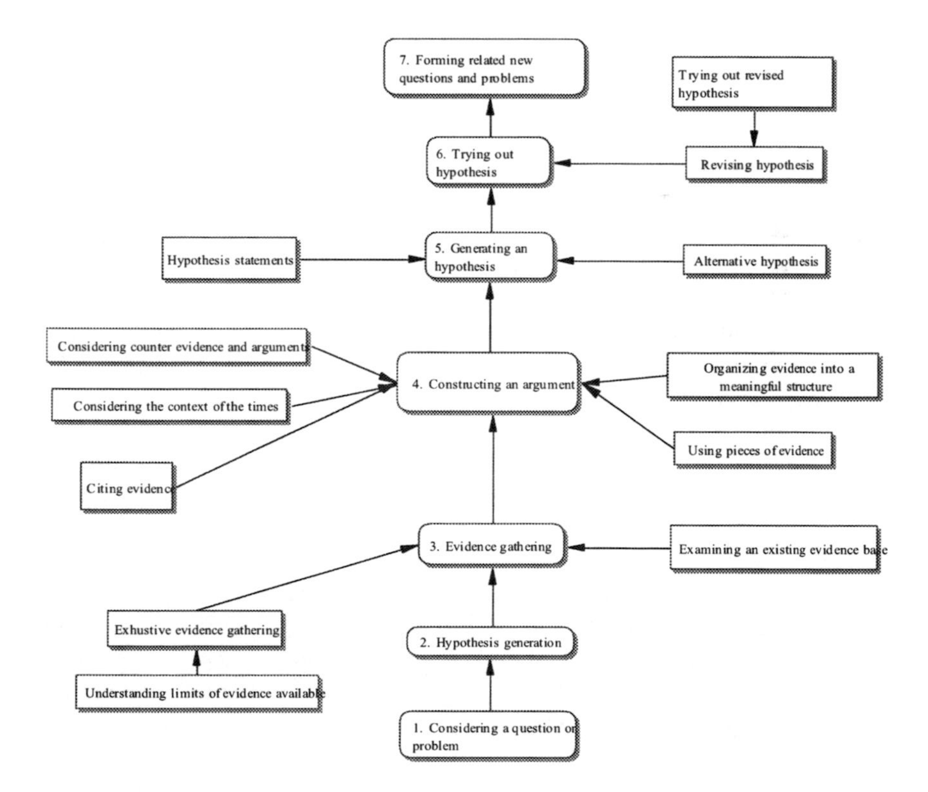

Figure 9.2. Web of argumentation.

testing and finally to related new questions or problems. A student or a researcher can and often does, enter the process at any step. Although Figure 9.2 portrays a linear process, an individual may jump over a step then return to it later. The process is a recursive one since a new question, problem, and hypothesis lead to evidence gathering, the construction of an argument, debate of the argument, and so on. Figure 9.2 suggests components of a process used to help students develop argumentation. As such, it is a model for curriculum and instruction aimed at using in science those cognitive processing skills well defined by historians. The application of the cognitive processing skills used by historians to students' scientific studies is complex. Therefore, the model is presented as a means of suggesting starting points and components.

It is necessary to determine the extent to which students construct cases when they address problems in science. The research base in science that has examined the acquisition of argument construction and its subcomponents has found a broad lack of ability among students and adults. However, students involved in experimental inquiry-based science in elementary and middle school may have proficiency in the main components of the model in Figure 9-2.

Since science-related social issues are an area where science, history and the social sciences intersect, an examination of students' consideration of such issues should give insights into their use of case building in argumentation of science ideas. Such investigation should use topics found in middle school science programs that incorporate evident science-related social issues such as "energy" (AAAS, 1993; NRC, 1996). An open-ended "What if?" question can be used to explore students' argumentation abilities (Driver & Newton, 1997). For example, What if we found a way to more exactly identify the location of oil and how much of it exists in rock layers underground? Such a question could be given at the end of a unit when students have access to resource materials and have investigated relevant concepts. Or, it could serve as a problem around which a unit is structured and evidence is collected. Students' argumentation of their ideas can be examined both in small groups and as a whole group.

Analysis of students' argumentation should consider the six clusters of ideas related to construction of an argument case. One can begin with how many claims are made, whether claims are supported with reasons, whether qualifiers are present, whether claims with reasons are rebutted and whether judgments integrate different arguments (Carretero et al., 1994; Giere, 1991; Kuhn, 1993; Yerrick, 2000). The case that is built by students can include examination of criteria related to the components identified in Figure 9.2. These would include whether hypothesis generation is attempted, evidence is gathered, the construction of an argument

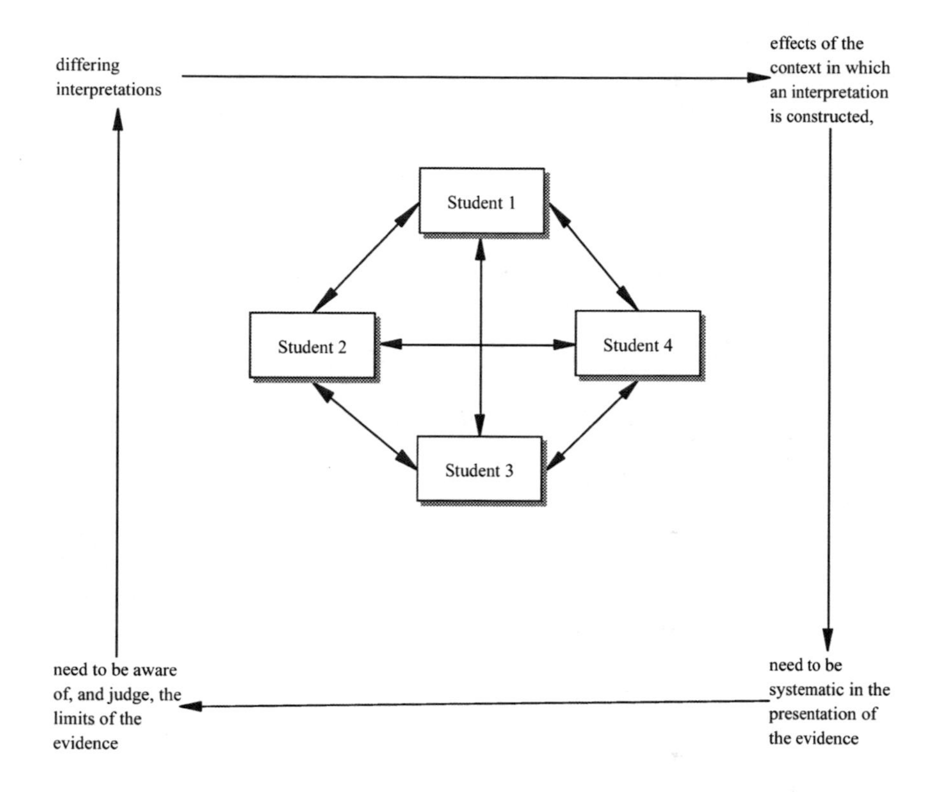

Figure 9.3. Collaboration, learning, reflection, and argumentation.

is attempted and new questions are raised. Group processes could also be examined since learning is socially situated within the classroom and context is an important part of an argument (Driver & Newton, 1997; Lee, Dickinson, & Ashby, 1994). Some of these are whether students check the coordination of evidence and claims, encourage the presentation of different ideas and attempt to coordinate different perspectives

Collaborative reflective learning occurs during argumentation (see Figure 9.3). The student is challenged, encountering differing interpretations, the effects of the context in which an interpretation is constructed, the need to be aware of, and judge, the limits of the evidence and the need to be systematic in the presentation of the evidence (Sunal & Sunal, 1999). Anomalous data can set up a challenge to students' thinking but Chinn and Malhotra (2002) report that students often do not modify their views as a consequence of evidence contradictory to their previously held beliefs. Contradictory claims can highlight limitations of the evidence in use. If students do not easily modify their views when encountering contradictory evidence, they may have difficulty identifying and considering limitations in the evidence they are using.

Students need to develop the skills necessary to ask questions and gather and analyze evidence. Then, they must interpret that evidence building a case they argue with their peers (Sunal & Sunal, 1999). Throughout this process, a key component is the estimation and comprehension of the limitations of the evidence.

AN ARGUMENTATION MODEL FOR STUDENT INQUIRY

A seven-step process can be described (see Figure 9-2) beginning with a question or problem, leading to hypothesis generation, then to evidence gathering, next to construction of an argument, on to further hypothesis development based on the reflection involved in argument construction, next to hypothesis testing, and finally to related new questions or problems (Sunal & Sunal, 1999). A student or a researcher can, and often does, enter the process at any step. The process is a recursive one since a new question, problem, and hypothesis lead to evidence gathering, the construction of an argument, debate of the argument, and so on. These are basic components of a process used to help students develop argumentation and is presented as a means of suggesting starting points and components.

An examination of students' consideration of science-technology-society issues and their argumentation of evidence in relation to an issue gives insights into how they use evidence in argumentation (Driver & Newton, 1997). Such investigations could use topics found in middle school science programs incorporating science-related social issues such as "energy" (AAAS, 1993; NRC, 1994). Analysis of students' evidence gathering involves three components. First, students examine the existing evidence base. Second, they have opportunities to work toward exhaustive evidence gathering. As part of the process they will need to be helped to understand that exhaustive evidence gathering is not likely to be possible because of the limitations of their own maturity and skills, time, money, etc. However, the understanding that exhaustive evidence gathering is a goal toward which scientists strive is important. Finally, students must come to understand the limits of the evidence available to them. They must comprehend that, not only are they probably unable to carry out exhaustive evidence gathering, but the evidence they gather has limitations. Such critical analysis is difficult. Group processes, when used within a classroom, have the potential to challenge students to recognize the limitations of evidence since learning is socially situated within the classroom and context is an important part of an argument (Driver & Newton, 1997; Lee, Dickinson, & Ashby, 1994). Group processes that may encourage the consideration of the limitations of available evidence include whether students check the coordination of evidence and claims, encourage the presentation of different ideas, and attempt to coordinate different perspectives (Driver & Newton, 1997).

INVESTIGATION OF SCIENTIFIC REASONING AND ARGUMENTATION IN SCIENCE CLASSROOMS

In order to determine the appropriateness of the call for inquiry using argumentation in the national science standards in current science classrooms, a study was conducted addressing students' use of evidence in scientific reasoning and argumentation in the context of a unit on a controversial science and society issue. The specific focus of the study was on students' understanding of the limitations of evidence and interpretation of the meaning of that evidence in terms of the arguments they constructed. The study sought to determine whether students' argumentation in small group discussion can be categorized based on their ability to use evidence and identify its limitations. Framed questions were; What characteristics of argumentation occur in small groups? What characteristics of evidence are found in students' argumentation? What group processes do students undertake when involved in interpreting evidence and understanding its limitations? What relationships exist between group ability level and the argumentation used (argument characteristic, use of evidence, and argumentation processes)? What relationship exists between group ability level and their identification of the limitations of argumentation with the characteristics of argumentation and group processes?

This study involved eighth- and ninth-grade students in the interpretation of evidence and consideration of its limitations as they explored nuclear fusion as a potential energy source and related science-technology-society issues through laboratory experiments on fundamental energy concepts in electricity and magnetism, the use of briefing papers, a panel presentation, discussion of the quality of evidence available and of the limitations of the evidence and finally, in voting by ballot as a member of a decision-making body considering whether nuclear fusion is the best and safest source of energy for our future needs.

Procedures

Seven eighth- and ninth-grade science classes (n = 175) at two sites were involved. All students worked in cooperative learning groups at least twice weekly. The age range was 13.6-15.6 with a mean of 14.4 among the

86 females and 89 males participating. At school A, two intact ninth-grade earth and physical science classes of 25 students each a mean age of 15.0, taught same teacher, were used. This high school served students from a rural and small town region. Table 9.2 describes data by class for each school including size, number by gender, age range and mean age. Students were considered by the school system to be achieving at an average or above average level based on grade point average and scores of the state-mandated Stanford Achievement Test (SAT) which categorized school A as average, scoring at the 50th–60th percentile. These students were 70% European American and 30% African American.

At school B, five class sections averaging 25 African American students, with a mean age of 14.2, were taught eighth-grade earth and physical science by the same teacher. These students lived in a small southeastern city with 150,000 residents and a mixed financial base adjacent to the county in which school A students lived. Class three was considered by school administration and the school system's special education department to contain 25 students mostly achieving at a gifted level. Class five had the oldest class mean at school B among its 24 students and was considered as achieving at a below-average level. Classes four, six, and seven were considered by the school system as achieving at an average or above average level. This middle school was categorized as low performing within the state, averaging in the 30th-40th percentile range on the SAT.

Based on school system designations, of the whole group, 24 students were considered as performing at a below average level constituting the

Table 9.2. Class Data

School Site	*Class/Number of Groups/Designation*	*Class Size*	*Gender*	*Age*
A (2 classes, 10 total groups)	one/5 groups/ average-above	25	13 females 12 males	14.7-15.6 mean = 14.9
	two/5 groups/ average-above	25	11 females 14 males	14.7-15.3 mean = 15.1
B (5 classes, 28 groups)	three/6 groups/ gifted	25	13 females 12 males	13.7-14.1 mean = 13.9
	four/6 groups/ average-above	26	14 females 12 males	13.6-14.5 mean = 14.2
	five/5 groups/ low	24	11 females 13 males	14.1-15.1 mean = 14.7
	six/5 groups/ average-above	25	11 females 14 males	13.8-14.4 mean = 14.2
	seven/6 groups/ average-above	25	13 females 12 males	13.7-14.1 mean = 13.9

Table 9.3. Characteristics of Argument: Summary

Class/School	*Characteristic, Number, and Percentage of Groups*					
Designation	*Level*					
Class	*1. Level 0*	*2. Level 0*	*3. Level 1*	*4. Level 3*	*5. Level 3*	*6. Level 4*
One, $n = 5$				2, 5%	3, 8%	
Two, $n = 5$				2, 5%	2, 5%	1, 3%
Three, $n = 6$				2, 5%	4, 11%	
Four, $n = 6$			2, 5%	1, 3%	3, 8%	
Five, $n = 5$	1, 3%	1, 3%	2, 5%	1, 3%		
Six, $n = 5$			1, 3%	2, 5%	3, 8%	
Seven, $n = 6$			3, 5%	1, 3%	2, 5%	
Low	1, 17%	1, 17%	2, 33%	1, 17%		
Average to above			6, 22%	8, 29%	13, 48%	1, 4%
School A				4, 15%	5, 19%	1, 4%
School B			6, 22%	4, 15%	8, 30%	
Gifted				2, 33%	4, 67%	
All Groups	1, 3%	1, 3%	7, 18%	11, 29%	16, 42%	1, 3%

Note: Percentages are rounded off.

membership of 5 of the groups. Another 126 students were at an average to above average level forming 27 groups. The remaining 25 students were performing at a gifted level forming 6 groups.

Observations over 2 months identified the types of science content taught and instructional strategies. Students worked in cooperative groups chosen through a stratified random selection procedure based on gender with a target group of four including two males and two females. Because of unequal male-female balances, 11 of the 38 groups had five students at the beginning of this study of whom at least two represented each gender (Table 9.3). In this study, students worked in cooperative groups over a 2-week period conducting experiments, teaching each other the content of briefing papers provided and preparing for whole group discussions. The group leader guided discussions asking; What are the limitations of the evidence you have about nuclear fusion? Where do you need more information, or different information, in order to make a decision about a question and give an answer you think is supported by evidence? Why can you not get all the information, or evidence, you think you need to give you a good answer? Where could you get more information, or evidence? and When you think about where you need more information, or different information, what does this tell you about how sure you can be of your answers to the questions we talked about?

Analysis

Small group discussion transcripts were analyzed using three schemes. The first scheme evaluated small group argumentation structure incorporating portions of hierarchical criteria devised by Driver and Newton (1997) drawing on the characteristics of argument proposed by Tolumin (1958). These characteristics and their scoring levels were; (1) single claim with no reasons, *level 0*, (2) competing claims with no reasons, *level 0*, (3) single claim with reason(s), *level 1*, (4) competing claims with reasons and qualifiers, *level 3*, (5) claim(s) with reasons responded to by rebuttal, *level 3*, and (6) making judgment(s) integrating different arguments, *level 4*.

The second analysis scheme, *Identification of Evidence and its Limitations in Argumentation*, used criteria to evaluate the ability of small groups to build a case for argumentation that incorporated the use of evidence and recognized its limitations. The criteria were scored as a *1* indicating presence of the criterion or *0* indicating absence, for a possible maximum score of 10. Three subgroups of criteria were scored with a total score possible of 1-10. Subgroup A; Examination of an existing evidence base, gathering of evidence beyond that provided by teacher, and exhaustive evidence gathering. *Subtotal (0-3)* Subgroup B; Identification of one limitation of evidence available, identification of more than one limitation of evidence available, and consideration of the context of the times. *Subtotal (0-3)* and Subgroup C; Citation of evidence, linkage of two pieces of evidence in an argument, organization of evidence into a meaningful structure, and consideration of counter evidence. *Subtotal (0-4)*.

The third scheme, derived from Driver and Newton (1997), evaluated the group processes students used; check information and evidence, question each other's reasons, build on each other's arguments, monitor involvement of group members, check coordination of evidence and claims, encourage presentation of different ideas, and distinguish between scientific claims and those based on other types of knowledge. These were scored 1 if present or 0 if absent, with a range of 0 – 7.

FINDINGS

The findings of the study first are reported in regard to each of the analysis schemes used. Findings for each analysis scheme are presented for the small groups then the groups are considered by school site. The findings are next considered by class designation within school B. Analysis with ANOVA is reported to examine relationships.

Characteristics of the Arguments

Arguments were examined in a progression of four levels demonstrating increasing complexity of thought beginning with making a single claim with no reasons given and ending with making judgment(s) integrating different arguments (Table 9.4). These students mostly fell in a range of three characteristics of argument at two levels. Eight groups

Table 9.4. Summary of Small Groups' Ability to Use Evidence and Identify Its Limitations

	Class, Criterion, Number of Groups and Percentage Meeting the Criterion									
	Subgroup A			*Subgroup B (4-6)*			*Subgroup C*			
	1	*2*	*3*	*4*	*5*	*6*	*7*	*8*	*9*	*10*
1	5, 100%	4, 80%	0, 0%	5, 100%	4, 80%	4, 80%	5, 100%	5, 100%	3, 60%	2, 40%
2	5, 100%	5, 100%	2, 40%	5, 100%	5, 100%	4, 80%	5, 100%	5, 100%	3, 60%	3, 60%
3	6, 100%	6, 100%	1, 17%	6, 100%	5, 83%	4, 67%	6, 100%	6, 100%	5, 83%	3, 60%
4	5, 83%	5, 83%	1, 17%	5, 83%	3, 60%	4, 67%	6, 100%	5, 83%	3, 60%	1, 17%
5	4, 80%	1, 20%	0, 0%	5, 100%	3, 60%	1, 20%	5, 100%	4, 80%	2, 40%	0, 0%
6	5, 100%	5, 100%	1, 20%	5, 100%	4, 80%	3, 60%	5, 100%	5, 100%	3, 60%	0, 0%
7	6, 100%	6, 100%	2, 33%	6, 100%	4, 67%	2, 33%	6, 100%	6, 100%	6, 100%	2, 33%
sub-total	36, 95%	32, 84%	7, 18%	37, 97%	28, 74%	22, 58%	38, 100%	36, 95%	25, 66%	11, 29%
low	4, 80%	1, 20%	0, 0%	5, 100%	3, 60%	1, 20%	5, 100%	4, 80%	2, 40%	0, 0%
aver-age	25, 93%	25, 93%	6, 22%	26, 96%	20, 74%	17, 63%	27, 100%	26, 96%	18, 67%	8, 30%
A	10, 37%	9, 33%	2, 7%	10, 7%	9, 33%	8, 30%	10, 37%	10, 37%	6, 22%	5, 19%
B	15, 88%	16, 94%	4, 24%	16, 94%	11, 65%	9, 53%	17, 100%	16, 94%	12, 71%	3, 18%
gifted	6, 100%	6, 100%	1, 17%	6, 100%	5, 83%	4, 67%	6, 100%	6, 100%	5, 83%	3, 50%

Note: Percentages are rounded off.

made level 1 arguments involving a single claim with reasons given. Level 3 argumentation characteristics were found among 27 groups with 11 making competing claims with reason(s) given and qualifier(s) and 16 making claim(s) with reason(s) responded to by rebuttal.

The five below average groups demonstrated characteristics spread across levels 0-3. School A groups, all of whom were designated as average to above average, showed level 3 and 4 characteristics with 50% making claim(s) with reason(s) given and qualifier(s) (Table 9.4). Another 40% made competing claims with reason(s) given and qualifier(s) while 10% made judgment(s) integrating different arguments. School B's average to above average groups demonstrated characteristics 3, 4, and 5, which were found at levels 1 and 3 with the exception of one group who demonstrated a level 4 characteristic. Most (47%) demonstrated level 3 characteristic 5, claim(s) with reason(s) given and qualifier(s), while 35% showed level 1 characteristic 3 and another 24% showed level 3 characteristic 5. The class designated as gifted demonstrated argumentation at level 3 with 67% displaying characteristic 5 and 33% characteristic 4.

Use of Evidence and Identification of its Limitations

The use of evidence by small groups and their identification of its limitations were examined with the second analysis scheme, using criteria in three subgroups. Subgroup A focused on evidence gathering, considering whether the small group had; (1) examined an existing evidence base, (2) gathered evidence beyond that provided by the teacher, and (3) carried out exhaustive evidence gathering (Table 9.3). A range exists in Subgroup A with criterion one considered the lowest level of operation and criterion three considered the highest level of operation. Subgroup B focused on identification of the limitations of evidence, considering whether the small group had; (4) identified one limitation of the evidence available, (5) identified more than one limitation of the evidence available, and (6) considered the context of the times. Within Subgroup B a range exists with criterion four at the lowest level of operation and criterion six at the highest level of operation. Subgroup C focused on the exhaustivity of the use of evidence in argumentation, considering whether the small group had; (7) cited evidence, (8) linked two pieces of evidence in an argument, (9) organized evidence into a meaningful structure, and (10) considered counter evidence. A range exists in Subgroup C with criterion seven considered the lowest level of operation and criterion ten the highest level.

Subgroup A

All but two groups (95%) examined the existing evidence base provided by their teacher (Table 9.3). Many (84%) gathered evidence beyond that provided. Just seven groups (18%) attempted exhaustive evidence gathering. The existing evidence base was examined by 67% of the low-performing designated groups. Gathering of evidence beyond that provided occurred among 17% of the low performing groups.

Among average to above average groups, 93% examined the existing evidence base provided and gathered evidence beyond that base while 22% demonstrated exhaustive evidence gathering (Table 9.3). At school A all groups examined the existing evidence base while 88% did so at school B. Exhaustive evidence gathering was noted in 20% of school A's groups and 24% of school B's groups. One of the gifted groups (17%) attempted exhaustive evidence gathering.

Subgroup B

All but one group of students, in school B's low performing class, identified one limitation of the evidence available (Table 9.3). Many groups (74%) identified more than one limitation. The context of the times in which the issues were set was considered by 58%.

Of the low-performing groups, 60% identified more than one limitation. Consideration of the context of the times occurred among 20%. Among average to above average groups, more than one limitation of the evidence was identified by 74% of the groups including 90% at school A and 65% at school B. The context of the times was considered by 63% including 80% at school A and 53% at school B. All but one of the gifted (83%) identified more than one limitation of the evidence and 67% considered the context of the times.

Subgroup C

The use of evidence in argumentation was examined using the criteria found in Subgroup C (Table 9.3). All groups cited evidence. Two pieces of evidence were linked in making an argument by 95% of the groups. The evidence was organized into a meaningful structure by 66% with counter evidence considered by 39%.

All low-performing groups cited evidence in their argumentation while two pieces of evidence were linked in their arguments by 80% and two groups (40%) organized evidence into a meaningful structure. All the average to above average groups cited evidence. At school A, 60% created a meaningful structure and 50% cited counter evidence. At school B, 71% created a meaningful structure while 18% considered counter evidence. All of the gifted-designated groups cited evidence in their argumentation

Table 9.5. Summary of Group Processes Used

	Process							
Class	*Check information/evidence*	*Question other's reasons*	*Build on other's arguments*	*Monitor members*	*Coordinate evidence/ claims*	*Encourage different ideas*	*Distinguish knowledge*	*Totals*
1 $n = 5$	5 100%	5 100%	3 60%	5 100%	4 80%	3 60%	2 40%	27 77%
2 $n = 5$	5 100%	5 100%	3 60%	5 100%	4 80%	4 80%	3 60%	29 83%
3 $n = 6$	6 100%	6 100%	3 50%	5 83%	4 67%	4 67%	1 17%	29 69%
4 $n = 6$	6 100%	6 100%	4 67%	5 83%	5 83%	5 83%	3 50%	34 81%
5 $n = 5$	3 60%	2 40%	1 20%	4 80%	2 40%	3 60%	1 20%	16 38%
6 $n = 5$	5 100%	4 80%	3 60%	4 80%	3 60%	2 40%	2 40%	23 55%
7 $n = 6$	6 100%	6 100%	3 50%	4 67%	5 83%	5 83%	3 50%	32 91%
subtotal	36 95%	34 89%	20 53%	32 84%	27 71%	26 68%	15 39%	190 73%
low (5)	3 60%	2 40%	1 20%	4 80%	2 40%	3 60%	1 20%	17 49%
average (27)	27 100%	26 96%	16 59%	23 85%	21 78%	19 70%	13 48%	124 66%
A (10)	10 37%	10 37%	6 22%	10 37%	8 30%	7 26%	5 19%	56 30%
B (17)	17 63%	16 59%	10 37%	13 48%	13 48%	12 44%	8 30%	89 47%
gifted (6)	6 100%	6 100%	3 50%	5 83%	4 67%	4 67%	1 17%	29 69%

while (83%) created a meaningful structure citing evidence and half considered counter evidence.

Use of Group Processes in Argumentation

Seven processes Driver and Newton (1997) identified as facilitative of the involvement of all members in small group argumentation were considered (Table 4). All but two groups checked information and evidence (95%). Most (89%) questioned each other's reasons during argumentation. Monitoring to insure the involvement of each group member occurred among 84%. The presentation of different ideas by group members was encouraged by 68%. Many (71%) checked the coordination of evidence and claims. Fewer (53%) built on each other's arguments. Only 39% tried to distinguish between scientific claims and those based on other types of knowledge.

Among low performing groups, 60% checked information and evidence while 40% questioned each other's reasons. Just one group (20%) built on each other's arguments. During discussions 80% monitored the involvement of members, encouraging each other to participate. As discussion proceeded, 40% checked the coordination of evidence with claims being made. Among 60% members were encouraged to present different ideas. In one group (20%) an effort was made to distinguish between scientific claims and those based on other types of knowledge.

All average to above average groups checked information and evidence. All but one group in school B questioned each other's reasons. In 16 groups members built on each other's arguments, 60% at school A and 59% at school B. Students often monitored each other's involvement in discussions, all at school A and 71% at school B. Checking of the coordination of evidence and claims occurred among 78%, including 80% at school A and 71% at school B. The presentation of different ideas was encouraged among 70% at both schools. Almost half at each school distinguished between scientific claims and those based on other types of evidence.

All gifted-designated groups checked information and evidence and questioned each other's reasons during argumentation. Half built on each other's arguments. Most (83%) monitored the involvement of each other. The coordination of evidence and claims was checked by 67%. The presentation of different ideas was encouraged among 67%. One group (17%) tried to distinguish between scientific claims and those based on other types of knowledge.

What Relationships Exist Within the Data?

Examination of the relationship between ability level and other characteristics demonstrated in group members' work with each other and their argumentation indicated that the difference between use of evidence and characteristics of argumentation was significant at $p < .001$ (F = 4.298) between the top quartiles and lowest quartile. As the use of evidence increased, so did the characteristics of argumentation. The use of group processes also increased as the characteristics of argumentation increased. Differences were found between the lowest quartile and the top quartiles (average to above average and gifted) using univariate analysis of variance with a significance level of $p < .001$ ($F = 4.312$) between the lowest group and average to above average group and p $< .004$ ($F = 4.106$) with the gifted group. There was a significant difference ($p < .000$, $F = 4.401$) between the lowest and top two quartiles indicating that argumentation processes increased as use of evidence increased. Greater usage of evidence was accompanied by concomitant increase in use of group processes among the top two quartiles. A significant difference at $p < .001$ ($F = 4.297$) was also found between group ability level and identification of the limitations of arguments with the characteristics of argumentation but no differences were found with group processes. The top two quartiles' identification of the limitations of argumentation increased at the $p < .001$ ($F = 4.168$) level with their characteristics of argumentation while the lowest quartile showed no increase and continued to demonstrate lower levels of characteristics of argumentation.

CONCLUSIONS AND IMPLICATIONS FOR CLASSROOM PRACTICE

The findings in the study on the characteristics of argument present in small group discussion indicated that many of the groups demonstrated level 3 argumentation with the majority of these making claim(s) with reason(s) responded to by rebuttal and the others making competing claims with reason(s) given and qualifier(s). When examined by school, a greater percentage of student groups at school A demonstrated level 3 argumentation characteristics than at school B. This difference may be partially explained by the higher mean age found at school A, however, some studies of reasoning and argumentation have found little change in argumentation ability after the end of middle/junior high school (Kuhn, 1993; Perkins, 1985; Voss & Means, 1991). The conclusion is that maturation alone does not enhance reasoning and argumentation abilities. Hence, if teachers are to address the inquiry components of the national standards, educational intervention such as that carried out by Shayer and Adey (1993) in fostering the development of specific formal reasoning skills appears to be necessary to facilitate students' development of reasoning and argumentation.

The findings indicate that class performance designation is related to the level of argumentation characteristics demonstrated. There is less difference between the average to above average and gifted designated groups than there is between these groups and those designated as low performing. The average to above average groups had a larger range of students, some of whom are very close in test scores and grade point average to those students designated as gifted. The findings suggest that the characteristics used in analyzing their argumentation reflect the level of development found in the students since the characteristics they demonstrate match to some extent their general designation.

Most of the student groups gathered evidence beyond that given them by their teacher with a few attempting exhaustive evidence gathering. There is more divergence between the low performing groups and the other groups when gathering of evidence beyond that provided by the teacher is used as a criterion. Exhaustive evidence gathering was done by very few groups but these were found in the average to above average and gifted designated class groups and not at all in the low performing class groups. Because of limitations of time in a 2-week unit, the small effort involving a few groups doing exhaustive evidence gathering could be expected. Such evidence gathering requires time to carry out, time to reflect on the available evidence and time to decide how to acquire further evidence. The low performing designated groups may have the most difficulty planning and carrying out a search for further evidence because of a lack of skills, a low experience base and/or a greater need for additional time for reflecting and organizing.

Almost all student groups identified one limitation in the evidence and most identified more than one. Although the low performing groups identified fewer limitations in the evidence overall, half identified more than one limitation and all identified one limitation. These student groups were able to identify limitations in evidence at a minimum level, and generally to a level that was at more than a minimum.

Since the topic investigated dealt with nuclear fusion and its possibilities as a source of commercial energy, the criterion of context of the times examined students' recognition and consideration of factors impacting the search for new energy sources in current society such as the cost of petroleum and the current ability of scientific research to overcome technical problems arising when trying to develop viable commercial nuclear fusion. Student groups met this criterion less often than they met the other criteria in this subgroup. A larger percentage of groups at school A met the criterion than at school B. Since students at school A were in classes one grade level higher than were students in school B this difference could be due to the additional social studies as well as additional science classes taken. The state course of study emphasized the use of primary source documents in social studies classes and considerations of the context of the times. So, additional participation in experiences with such a focus may have impacted school A's students. Consideration of the context of the times requires information, the ability to weigh the influence of factors and reflection (Lee, Dickinson, & Ashby, 1994). Research on students' understanding of history, however, indicates that the context of the times is difficult for middle school students (VanSledright, 1994).

Because some low performing designated groups either did not consider all of the evidence provided and/or did not gather some evidence on their own, they had a weaker information base than did other groups. A weaker information base provides less opportunity for identifying potentially important factors relating to the questions under investigation and for weighing the differential influence of factors. The lower ability of some of these groups to identify more than one limitation of the evidence may result from their weaker information base and may indicate less ability at reflection and weighting factors. Further investigation is needed to establish patterns that may be occurring within a group.

All groups cited evidence in their argumentation and all but two linked two pieces of evidence in making an argument. While groups may link evidence in a multitude of ways, the criterion was whether their final structure could be considered meaningful. Thus, pieces of evidence had to support each other and some recognition of limitations of the evidence was needed in order for the argument to be meaningful. Because earlier criteria were not attained, it was not possible for some groups to develop a meaningful structure for their argumentation. The consideration of counter evidence follows a pattern similar to that found with other criteria in this portion of the study: attainment of the criteria more often by the groups designated as gifted, somewhat less often by those designated as average to above average and least often by the low performing designated groups. A second, less frequent pattern may also be occurring: School A's somewhat older student groups are more successful than are school B's student groups.

The last analysis scheme examined internal group usage of seven processes facilitative of the involvement of all members. The findings indicate that most of these student groups demonstrate the ability to check information and evidence being presented. Within most groups, students questioned each other's reasons during argumentation.

Two group processes that foster the participation of all in discussion were examined. Monitoring to insure each member's involvement is a group process skill found widely among these groups. Facilitation of this process in groups not demonstrating it appears to have potential for success since it has already developed among the majority of the groups. A second group process skill fostering participation is encouragement of the presentation of different ideas. The rate of usage of this skill was less among low performing designated groups than among others.

Internal group processes supporting argumentation included checking the coordination of evidence and claims made during discussion. The usage of checking by low performing groups was much lower than by other groups. Building on each other's arguments is a group process found in fewer groups than were the prior group process skills investigated. A similar pattern was found with the low performing students again utilizing this process less often than did others. Of all the group processes discussed so far, this process is most in need of development by students. The findings also suggest there is potential for the development of the skill of building on each other's arguments since about half of the groups demonstrated its usage.

A final group process examined, distinguishing between scientific claims and those based on other types of knowledge, was the least used. Low performing groups used this process much less frequently than did the other groups. Although all of the group processes have relevance in consideration of students' use of evidence in argumentation and recognition of its limitations, this group process can be considered to most strongly impact such argumentation. Evidence with abundant limitations has restricted applicability. As the applicability is increasingly restricted, the use of the evidence in argumentation also is restricted. Distinguishing scientific claims requires students to have an appropriate information base, to have developed a range of skills in processing such information and to reflect on the claims made. This is a complex set of requirements that may have impacted student groups' usage of this process.

When examining relationships between ability level, argumentation characteristic, and use of evidence, the primary pattern found was a lack of difference between the top two quartiles accompanied by a difference between those two top quartiles and the bottom quartile. The top two quartiles consisted of groups designated as performing at gifted and average to above average levels while the bottom quartile consisted of groups identified as low performing.

When taken together, all of the group processes, which examine students' interaction with each other within a group, showed no significant differences between ability groups. There was more frequent usage of some of the processes. Students often checked information and evidence and questioned each other's reasons but least often built on each other's arguments or distinguished between scientific claims and those based on other types of knowledge. The range found in the frequency of the group processes examined indicates generalized accomplishment of several of the processes. The findings indicate that facilitation of the two much less used group processes may be most needed but do not explain why their usage is low. Further study should explore whether the group processes findings constitute a pattern found among other student populations. The findings also suggest that the least-used group processes may require greater reflection from students, better comprehension of the nature of evidence and more ability to analyze claims.

The lack of significant differentiation between the two top quartiles suggests that these students can accomplish similar levels of argumentation. The gifted groups' designation is not an indicator of difference in this study so perhaps it is not a relevant designation when variables such as those studied here are considered. These findings indicate teachers should begin with an expectation that students designated as average through gifted can accomplish scientific reasoning including consideration of the limitations of evidence. Hence, this study indicates that the national science standards' emphasis on inquiry, reasoning and argumentation is appropriate for many students at the eighth and ninth grade levels.

The findings further indicate that students designated as low performing may be of special concern. A range of accomplishment was found among these student groups yet that range was significantly lower than those of the other groups. Since there is a range of accomplishment, teachers should consider which characteristics and processes students are demonstrating and focus on further development. Least-used characteristics and processes appear to be those requiring higher levels of thinking skills and reflection. These students may need to first build the foundation for such higher skills. Because argumentation as a generalized capacity is found among these low performing groups, the national science standards emphasis on inquiry, reasoning and argumentation is also appropriate for these lower performing students. However, these students need facilitation and support if they are to be able to meet the standards.

Recognizing and understanding the limitations of evidence are key components of argumentation. This study's findings indicate these student groups have some facility with incorporating the limitations of evidence into their argumentation. While differences are found in the lower performance of the bottom quartile as compared to the top quartiles, there is some consideration of the limitations of evidence across the groups. As teachers focus on argumentation and the use of evidence, this study's findings suggest young adolescent students may be generally ready to work with the limitations of the evidence used.

Further, the findings of this study indicate that these student groups had developed some of the characteristics of argumentation and group process skills used in building and arguing an evidentiary case that recognizes the limitations of the evidence. The national science standards' emphasis on inquiry, reasoning and argumentation is appropriate for, and can be accomplished by middle school and secondary students. However, students need guidance and facilitation if they are to develop and expand their use of evidence and recognition of its limitations in argumentation.

SUMMARY

National science standards emphasize inquiry by students that uses reasoning skills and involves them in argumentation of the results with their peers. Much research is occurring in science education to examine how students become inquirers. Researchers have concluded that inquiry is a complex, interconnected web of skills. Hence, researchers are examining how to ask questions, how to gather evidence, how to analyze evidence, and how to argue the interpretation of that evidence with one's peers. Such research also is occurring within a variety of other disciplines. This chapter has examined the intersection of argumentation skills used by scientists and by historians, taking the perspective that each group has developed skills using methods that can inform the other group. A model for argumentation resulting from this combination of perspectives is described. Then, the role of argumentation in science and the web of skills students develop to construct an argument that recognizes the limitations of the evidence being utilized in the argumentation are examined in some depth. Finally, a study is discussed that reports argumentation among 13-15 years olds with a strong consideration of their ability to recognize the limitations of the evidence argued. The results of the study reported indicate that young adolescents can meet the national standards' call for the use of inquiry, reasoning and argumentation. However, the standards are not likely to be met through traditional science teaching, nor by a focus on the best performing students. This study finds that students designated by their school systems as low performing have several skills required for inquiry and argumentation while students who are considered average in performance are able to utilize a wide range of skills. With facilitation and scaffolding of inquiry, reasoning and argumentation skills, the study indicates most young adolescents can argue evidence incorporating the recognition of its limitations. The national standards' emphasis on inquiry is attainable with teaching that recognizes inquiry as an important educational and personal goal.

REFERENCES

Alvermann, D. E., & Hynd, C. R. (1989). The role of subject matter knowledge and interest in the processing of linear and nonlinear texts. *Review of Educational Research, 64*, 201-252.

American Association for the Advancement of Science. (1993). *Benchmarks for scientific literacy.* New York: Oxford University Press.

Beeth, M. E., & Hewson, P. W. (1999). Learning goals in an exemplary science teacher's practice: Cognitive and social factors in teaching for conceptual change. *Science Education, 83*,738-760.

Booth, M. (1993). A modern world history course and the thinking of adolescent pupils. *Educational Review, 32*(3), 245-257.

Brophy, J. VanSledright, B. & Bredin, N. (1993). What do entering fifth graders know about U.S. history? *Journal of social Studies Research, 16 & 17*, 2-19.

Carreterro, M., Jacott, L., Limon, M., Lopez-Manjon, S., & Leon, J. (1994). Historical knowledge: Cognitive and instructional implications. In M. Carretero & J. Voss (Eds.), *Cognitive and instructional processes in history and the social sciences* (pp. 357-376). Hillsdale, NJ: Erlbaum.

Chinn, C., & Malhotra, B. (2002). Epistemologically authentic inquiry in schools: A theoretical framework for evaluating inquiry tasks. *Science Education, 86*, 175-218.

Driver, R., Leach, J., Miller, R., & Scott, P. (1996). *Young peoples images of science.* Philadelphia: Open University Press.

Driver, R., Leach, J., Scott, P., & Wood-Robinson, C. (1994). Young people's understanding of science concepts: Implications of cross-age studies for curriculum planning. *Studies in Science Education, 24*, 75-100.

Driver, R., & Newton, P. (1997, October). *Establishing the norms of scientific argumentation in classrooms.* Paper presented at the European Science Education Research Association Conference, Rome.

Driver, R., Newton, P., & Osborne, J. (2000). Establishing the norms of scientific argumentation in classrooms. *Science Education, 84*(3), 287-312.

Duschl, R. (1990). *Restructuring science education: The importance of theories and their development.* New York: Teachers College Press.

Epstein, T. (1994). *Makes no difference if you're black or white? African American and European-American working-class high school students' beliefs about historical significance and historical authority.* Paper presented at the annual meeting of the American Educational Research Association, New Orleans, LA.

Fehn, B. (1997). *Historical thinking ability among talented math and science students: An exploratory study.* Paper presented at the annual meeting of the American Educational Research Association, Chicago, IL.

Germann, P. J., Haskins, S., & Ails, S. (1996). Analysis of nine high school biology laboratory manuals: Promoting scientific inquiry. *Journal of Research in Science Teaching, 33*, 475-499.

Giere, R. N. (1991). *Understanding scientific reasoning* (3rd. ed.) Ft. Worth, TX: Holt, Rinehart and Winston.

Gilles, D. (1992). *Philosophy of science in the twentieth century.* Oxford: Blackwell.

Hewson, P. W., & Thorley, R. (1989). The conditions of conceptual change in the classroom. *International Journal of Science Education, 11*, 541-553.

Kuhn, T. S. (1962). *The structure of scientific revolutions* (2nd ed.). Chicago: University of Chicago Press.

Kuhn, D. (1993). Science as argument: implications for teaching and learning scientific thinking. *Science Education, 77*(3), 319-337.

Lee, P., Dickinson, A., & Ashby, R. (1994, October). *Researching children's ideas about history.* Paper presented at Madrid conference, Cognitive and Instructional Processes in History, Madrid, Spain.

Leinhardt, G., Beck, I., & Stainton, C. (Eds.). (1994). *Teaching and learning history.* Hillsdale, NJ: Erlbaum.

Leinhardt, G., Stainton, C., Virji, S., & Odoroff, E. (1992). *Learning to reason in history: Mindlessness to mindfulness.* Pittsburgh, PA: University of Pittsburgh, Learning Research and Development Center.

Levstik, L. (1992). New directions for studying historical understanding. *Theory and Research in Social Education 20*(4), 369-385.

Levstik, L., & Pappas, C. (1987). Exploring the development of historical understanding. *Journal of Research and Development in Education, 21,* 1-15.

McKeown, M., & Beck, I. (1990). The assessment and characterization of young learners' McKeown, M. & Beck, I. (1990). The assessment and characterization of young learners' knowledge of a topic in history. *American Educational Research Journal, 27,* 688-726.

McKeown, M., & Beck, I. (1994). Making sense of accounts of history: Why young students don't and how they might. In G. Leinhardt, I. Beck, & C. Stainton (Eds.), *Teaching and learning in history* (pp. 1-27). Hillsdale, NJ: Erlbaum.

National Council for the Social Studies, (1994). Expectations of excellence: Curriculum guidelines for the social studies. Washington, DC: author.

National Research Council. (1996). *National science education standards.* Washington, DC.: Author.

National Research Council. (2000). *How people learn: Brain, mind, experience, and school.* Washington, DC: National Academy Press.

Norris, S. P., & Phillips, L. M. (1994). Interpreting pragmatic meaning when reading popular reports of science. *Journal of Research in Science Teaching, 31*(9), 947-967.

Novak, J. E., & Gowan, D. B. (1984). *Learning how to learn.* Cambridge, New York: Cambridge University Press.

Osbourne, J. (November, 1997). *The structure and organization of the science curriculum.* Paper presented at the Science Education Seminar, King's College: London.

Osborne, J., Erduran, S., Simon, S., & Monk, M. (2001, March). *Enhancing the quality of argument in science lessons*. Paper presented at the annual meeting of the National Association for Research in Science Teaching, St. Louis, MO.

Pearsall, N. R., Skipper, J. J., & Mintzes, J. J. (1997). Knowledge restructuring in the life sciences: A longitudinal study of conceptual change in biology. *Science Education, 81,* 193-215.

Perkins, D. (1985). Postprimary education has little impact on informal reasoning. *Journal of Educational Psychology, 77,* 562-571.

Pintrich, P. R., & Schunk, D. H. (1996). *Motivation in education: Theory, research and applications.* Englewood Cliffs, NJ: Prentice-Hall.

Popper, K. R. (1934). *Logik der forschung [The logic of scientific discovery].* London: Hutchinson.

Posner, G. J., Strike, K. A., Hewson, P. W., & Gertzog, W. A. (1982). Accommodation of a scientific conception: Toward a theory of conceptual change. *Science Education, 66,* 211-227.

Salter, D. (1997). *Evaluation of evidence in historical text and recognition of historical authorial intention.* Paper presented at the annual meeting of the American Educational Research Association, Chicago, IL.

Shayer, M., & Adey, P. (1993). Accelerating the development of formal thinking in middle and high school students IV: Three years after a two-year intervention. *Journal of Research in Science Teaching, 30*(4), 351-366.

Shymansky, J. A., Yore, L. D., Treagust, D. F., Thiele, R. B., Harrison, A., Waldrip, B. G., Stocklmayer, S. M., & Venville, G. (1997). Examining the construction process: A study of changes in level 10 students' understanding of classical mechanics. *Journal of Research in Science Teaching, 34,* 571-593.

Solomon, J. (1991). Group discussion iun the classroom. *School Science Review 72,* 29-34.

Sunal, C. (1998). *The intersection of historical reasoning and scientific reasoning: Exploring cognitive intervention.* Paper presented at the annual meeting of The Association for Science Education, Liverpool, England.

Sunal, C., & Haas, M. (1993). *Social studies and the elementary/middle school student.* Ft. Worth, TX: Harcourt Brace.

Sunal, C., Karr, C., & Smith, C. (1998, November). *Fuzzy logic, neural networks, genetic algorithms: Adults' views of three artificial intelligence concepts used in modeling scientific systems.* Paper presented at the annual meeting of NOVA-NASA, College Park, MD.

Sunal, C., & Sunal, D. (1999, April). *Scientific reasoning: exploring characteristics of middle school students' argumentation.* Paper presented at the annual meeting of the American Educational Research Association, Montreal, CA.

Sunal, D., & Sunal, C. (2003). *Teaching elementary and middle school science.* Columbus, OH: Prentice Hall.

Tolumin, S. (1958). *The uses of argument.* Cambridge, United Kingdom: Cambridge University Press.

VanSledright, B. (1994). *I don't remember—the ideas are all jumbled in my head. Eighth graders' reconstructions of colonial American history.* Paper presented at the annual meeting of the American Educational Research Association, New Orleans, LA.

VanSledright, B., & Brophy, J. (1992). Storytelling, imagination, and fanciful elaboration in children's historical reconstructions. *American Educational Research Journal, 29*(4), 837-859.

Voss, J., & Means, M. (1991). Learning to reason via instruction in argumentation. *Learning and Instruction, 1,* 337-350.

Wegener, A. (Trans.). (1966). *The origin of continents and oceans* (4th ed.). New York: Dover.

Windschitl, M. (2004). Folk theories of "inquiry:" How preservice teachers reproduce the discourse and practices of an atheoretical scientific method. *Journal of Research in Science Teaching, 41*(5), 481-512.

Wineburg, A, (1998). *Historical thinking and other unnatural acts.* Paper presented at the annual meeting of the American Educational Research Association, San Diego, CA.

Wineburg, S. (1991). Historical problem solving: A study of the cognitive processes used in the evaluation of documentary and pictorial evidence. *Journal of Educational Psychology,* 83, 73-87.

Wineburg, S., & Fournier, J. (1994). Finding home in a foreign country: The nature of contextualized thinking in history. In M. Carretero & J. Voss (Eds.),

Cognitive and instructional processes in history and the social sciences (pp. 285-308). Hillsdale, NJ: Erlbaum.

Yerrick, R. K. (2000). Lower track science students' argumentation and open inquiry instruction. *Journal of Research in Science Teaching, 37*(8), 807-838.

CHAPTER 10

IMPROVING THE ALIGNMENT OF CURRICULUM AND ASSESSMENT TO NATIONAL SCIENCE STANDARDS

Luli Stern and Jo Ellen Roseman

This chapter focuses on the continuing efforts of Project 2061 of the American Association for the Advancement of Science (AAAS) to develop and validate tools for evaluating science curriculum and assessment materials in light of *Benchmarks for Science Literacy* (AAAS, 1993) and the *National Science Education Standards (NSES)* (National Research Council, NRC, 1996). It describes a procedure and research-based criteria for judging the alignment of curriculum materials and their assessments to specific ideas in these documents and for evaluating the quality of instructional support tied to those ideas. It then describes findings from one of Project 2061 curriculum evaluation studies, pointing to major strengths and weaknesses of today's curriculum materials and assessment tasks included in them. Finally, it proposes how the same criteria can be used to guide the development of materials that are better aligned with important science ideas.

The Impact of State and National Standards on K-12 Science Teaching, 301–324

INTRODUCTION

During the past 3 decades, research studies on student learning have consistently shown that students of all ages—from middle school to college and beyond—have difficulties understanding many key ideas in science and hold naïve ideas instead. Even after instruction, few students understand that plants make their own food from water and air (rather than take their food from the soil), that matter is particulate (rather than continuous), and that populations change over time by the inadvertent increase in the proportion of individuals that have advantageous characteristics (rather than by the deliberate change of all individuals). When U.S. students are compared internationally, their science performance is favorable in the elementary grades but becomes increasingly poor up through the secondary grades.

Improving student learning is an enormous challenge, requiring reform of many different parts of a complex and inertia-prone educational system. Such reform involves creating more effective assessment, developing sound curriculum materials, improving the quality of teaching, and allocating adequate financial and conceptual resources in school settings. Standards-based reform is founded on the premise that all changes in the education system need to be grounded in a coherent, well-articulated set of specific learning goals. Project 2061, through *Science for All Americans* (AAAS, 1989) and *Benchmarks* (AAAS, 1993), and the NRC, through the *NSES* (NRC, 1996), have provided such a vision for reform of K-12 science education. In general terms, these documents emphasize the understanding and application of ideas and skills by students rather than the memorization of vocabulary and procedures. The convergence of the two sets of recommendations—*Benchmarks* and the *NSES*—demonstrates a substantial consensus on what all students should know and be able to do at specific K-12 grade levels and what topics can be postponed or excluded to provide the time to teach core topics well. Although all states have developed their own standards or frameworks to guide their reform efforts to improve science curriculum and instruction, many have used the national documents in crafting their frameworks, as SRI International found in an external evaluation of Project 2061 in 1996:

> A reading of 43 state curriculum (or other standards-type) documents showed that, overall, such documents do exhibit evidence of influence from Project 2061, as apparent in bibliographic references to *Science for All Americans* and *Benchmarks*, quotations establishing science literacy visions, and organizational schemas similar to that of Project 2061. (SRI International, 1996a, p. 3)

Many states have organized their recommendations according to topics found in *Science for All Americans* and the *NSES* (Blank, Langesen, Sardina, Pechman, & Goldstein, 1997) and some have chosen to adopt statements from these documents and AAAS's *Benchmarks* verbatim (SRI International, 1996b).

Crafting K-12 science standards and building consensus for them among educators, scientists, and the public has been a complex undertaking; however, the implementation of them in a coherent and consistent fashion is a much more daunting task. For students to progress toward attaining the literacy goals found in the standards, it is essential that all aspects of the education system—curriculum materials, instruction, and student assessment—are aligned with these goals. During the past decade, Project 2061 has devoted significant efforts to designing tools that evaluate the alignment of existing curriculum materials with benchmarks and standards and that inform the development of better aligned materials.

Textbooks, for better or for worse, are the de facto curriculum of most schools: they largely determine *what* topics and ideas are taught in the classrooms and *how* these topics are taught (Association for Supervision and Curriculum Development, 1997; Tyson, 1997). A study conducted 20 years ago found that 90% of all science teachers use a textbook 95% of the time (Harmes & Yager, 1981, cited in Renner, Abraham, Grzybowski, & Marek, 1990). More recent studies indicate that many teachers rely on curriculum materials to provide them with some or all of the content as well as the knowledge they need to help their students grasp that content (Ball & Feiman-Nemser, 1988; National Educational Goals Panel, 1994). Reliance on curriculum materials is more apparent when teachers are teaching outside their own area of expertise. Poor curriculum materials can deprive both students and teachers of the needed understanding. In contrast, when used as intended, good curriculum materials can be a powerful catalyst for improving teaching and learning (Ball & Cohen, 1996; Schmidt, McKnight, & Raizen, 1997). Indeed, some studies have suggested that textbooks that incorporate effective teaching strategies improve student learning and provide good models for teaching (e.g., Bishop & Anderson, 1990; Lee, Eichinger, Anderson, Berkheimer, & Blakeslee, 1993). While better curriculum materials alone are unlikely to improve student learning, it is our belief that high-quality curriculum materials can positively influence student learning both directly and through their influence on teachers. Therefore, valid identification of curriculum materials that actually support learning of worthwhile ideas and help teachers build their own content and related pedagogical knowledge is essential.

Over a period of 4 years, with input from hundreds of K-12 teachers, teacher educators, materials developers, scientists, and cognitive researchers nationwide, Project 2061 developed and field-tested a rigorous procedure and criteria for evaluating science and mathematics curriculum materials and the assessment tasks included in them (Kulm, 1999; Roseman, Kesidou, & Stern, 1996). Project 2061 then used the procedure to conduct a series of evaluations of widely used textbooks, including those for middle grades science and high school biology. What makes this procedure unique is that it examines the content and instructional strategies used in curriculum materials for how well they appear to support students achieving specific learning goals, such as the AAAS' benchmarks and the NRC's science standards.

Alignment of curriculum and assessment materials to benchmarks and content standards means more than using them as topic headings. It entails being clear about their precise meaning (i.e., which key ideas are included and which are not), delving into learning research on related student learning difficulties, and holding activities and tasks accountable for aligning content with the ideas and supporting instruction of them. But even alignment of content in curriculum materials to specific ideas in benchmarks and standards is not sufficient to promote meaningful learning.

For learners to develop a coherent understanding of science (as opposed to fragmented knowledge consisting of bits and pieces), curriculum materials need to make explicit connections among the ideas they present. Sound instructional features in curriculum materials are essential, too. The Project 2061 criteria for making the judgments about adequacy of content and instructional support were derived from the past 3 decades of published research on learning and effective teaching, which highlights the importance of establishing a sense of purpose for students, taking account of students' prior ideas (both troublesome and helpful) and attempting to address these ideas, presenting sufficient relevant phenomena to make the scientific ideas plausible, helping students to conceptualize ideas by including helpful representations, providing varied opportunities to apply the ideas, and guiding students' thinking about connections between their experiences and the ideas (e.g., Lee et al., 1993; NRC, 2000; Posner, Strike, Hewson, & Gertzog, 1982; Smith, Blakeslee, & Anderson, 1993). The criteria are consistent with equity concerns and take into consideration both individual and social aspects of learning (Kesidou & Roseman, 2002). Each criterion is defined by a unique set of indicators that further specify what constitutes evidence for meeting it. The complete set of criteria and indicators used in the middle school science textbook evaluation is available on Project 2061's Website (see AAAS, 2002).

An underlying premise of Project 2061 is that curriculum materials can and should play an important role in improving teaching and learning. Several well-respected education researchers have argued that curriculum materials have the potential to serve as sources of reflection for teacher learning as well as for student engagement with important mathematical ideas (Acquarelli & Mumme, 1996; Ball & Cohen, 1996; Smith, 2001). The Project 2061 procedure assumes that textbooks can support effective teaching by providing idea-specific resources—in both the student text and the teacher guide—as defined by the set of criteria. We do not advocate *teacher-proof* materials, which is neither possible nor desirable. Instead, we maintain that materials that rate highly according to the Project 2061 criteria can support teachers in building their own content and pedagogical knowledge and in incorporating that knowledge into their teaching.

The following sections offer some insights and practical directions for evaluating and designing materials that support effective teaching and learning of important ideas in *Benchmarks* and the *NSES*.

FROM TOPIC HEADINGS TO SPECIFIC IDEAS

Benchmarks and *NSES* are very similar in terms of philosophy, language, difficulty, and grade placement, which is highlighted by the very small number of differences found between the two documents (AAAS, 1997). Both *Benchmarks* and *NSES* are based on the premise that the time available in 13 years of schooling is insufficient to teach meaningfully all of the topics and ideas within each topic that are included in today's textbooks. Considerable care went into selecting a set of the most important ideas for all students to understand and be able to use (AAAS, 1993, Chapter 14: Issues and Language; NRC, 1996, Chapter 1). Alignment merely to a topic heading undermines the intent and effort involved in this careful selection.

By reviewing topic headings or a textbook's table of contents, any textbook can appear to be aligned with *Benchmarks* or *NSES*. Nearly all middle and high school textbooks address the topics included in these documents—topics like the structure of matter, processes that shape the Earth's surface, food webs, heredity, and evolution. However, materials may differ considerably in terms of which ideas within these topics are addressed. To be considered aligned, an activity, section of text, or assessment task needs to address the specific idea or ideas in a benchmark or standard—not just its topic.

Yet more than a decade of work with scientists and educators has shown us that expecting alignment to be considered in terms of specific ideas is more radical than we had anticipated. Readers often mentally substitute topic headings for the precise language used to characterize the important ideas in *Benchmarks*. It is common for casual readers to either ignore intended meanings or read unintended meanings into benchmarks (Kesidou & Roseman, 2003; Roseman, 1997; Stern & Roseman, 2001). Our experience has convinced us that before examining the alignment of curriculum and assessments to particular standards or developing aligned materials, it is essential to clarify the meaning of each included idea. Only then do reviewers reach consistent judgments (Roseman, 2004).

Clarifying a benchmark or a standard means identifying which specific ideas are included and which are not. Some ideas may be included in another benchmark or standard at the same grade level or in one at a higher or lower grade level, while other ideas may be considered by both AAAS and the NRC to be outside the scope of science literacy.

For example, *Benchmarks* recommends that by the end of grade 8, students should know that

> Food provides molecules that serve as fuel and building materials for all organisms. Plants use the energy in light to make sugars from carbon dioxide and water. This food can be used immediately or stored for later use. Organisms that eat plants break down the plant structures to produce the materials and energy they need to survive. Then they are consumed by other organisms. (AAAS, 1993, p. 120)

The ideas included in this benchmark are related to several topics, including photosynthesis, cellular respiration, and digestion in organisms and food webs, and focus on the overall transformations of matter and energy. Specifically, the benchmark intends that students should know that plants make their own food by breaking down substances and reassembling the ingredients into other substances using light energy and that some of the light energy is stored in the food. The idea that plant leaves contain different pigments—an idea that is relevant to the general topic of photosynthesis—is not included in this benchmark because it does not deal with matter or energy transformation. On the other hand, not all ideas about matter and energy transformation are included in the benchmark. For example, this benchmark does not include the idea that different energies are associated with the different configurations of atoms in carbon dioxide and sugar molecules or the idea that chlorophyll molecules can be excited to a higher-energy configuration by sunlight and that this energy transformation can be used to excite molecules of carbon dioxide and water so they can link. These more sophisticated ideas are recommended for grade 12 students. Similarly, the idea that organisms need energy to stay alive and grow is not included but is recommended instead for grade 2 students.

Consider another example. The benchmarks related to the kinetic molecular theory recommend that by the end of grade 8, students should know that:

- All matter is made up of atoms, which are far too small to see directly through a microscope;
- Atoms and molecules are perpetually in motion. Increased temperature means greater average energy of motion, so most substances expand when heated. In solids, the atoms are closely locked in position and can only vibrate. In liquids, the atoms or molecules have higher energy, are more loosely connected, and can slide past one another; some molecules may get enough energy to escape into a gas. In gases, the atoms or molecules have still more energy and are free of one another except during occasional collisions. (AAAS, 1993, p. 78)

These benchmarks expect middle school students to know that matter is particulate, rather than being continuous or just including particles, and that these particles are in continuous motion. Students are also expected to know the differences in arrangement and motion of particles in solids, liquids, and gases and a molecular explanation of thermal expansion. The benchmarks expect more than the idea that substances (e.g., mercury in a thermometer or the helium in a balloon) expand upon heating, which stops short of a molecular explanation and is therefore less sophisticated. But the benchmarks stop short of requiring knowledge of types of molecular motion or the ideal gas law, which go beyond reasonable expectations for all middle school students.

The process of clarifying the precise meaning of benchmarks and standards involves examining relevant research on student learning, other benchmarks on the topic, and the adult knowledge that the benchmark anticipates (see Roseman, 1997 for a detailed description of the process). A common and precise understanding of the meaning of a learning goal among reviewers of curriculum and assessment materials is essential for obtaining consistent analysis judgments and it is no less critical for materials developers. Painstaking as this clarification process might seem, experience has shown that it needs to be applied to any benchmark or standard that is used as the basis for the evaluation or design of curriculum and assessment materials (Heller, 2001).

JUDGING ALIGNMENT OF ACTIVITIES AND ASSESSMENT

In principle, a variety of activities—from hands-on investigations to readings and discussions—can align with a particular benchmark. Indeed, variety is often needed for students to find an idea plausible. For example, to help middle school students understand that plants make their own food from carbon dioxide and water, a well-aligned activity could have the students use diabetic test strips to show that sugar is present in iris leaves that have been grown in the presence of CO_2_but not in its absence. An experiment in which sugar (or starch) was detected in leaves grown in open jars, but was detected only in the first few hours on similar leaves grown in closed jars would also align with this middle school benchmark. Other helpful activities might include having middle school students observe the formation of sugars in leaves grown in the light but not in the dark or discussing how a plant's leaf compares to a factory in terms of inputs, outputs, and energy source. Moreover, for middle school students to make important generalizations, they will need to encounter ideas in multiple instances and in a variety of contexts. For example, students who observe that a potted plant makes sugars only in the presence of carbon dioxide may not appreciate that the same holds true for trees, bushes, grass, flowering plants, and algae, but could be helped to make the generalization by guided discussions about other types of plants.

In a similar way, a variety of assessment tasks ranging from True/False and multiple-choice questions to essays and hands-on performance can align with a benchmark idea. To be truly aligned, assessment tasks must probe student understanding of specific ideas in benchmarks and standards at the desired grade range. For example, the following task would not be considered aligned to middle school benchmarks on the kinetic molecular theory, because the task targets less sophisticated ideas:

Would you expect a solid iron ball heated on the stove to:

(a) be a little smaller than before?
(b) be a little larger than before?
(c) stay exactly the same size as before?
(d) I don't know.

The task, as written, does not require knowledge of molecular motion but only knowledge of how the properties of a substance change with temperature (a less sophisticated idea that is expected of upper elementary school students). Students who are familiar only with the macroscopic phenomenon of thermal expansion will be able to respond correctly. To assess students' understanding at the level of sophistication intended by the middle school benchmarks, students could be asked to explain their answers in molecular terms. Alternatively, the choices in the original question could be modified to reflect the molecular level and the research literature on common student misconceptions:

When you heat a solid iron ball on the stove:

(a) the number of molecules increases.
(b) molecules expand or get larger.
(c) molecules stay the same size but move farther apart.
(d) molecules contract or get smaller.

As the examples above indicate, reviewers using Project 2061's analysis procedure are required to judge textbooks and their assessments on the basis of what is explicitly included in them—not by what good teachers might be able to do to improve them. For example, an activity that demonstrates that a plant deprived of light does not grow well would be judged *aligned* with the middle school benchmark only if an explicit connection is made (in prose or questions) to the important role of light energy in the production of food in plants. Without this connection, students can conclude only that *plants need light to grow,* a less sophisticated idea.

To reflect the rigor involved in attending to a benchmark's meaning, Project 2061 reviewers represent alignment findings by bolding only the part(s) of a benchmark statement for which evidence of alignment was found. For example, alignment of a textbook that treated only food making in plants and food breakdown in animals might be represented as shown below:

> Food provides molecules that serve as fuel and building materials for all organisms. *Plants use the energy in light to make sugars from carbon dioxide and water.* This food can be used immediately or stored for later use. Organisms [animals] that eat plants break down the plant structures to *produce* the materials and *energy they need to survive. Then they are consumed by other organisms.*

The above representation is used to make clear that the particular textbook does not clarify the role of food, what plants do with the sugars they make, or the role of food as building materials in consumers.

FINDINGS FROM PROJECT 2061'S MIDDLE SCHOOL CURRICULUM REVIEW STUDY

A closer look at Project 2061's evaluation of nine comprehensive middle school science curriculum materials provides additional insights into what's needed to improve the alignment of materials to benchmarks and standards. After a brief discussion of the evaluation's design and methods, this section summarizes the major findings of the study.

Analysis of each curriculum material, including the assessment tasks provided by materials developers, was carried out by independent teams of trained reviewers (including both experienced middle school teachers and university faculty). For each criterion, reviewers cited all relevant evidence to support their judgments that they found in both the student text and teacher's guide, and these citations served as the basis for detailed analysis reports on each material (AAAS, 2002). The design of the analysis process is described in detail elsewhere (Kesidou & Roseman, 2002; Roseman, Kesidou, & Stern, 1996; Stern & Roseman, 2004).

The rigorous nature of the Project 2061 analysis makes it impractical to examine how well all the ideas specified in national standards are treated in each curriculum material. Preliminary experience in examining a variety of curriculum materials had shown that analyzing a small but carefully chosen set of key ideas could be used to estimate the instructional strengths and weaknesses of a material as a whole. Therefore, key ideas from three important topics—the kinetic molecular theory (physical science), flow of matter and energy in ecosystems (life science), and processes that shape the earth (earth science)—were used as the basis for the middle school analysis. These ideas are examples of the core science content likely to appear in any middle grades material and are common to benchmarks, national standards, and most state frameworks. In addition, the topics selected are the basis for learning other, more complex ideas in middle school as well as in high school (AAAS, 2001a). Furthermore, considerable research has been conducted about learning ideas related to matter and energy transformations and the kinetic molecular theory. The similarity in findings across the topics chosen confirmed our expectation that a well chosen sample could be used to estimate characteristics of instructional support in the material as a whole.

Since much of the knowledge needed for effective teaching is specific to a particular idea (Shulman, 1986), research findings could be used to inform both Project 2061 curriculum analysis and subsequent revision of the materials. Regrettably, we were unable to use ideas about the nature of science or critical response skills as a basis for the textbook evaluation even though both topics are essential for science literacy. Training reviewers requires being able to illustrate the application of the criteria and indicators to the ideas being analyzed, and existing materials did not provide us with good examples. We hope that the new, research-based materials that have explicitly targeted some of these ideas and skills will enable the future analysis and empirical studies needed to increase our knowledge of what it takes to help all students achieve them.

Adhering closely to the clarifications and examples provided by Project 2061 for each key idea and analysis criterion enabled reviewers to reliably judge how well the content in each of the middle school curriculum materials aligns with each key idea and how well the instructional strategies in the student text and teacher's guide can support students' learning of this content. The major findings are outlined below.

Alignment of Content and Key Ideas

For the most part, middle school materials include content that aligns with key ideas in benchmarks and standards. On all three sets of ideas used to probe the conceptual development of important science ideas, curriculum materials reviewed did not differ greatly in their *inclusion* of the specific ideas. Each curriculum material addressed all (or nearly all) key ideas that served as the basis for the analysis. Only a few of the specific benchmark ideas were consistently absent in middle school materials. For example, there was often not a match to the idea that food serves as fuel and building blocks for all organisms. Materials typically presented food only as an energy source while ignoring its important role as building material. When materials did focus on the dual role of food, they did so mostly in the context of humans, rarely generalizing to other organisms.

However, while most key ideas were presented in the curriculum materials, they were often buried between detailed, conceptually difficult, or even unrelated ideas, making it difficult for students to focus on the main ideas. This was true particularly for the life and earth science topic areas. For instance, curriculum materials presented *photosynthesis* and *respiration* repeatedly (but briefly) in the context of detailed descriptions of cells, organisms, or ecosystems, but never developed them as instances of matter and energy transformation. This was true even in curriculum materials that listed the key ideas as main learning outcomes.

Experiences and Key Ideas

Different experiences with the key ideas are usually not tied to one another. While *Benchmarks* and *NSES* specify what students ought to know by the end of particular grade ranges, they do not dictate *when* during a particular grade range (e.g., grades 6-8) a given benchmark (or standard) should be taught and whether experiences related to specific benchmarks should be distributed among several chapters or concentrated in a single chapter. In most middle school textbooks, key life science ideas are taught in different chapters and sometimes in different grades. For example, ideas about matter and energy transformation involve consideration of different levels of biological organization—cells, organisms, and ecosystems—and textbooks typically present parts of the matter and energy story in chapters treating the different levels. This presents a challenge to ensure that students see all these separate treatments as instances of matter and energy transformation. However, the curriculum materials we reviewed rarely related treatments of matter and energy transformation in different chapters. For example, a chapter on plants might indicate that plants are producers; a chapter on cells might indicate that chloroplasts are the sites where photosynthesis occurs and that mitochondria are the powerhouses of cells; and a chapter on ecosystems might show an energy pyramid with plants on the bottom. Yet nowhere was it made clear, for example, that the reason that the amount of energy available decreases at each level in the pyramid is because producers and consumers use some of the material as fuel, which decreases the available biomass. Moreover, only one curriculum material even referenced the different units in which related key ideas were treated so that teachers could design their own ways of making connections for students.

Instructional Support

Instructional support for effectively teaching the key ideas is minimal. Even though materials include content that aligns with the key ideas, the instructional support tied to these ideas is minimal. Most notably, materials do not take student ideas into account, do not provide appropriate relevant phenomena, and lack helpful representations of abstract ideas.

Much of the point of science is explaining phenomena in terms of a small number of principles or ideas. For students to appreciate this explanatory power, they need to have a sense of the range of phenomena that science can explain. Appropriate phenomena—introduced directly through hands—on activities or demonstrations or indirectly through the use of videos or text—can help students view scientific ideas as plausible (Anderson & Smith, 1987; Champagne, Gunstone, & Klopfer, 1985; Strike & Posner, 1985). For scientists to find ideas powerful, they need to see that they can explain a wide variety of phenomena and we should not expect students to settle for less. Unfortunately, middle school materials do not include a sufficient number and variety of phenomena relevant to the key ideas chosen for the Project 2061 study. The problem was not that middle school materials lacked hands-on activities, but that the activities included were not focused on the key ideas that served as the basis of the analysis.

Fostering understanding requires taking the time to attend to the ideas students already have, for example, to inform teachers about prerequisite ideas and about students' likely conceptions, and to incorporate appropriate strategies to address students' ideas (Bishop & Anderson, 1990;

Eaton, Anderson, & Smith, 1984; Lee et al., 1993). As described in the introduction of this chapter, research studies conducted over the past 3 decades indicate that students have difficulties understanding ideas about the particulate nature of matter and about matter and energy transformations in living things. Furthermore, these particular misconceptions are especially resistant to change (Bell & Brook, 1984; Anderson, Sheldon, & Dubay, 1990; Roth & Anderson, 1987). For meaningful learning to occur, teachers need to be informed about commonly held ideas students might have and about what is likely to work. However, despite the extensive, well-documented research on students' difficulties regarding food, matter and energy transformations, matter conservation in living systems, plant and animal nutrition, and the particulate nature of matter, middle school textbooks do not take these difficulties into account. Furthermore, text statements and representations in materials can often reinforce or even induce students' naive conceptions. What we already know about student learning does not seem to influence materials design.

Assessing Students' Understanding

Students' understanding of key ideas is not adequately assessed. Like the text and activities, the assessment tasks included in textbooks were judged for their alignment to specific key ideas. Materials were examined first for their inclusion of sufficient goal-relevant assessments and those found to align with the specific ideas used in the analysis were further examined for whether they merely require rote memorization or actually attempt to probe students' understanding. Rather than memorizing bits of information, science literate adults should be able to use knowledge to describe, explain, and predict real world phenomena, consider alternative positions on issues, and thoughtfully consider solutions to practical problems. Meaningful assessment tasks should reflect these expectations (White & Gunstone, 1992).

In general, our study found assessments included in curriculum materials lacking in value. End-of-unit tests, as well as questions embedded throughout instruction, provide little or no assistance for the teacher in finding out what students actually know about important science literacy ideas. This is consistent with other studies that argued that classroom assessment often encourages superficial learning and that its grading function is overemphasized (Black, 1998; Rudman, 1987).

In analyzing assessment tasks in light of the key ideas we found that the tasks often confound attempts to determine what students actually do or do not understand. We found that:

- *Many tasks could be answered successfully by general intelligence alone or some "test wiseness."* Successful responses could result without knowing the benchmark. For example, in the following task, options c and d are relevant opposite statements, leading a thoughtful student to narrow the options without any knowledge of the kinetic molecular theory:

  Which of the following does *not* happen as the temperature of a gas increases?

  (a) Molecules move faster.
  (b) Molecules have more collisions.
  (c) Kinetic energy increases.
  (d) Kinetic energy decreases.

  Other assessment tasks merely asked students to unscramble relevant terms (e.g., "qateekuhar" for "earthquake"), solve crossword puzzles when definitions are given (in which often relevant terms can be deciphered simply based on letters of irrelevant terms solved in the puzzle), or crack the code of terms written in a "different language." Many of these tasks can be answered successfully just by general intelligence, without needing the scientific ideas.

- *Many tasks could be answered successfully using ideas less sophisticated than those in benchmarks or standards.* Again, successful responses could result without knowing the benchmark. For example, according to benchmarks, middle school students are expected to know that the particles in solids are closely packed and attract one another. In the example below, a student familiar with only the macroscopic phenomena will be able to respond successfully.

  Solids have

  (a) A definite shape but no definite volume.
  (b) A definite shape and a definite volume.
  (c) A definite volume but no definite shape.
  (d) Neither a definite shape nor a definite volume.

- *Tasks were found for which ideas in benchmarks and standards are necessary, but not sufficient for an adequate response.* Unsuccessful responses could therefore result from not knowing the benchmark or from not knowing something else. For example, consider the task:

  A microwave oven works by heating water molecules inside of a substance. Use what you have learned about the particle model of matter to answer the following questions:

  - Why do potatoes sometimes explode in microwave ovens?

- What can a cook do to avoid having a potato explode in a microwave oven? Why would this work?

The answer provided by the Answer Key states that:

> As the water molecules inside a potato are heated, they move faster and eventually become water vapor (a gas). Because the gas expands quickly, it can cause the potato to explode. By poking a few holes in the potato before heating, a cook can provide an opening for the water vapor to escape through, preventing the gas from building up inside the potato.

To respond successfully, students must know that substances expand when heated due to increased molecular motion—an idea that is included in a middle school benchmark—but they must also know that cooks make holes in potatoes to avoid their *explosion* (which goes beyond benchmarks). Since it would probably be unreasonable to expect students who had never seen someone pricking a potato to arrive at the correct answer, perhaps a better way is to provide this information in the question: "Some cooks poke a few holes in the potato before heating. How would this help to avoid having a potato explode in a microwave oven?"

The poor quality of assessments found in curriculum materials has implications for improving teaching and learning. Given that the current assessment in most commercially available and teacher generated materials does not effectively probe for understanding, high student scores on such tests would likely mislead teachers into thinking that their students understand more than they actually do (and might discourage teachers from seeking better curriculum materials). As research has shown us repeatedly, when understanding is probed in effective ways, students do not perform well. Until teachers, parents, and scientists see how little students understand when their understanding is probed in depth, their reluctance to change is understandable.

Because assessment of student performance exerts such influence on the lives of children and on every level of the education system, high-quality assessment can be a powerful catalyst for improving both curriculum and instruction. Poor assessment practices, on the other hand, can impoverish our expectations for learning science, focusing teachers' and students' efforts on less important concepts and skills or on test-taking as an end in itself.

IMPLICATIONS FOR MATERIALS DEVELOPMENT

In the ideal, what students are taught and what they are tested on should be aligned with the same set of specific learning goals. For this to happen, valid interpretation of the learning goals needs to drive the development of both assessment and curriculum materials. But as argued above, alignment to learning goals is necessary but not sufficient; also needed are sound instructional features and explicit connections among key ideas to help students gain a coherent understanding of the learning goals.

Creating Effective Assessment

According to the standards-based reform movement, science literate adults should know the ideas specified in the national science standards documents and draw upon them in a variety of contexts. But what does it mean to assess a benchmark or standard effectively? The following list, assembled from examples found in curriculum materials and the cognitive research literature, attempts to illustrate possible kinds of tasks that require students to apply the ideas in benchmarks and standards (a more comprehensive list is presented in Stern & Ahlgren, 2002). These examples do not dictate an assessment format. They could be set up as open-ended or selected-response tasks (such as multiple-choice); they could be administered as written or oral assignments; and they could be used to track individual or group progress. Tasks that probe understanding and require application of science ideas may include (but are not restricted to) tasks that have students:

- *Decide whether naïve explanations of phenomena are correct and explain the basis for their decision.* For example, relevant to the idea that increased temperature means greater molecular motion, the following task anticipates students' naive attribution of observable properties to the invisible molecules:

  My friend says that when water freezes, the molecules get cold and turn hard. Do you agree? Explain.

- *Explain phenomena.* For example, relevant to the idea that increased temperature means increased molecular motion:

  Explain [in terms of molecules] why you can smell apple pie just taken out of the oven but can't smell apple pie just taken out of the refrigerator.

  Or, relevant to the idea that plants assemble some of the sugars they have synthesized from carbon dioxide and water into the plants' body structures:

  A maple tree weighs so much more than a maple seed. Where does this added mass come from? (Schneps & Sadler, 1988)

- *Predict new phenomena.* For example, relevant to the idea that increased temperature means greater molecular motion:

  What would happen to a solid chunk of steel that sits outside on a very hot summer day? Explain in terms of particles.

  Relevant to the idea that matter is made up of particles:

  Consider a piece of copper wire. Divide it into two equal parts. Divide one half into two equal parts. Continue dividing in the same way. Will this process come to an end? Explain your answer.

- And relevant to the idea that plants assemble some of the sugars they have synthesized into the plants' body structures:

  Assuming some animal eats the carrots' leaves (but not carrots), do you think it will affect the carrots' size this year? Explain your answer.

- *Decide whether certain phenomena are instances of a generalization* (or identify phenomena that could be explained by the generalization). For example:

  Which of the following statements are explained by the idea that the solid crust of the Earth consists of separate plates that move constantly? (Circle all possibilities)

  - The Atlantic Ocean is getting wider each year.
  - The center of the Great Rift Valley in Africa is spreading.
  - Sand dunes in African deserts migrate.
  - Most volcanoes are found in certain locations.

- *Represent the benchmark's idea.* For example, relevant to the idea that matter is made up of particles:

  Imagine that you could see everything magnified by many folds, down to their molecular level. Draw what you would observe in this flask before and after half of the air in it is removed.

- *Consider what would happen if the generalization is violated or modified.* For example, related to the idea that several processes contribute to building up and wearing down the Earth's surface:

  - What would the Earth's surface be like if erosion ceases?
  - What would the Earth's surface be like if there were 20 smaller crustal plates instead of the existing major ones?

The above examples, and the commentary on additional examples identified in the Project 2061 study (Stern & Ahlgren, 2002), can guide teachers and others interested in designing assessments to probe student understanding of benchmarks and standards. In addition, questions used in students' tests and interviews in high-quality research projects can serve as a source of good examples of assessment questions (Black, 1998; Black & Wiliam, 1998; Mintzes, Wandersee, & Novak, 2000; Sadler, 1998; White & Gunstone, 1992). Development of effective assessment tasks requires iterative research and development cycles (including student interviews) in which students' responses are used as the basis for revision and refinement of assessment questions (Gallagher, 1996; Treagust, 1988).

The underlying premise behind the Project 2061 assessment analysis criteria is that assessment should be used mainly to promote student learning and improve instruction (Black, 1998; Black & Wiliam, 1998; Neill, 1997). Effective assessment tasks are essential for judging the impact of curriculum and instruction on learning. But to help students progress toward literacy goals, new science curriculum materials are needed.

Creating Effective Curriculum Materials

Project 2061's interest in analysis goes well beyond materials evaluation as an end in itself. We see the analysis as a powerful means to understand how to design more effective curriculum materials and to play a critical role in the design process itself. *Science for All Americans* (AAAS, 1989), *Benchmarks*, and the *NSES* provide materials developers with a coherent set of learning goals on which to focus their materials design; *Atlas of Science Literacy* (AAAS, 2001a) points to conceptual prerequisites and other related ideas to help developers present a coherent story. The instructional criteria used to analyze curriculum materials can provide further guidance for curriculum development by defining and illustrating research-based qualities of instructional support. Concrete examples from curriculum materials, identified and described as a part of this study (AAAS, 2001b; AAAS, 2002; Stern & Roseman, 2004) can provide additional insights for curriculum development by illustrating characteristics of materials that are consistent with the learning research. The criteria and indicators, along with the standards to be targeted, constitute a coherent set of design specifications that are subject to modification as empirical studies reveal which have the greatest payoff for student learning.

Several middle school materials development groups have recently committed themselves to designing curriculum materials that not only align with important ideas in *Benchmarks* and *NSES* but also attempt to promote student learning of those ideas (Heller, 2001; Reiser, Krajcik,

Moje, & Marx, 2003). These groups are taking advantage of the above resources and consulting with Project 2061 in clarifying the meaning of specific benchmarks and standards. In addition, Project 2061 is clarifying what alignment and instructional support for them entails and analyzing draft materials to inform their revision.

It should be pointed out that while *Benchmarks* and *NSES* offer reasonable hypotheses based on both idea-specific learning research and findings of more general cognitive studies, neither purports to be perfect or final. As research on student science learning expands, more informed decisions will be possible regarding grade-level placements of science concepts. In the interim, we expect that as educators and scientists work together to carefully evaluate curriculum materials or to design standards-based materials, they will reconsider the appropriateness of the recommendations in the national documents and perhaps suggest slight modifications. And as research findings become available on the learning that results from use of highly-rated curriculum materials—those that narrow the number of topics, focus on a coherent set of age appropriate key ideas, and provide sound instructional support for these ideas—it will be possible to better estimate the extent to which the Project 2061 and NRC vision of reform is attainable. It may well be that taking standards and the criteria seriously, while improving student understanding, will take more time per topic or idea than is currently available in today's science curriculum. If so, then a further reduction in the number of topics and ideas within topics may be needed: Even though *Benchmarks* and *NSES* represent a considerable reduction over what is covered in today's textbooks, it is possible that this set is still be too large to be learned with understanding in the time available in 13 years of schooling. However, before making further reductions in recommended science learning for all, we need to acknowledge that we do not yet have the needed materials to implement a K-12 science literacy curriculum. For example, if K-5 students were to have adequate experiences with relevant phenomena, they might be ready to tackle and grasp some of the more abstract ideas in middle school. And if middle school students actually achieved the grade 8 learning expectations, then high school teaching would have a set of important ideas on which to build. At present, middle school and high school materials developers cannot assume that students have all the needed prerequisite understandings.

Will better curriculum materials necessarily make a difference in student learning? Clearly, teachers play a crucial role in mediating even the best available materials and in tailoring them as needed to meet their particular students' learning needs. In this regard, the procedure and detailed analysis reports can also be helpful to science teachers, by providing case studies to promote reflection on teaching practices. Highly rated materials can model attributes of exemplary teaching (Ball & Cohen, 1996). Indeed, Project 2061 is examining the relationship between teaching that meets the criteria and student mathematics learning and is using teacher analysis of case studies to inspire and guide changes in their practice (DeBoer et al., 2004).

Research is at the core of our work with curriculum materials. The criteria and indicators used in the analysis procedure are based on available research on how students learn specific concepts. We have hosted conferences to introduce materials developers and publishers to these criteria and to consider how research agendas and policies could support the development, selection, and use of highly rated materials. We are currently partnering with the University of Michigan, Michigan State University, and Northwestern University in a National Science Foundation-funded Center for Learning and Teaching dedicated to research on science materials development and to fostering a new generation of leaders in materials development, evaluation, selection, and successful implementation. More studies are needed, of course, and we encourage others to undertake them. It will not be easy, nor will it necessarily lead quickly to readily applicable findings. But research dedicated to the careful alignment of curriculum and assessment with national science standards can contribute significantly to tools that help educators overcome students' persistent difficulties with key science ideas and fulfill the vision of science literacy for all.

SUMMARY

Designing effective curriculum materials involves time consuming cycles of designing, testing, redesigning, and retesting the materials to ensure their effectiveness in helping all students achieve science literacy. Project 2061 continues to develop tools and instructional components to serve the needs of materials developers. This work involves designing model curriculum components, including examples of natural phenomena, representations, sets of questions, and research summaries that are well aligned to specific science standards and can be incorporated into curriculum materials or classroom lessons and empirically tested.

REFERENCES

Acquarelli, K., & Mumme, J. (1996). A renaissance in mathematics education reform. *Phi Delta Kappan*, 77, 478-484.

American Association for the Advancement of Science. (1989). *Science for all Americans*. New York: Oxford University Press.

American Association for the Advancement of Science. (1993). *Benchmarks for science literacy*. New York: Oxford University Press.

American Association for the Advancement of Science. (1997). *Resources for science literacy: Professional development*. New York: Oxford University Press.

American Association for the Advancement of Science. (2001a). *Atlas of science literacy*. Washington, DC: Author.

American Association for the Advancement of Science. (2001b). *AAAS science textbooks conference CD-ROM resource*. AAAS conference on developing textbooks that promote science literacy. Retrieved August 16, 2004, from http://www.project2061.org/meetings/textbook/literacy/cdrom/index.htm

American Association for the Advancement of Science. (2002). *Middle grades science textbooks: A benchmarks-based evaluation*. Retrieved October 20, 2004, from http://www.project2061.org/tools/textbook/mgsci/index.htm

Anderson, C., Sheldon, T., & Dubay, J. (1990). The effects of instruction on college nonmajors' conceptions of respiration and photosynthesis. *Journal of Research in Science Teaching, 27*, 761-776.

Anderson, C. W., & Smith, E. (1987). Teaching science. In V. Richardson-Koehler (Ed.), *The educator's handbook: A research perspective* (pp. 84-111). New York: Longman.

Association for Supervision and Curriculum Development. (1997). Education update, Vol. 39, No. 1.

Ball, D. L., & Cohen, D. K. (1996). Reform by the book: What is—or might be—the role of curriculum materials in teacher learning and instructional reform? *Educational Researcher, 25*, 6-8, 14.

Ball, D. L., & Feiman-Nemser, S. (1988). Using textbooks and teachers' guides: A dilemma for beginning teachers and teacher educators. *Curriculum Inquiry, 18*, 401-423.

Bell, B., & Brook, A. (1984). *Aspects of secondary students understanding of plant nutrition*. Leeds, United Kingdom: University of Leeds, Centre for Studies in Science and Mathematics Education.

Bishop, B., & Anderson, C. (1990). Student conceptions of natural selection and its role in evolution. *Journal of Research in Science Teaching, 27*, 415-427.

Black, P. (1998). Assessment by teachers and the improvement of students' learning. In B. J. Fraser & K. G. Tobin (Eds.), *International handbook of science education* (pp. 811-822). Boston: Kluwer Academic.

Black, P., & Wiliam, D. (1998). Assessment and classroom learning. *Assessment in Education, 5*, 7-74.

Blank, R. K., Langesen, D., Sardina, S., Pechman, E., & Goldstein, D. (1997). *Mathematics and science content standards and curriculum frameworks: States' progress on development and implementation*. Washington, DC: Council of Chief State School Officers.

Champagne, A., Gunstone, R., & Klopfer, L. (1985). Instructional consequences of students' knowledge about physical phenomena. In L. West & A. L. Pines (Eds.), *Cognitive structure and conceptual change* (pp. 61-90). Orlando, FL: Academic Press.

DeBoer, G., Morris, K., Roseman, J. E., Wilson, L, Capraro, M. M., Capraro, R., Kulm, G., Willson, V., & Manon, J. (2004, April). *Research issues in the improvement of mathematics teaching and learning through professional development.* Paper presented at the American Educational Research Association, San Diego, CA.

Eaton, J. F., Anderson, C. W., & Smith, E. L. (1984). Student preconceptions interfere with learning: Case studies of fifth-grade students. *Elementary School Journal, 64*, 365-379.

Gallagher, J. J. (1996). *Structure of matter.* East Lansing, MI: Michigan State University.

Heller, P. (2001, February 27-March 2). *Lessons learned in the CIPS curriculum project.* Paper presented at the American Association for the Advancement of Science Conference on Developing Textbooks That Promote Science Literacy, Washington, DC. Retrieved October 13, 2004, from http://www.project2061.org/meetings/textbook/literacy/heller.htm

Kesidou, S., & Roseman, J. E. (2002). How well do middle school science programs measure up? Findings from Project 2061's curriculum review. *Journal of Research in Science Teaching, 39*(6), 522-549.

Kesidou, S., & Roseman, J. E. (2003). Project 2061 analyses of middle-school science textbooks: A response to Holliday. *Journal of Research in Science Teaching, 40*, 535-543.

Kulm, G. (1999). Evaluating mathematics textbooks. *Basic Education, 43*(9), 6-8.

Lee, O., Eichinger, D. C., Anderson, C. W., Berkheimer, G. D., & Blakeslee, T. D. (1993). Changing middle school students' conceptions of matter and molecules. *Journal of Research in Science Teaching, 30*, 249-270.

Mintzes, J. J., Wandersee J. H., & Novak, J. D. (Eds.). (2000). *Assessing science understanding: A human constructivist view.* Boston: Academic Press.

National Education Goals Panel. (1994). *Data volume for the National Education Goals Report* (Vol.1). Washington, DC: Author.

National Research Council. (1996). *National science education standards.* Washington, DC: National Academy Press.

National Research Council. (2000). *How people learn. Brain, mind, experience, and school.* Washington, DC: National Academy Press.

Neill, D. M. (1997). Transforming student assessment. *Phi Delta Kappan, 79*, 34-40.

Posner, G. J., Strike, K. A., Hewson, P. W., & Gertzog, W. A. (1982). Accommodation of a scientific conception: Toward a theory of conceptual change. *Science Education, 66*(2), 211-227.

Reiser, B., Krajcik, J., Moje, E., & Marx, R. (2003, March). *Design strategies for developing science instructional materials.* Paper presented at the annual meeting of the National Association for Research in Science Teaching, Philadelphia.

Renner, J. W., Abraham, M. R., Grzybowski, E. B., & Marek, E. A. (1990). Understanding and misunderstandings of eighth graders of four physics concepts found in textbooks. *Journal of Research in Science Teaching, 27*, 35-54.

Roseman, J. E. (1997). Lessons from Project 2061: Practical ways to implement benchmarks and standards. *The Science Teacher, 64*(1), 26-29.

Roseman, J. E. (2004, April). *Mapping for curriculum coherence*. Paper presented at the annual meeting of the National Association for Research in Science Teaching, Vancouver, Canada.

Roseman, J., Kesidou, S., & Stern L. (1996, November). *Identifying curriculum materials for science literacy: A Project 2061 evaluation tool*. Paper presented at the National Research Council colloquium "Using standards to guide the evaluation, selection, and adaptation of instructional materials," Washington, DC.

Roth, K., & Anderson, C. (1987). *The power plant: Teacher's guide to photosynthesis*. Occasional Paper no. 112. Institute for Research on Teaching. East Lansing, MI: Michigan State University.

Rudman, H. C. (1987). Testing and teaching: Two sides of the same coin? *Studies in Educational Evaluation, 13*, 73-90.

Sadler, P. M. (1998). Psychometric models of student conceptions in science: Reconciling qualitative studies and distractor-driven assessment instruments. *Journal of Research in Science Teaching, 35*, 265-296.

Schmidt, W., McKnight, C., & Raizen, S. (1997). *Executive summary: A splintered vision: An investigation of U.S. science and mathematics education*. Lansing, MI: U.S. National Research Center, Michigan State University.

Schneps, M. H., & Sadler, P. M. (1988). *A private universe. Program 2. biology: Lessons pulled from thin air* [Videotape]. New York: Annenberg/CPB.

Shulman, L. (1986). Knowledge and teaching: Foundation of the new reform. *Harvard Educational Review, 57*, 1-22.

Smith, E. L., Blakeslee, T., & Anderson, C. W. (1993). Teaching strategies associated with conceptual change learning in science. *Journal of Research in Science Teaching, 30*, 111-126.

Smith, M. S. (2001). *Practice-based professional development for teachers of mathematics*. Reston, VA: National Council of Teachers of Mathematics.

SRI International. (1996a). *Evaluation of the American Association for the Advancement of Science's Project 2061, executive summary*. Menlo Park, CA: Author.

SRI International. (1996b). *Evaluation of the American Association for the Advancement of Science's Project 2061, Vol. I: Technical report*. Menlo Park, CA: Author.

Stern, L., & Ahlgren, A. (2002). Analysis of students' assessments in middle school curriculum materials: Aiming precisely at benchmarks and standards. *Journal of Research in Science Teaching, 39*(9), 889-910.

Stern, L., & Roseman, J. E. (2001, October). Textbook alignment. The Science Teacher, *68*(3), 52-56.

Stern, L., & Roseman, J. E. (2004). Can middle-school science textbooks help students learn important ideas? Findings from Project 2061's curriculum evaluation study: Life science. *Journal of Research in Science Teaching, 41*, 538-568.

Strike, K., & Posner, G. S. (1985). A conceptual change view of learning and understanding. In L. West & A. L. Pines (Eds.), *Cognitive structure and conceptual change* (pp. 211-231). Orlando, FL: Academic Press.

Treagust, D. F. (1988). Development and use of diagnostic tests to evaluate students' misconceptions in science. *International Journal of Science Education, 10*, 159-169.

Tyson, H. (1997, July). *Overcoming structural barriers to good textbooks.* Paper commissioned by the National Education Goals Panel for release at its meeting July 30, 1997, Washington, DC.

White, R., & Gunstone R. (1992). *Probing understanding*. London and New York: The Falmer Press.

PART IV

IMPACT OF SCIENCE STANDARDS ACROSS THE EDUCATION CONTINUUM

Part four examines science standards across the education continuum, promising practices in teacher preparation, school systems, schools, classrooms, and impact on student science content learning. Chapter eleven discusses the impact of the science standards on the teacher education continuum. In this chapter Gail Shroyer, Teresa Miller, and Cecilia Hernandez document two case studies of systemic reform in response to the science standards to improve teaching and student learning by enhancing teacher education across the continuum of teacher growth. In chapter twelve, a case study is reported that examines how national and state standards have affected middle school science education and student content learning. In order to interpret the influence of standards-based instruction and its impact on student performance Christy MacKinnon, Judith Fowles, Edward Gonzales, Bonnie McCormick, and William Thomann investigated the interaction of roles played by a university initiated professional development program and a school district in efforts to support changes in teacher practice. Chapter thirteen discusses the impacts in K-12 earth science education of the national standards against the backdrop of the intensions that guided their development. Larry Enochs and Fred Finley conclude that the impact of the national standards on earth science education has been limited. An international perspective is provided in chapter fourteen concerning the influence of science standards on teaching science in Australian schools. Warren Beasley discusses issues of professional teaching, curriculum, and assessment standards, their development and ownership, and the competing models from professional associations and business groups.

CHAPTER 11

TRANSLATING SCIENCE STANDARDS INTO PRACTICE ACROSS THE TEACHER EDUCATION CONTINUUM

A Professional Development School Model

Gail Shroyer, Teresa Miller, and Cecilia Hernandez

This chapter documents an example of systemic reform in response to the science standards, based on a professional development school (PDS) model, to improve teaching and learning by enhancing teacher education across the continuum of teacher growth—from initial preparation through continuous professional development. Two case studies of teacher development are presented to trace the impact of the science standards on K-16 science teachers, science teaching practices, and K-12 student learning in science. The evidence from these case studies suggests substantial impact of the standards on teaching and learning across the PDS Partnership. This chapter documents changes in teaching practices for K-16 teachers, continuous renewal in the Kansa State University (KSU) teacher education program, and expanded opportunities and learning for K-12 and teacher education students. Four conclusions have emerged from this analysis of K-

The Impact of State and National Standards on K-12 Science Teaching, 327–362

16 systemic reform in response to the standards: (1) significant educational change requires extensive and continuous time, resources, professional development, and implementation support across the system; (2) deeper understanding and implementation of the standards are developed through extensive and meaningful work with the standards; (3) the influence of teacher development initiatives on teachers, teaching practices, and student learning can be minimized or enhanced by curriculum, assessment, and accountability measures; and (4) teacher development efforts need to be centered on student learning.

INTRODUCTION

Since *A Nation At Risk: The Imperative of Educational Reform* (National Commission on Excellence in Education, 1983) numerous commissions, committees and foundations have documented an array of statistics indicating that K-12 education in the United States is not preparing students with the scientific literacy needed for success in the twenty-first century (American Association for the Advancement of Science, AAAS, 1989; Carnegie Forum on Education and the Economy, 1986; National Science Board, 1983). This era of reports was followed by the national science standards movement that identified what K-12 students and teachers must know and be able to do (AAAS, 1993; National Research Council, NRC, 1996). Twenty years after *A Nation At Risk* we wonder, what, if any, change has occurred? What was the response to the reform movement and the standards? How have the science standards been translated into practice? What are the results for teachers, teaching practices, and student learning (NRC, 2002)? This chapter documents one example of program change in response to the standards, based on a PDS model, to improve K-16 teaching and learning by enhancing teacher education across the continuum of teacher growth—from initial preparation through continuous professional development.

Over 70 faculty from KSU's colleges of arts and sciences and education and 100 teachers from KSU's PDSs have been engaged in standards-based reform in K-12 schools and teacher education for 15 years. The vision of this continuing partnership is to collaboratively restructure teacher education while simultaneously reforming K-12 schools to enhance the quality of teaching and learning at all levels of schooling for all students and educators. This K-16 systemic reform model is based on: (1) state and national K-12 science content standards (AAAS, 1993; Kansas State Department of Education, KSBE, 2002b; NRC, 1996); (2) program standards for teacher preparation and licensure, (KSBE 2002a; National Council for Accreditation of Teacher Education, NCATE, 2002); (3) national standards for beginning teachers, (Interstate New Teachers

Assessment and Support Consortium, INTASC, 1995); and (4) standards for professional teachers (National Board for Professional Teaching Standards, NBPTS, 1998). This chapter tells the story of K-16 standards-based reform in science teacher education and its impact on science teachers, science teaching practices, and student science learning.

The framework developed by the NRC (NRC, 2002) for investigating the influence of nationally developed standards in mathematics, science, and technology education guides the presentation of this story. The theme is teacher development (specifically teacher preparation and ongoing professional development) as a channel of influence of the science standards. The story will be presented as two case studies of change: (1) Initial Standards-Based Teacher Development Projects (from 1989-1995) and (2) Recent Standards-Based Teacher Development Projects (from 1999 to 2004). Both cases demonstrate how science standards have influenced teacher development and how teacher development impacts the educational system and student learning.

THE SETTING

Kansas State University is a land grant institution located in a rural/agrarian region of Kansas. The College of Education has the largest teacher preparation program in the state and is one of the largest in the nation, graduating over 400 elementary and secondary teacher education students each year. The College of Arts and Sciences includes 25 departments that educate teachers in the core academic subject areas in which they teach. The KSU PDS Partnership includes over 30 College of Education faculty members, 30 College of Arts and Sciences faculty members, and hundreds of teachers and administrators from 27 schools in five diverse districts across Kansas. These partners have been engaged in collaborative and simultaneous K-16 reform since 1989. This reform involved restructuring KSU's teacher education program while simultaneously improving teaching and learning in K-12 schools. Reform efforts in science and mathematics were supported by external funding during 2 periods of time, 1989–1995 and 1999–2004. Professional development in science was conducted between these periods, but new simultaneous K-16 reform efforts, involving both teacher preparation and professional development, were initiated only within these time frames. Our case studies of teacher development are set within this context of K-16 simultaneous reform, involving both teacher preparation and professional development.

The case studies will focus on just one of the KSU PDS districts. The Manhattan-Ogden Unified School District is the only district that has been a member of the KSU partnership since it began in 1989 and has been involved in all the partnership's teacher development projects. All schools in Manhattan-Ogden are PDSs, thus allowing for a more holistic examination of the systemic impact of standards-based teacher development initiatives. Manhattan-Ogden Unified School District is located adjacent to KSU and the Fort Riley Military Installation. At the beginning of the study the district had 13 schools, but lost 2,000 students between 1994 and 2004 and was forced to close two elementary buildings. By 2004, the district's student enrollment was 5,369 students with 500 teachers employed in eight elementary schools, two middle schools, and one high school. Thirty percent of these students were on free and reduced lunches, 22% were classified as minority, and 17% were identified as special needs. Students represented over 50 countries and spoke over 20 languages.

The Manhattan and Ogden communities are culturally, linguistically, and economically diverse. Manhattan is a university town, and many parents in the community are university faculty, support staff, and students. A growing number of KSU students and faculty are international. Ogden is a small community on the outskirts of Manhattan adjacent to the military installation with a large number of highly mobile military families. It has a 71% low income, 27% minority population, and 52% of the students are on free lunch subsidies.

BACKGROUND

The educational crisis proclaimed through reports of the 1980s was, and still is, based on a complex web of social, economic, political and educational factors. Complex problems require complex systemic solutions (Fuhrman & Massell, 1992). "Schools and universities must be willing to reexamine everything. They must creatively build different kinds of schools and preparation programs that bridge the gap between what is learned and what people need to understand and be able to do in order to be productive in the future" (Richardson, 1994, p. 1).

School reform advocates recognize that high quality career-long teacher education is the heart of educational reform (Darling-Hammond 1999; Holmes Group, 1986; National Commission on Teaching and America's Future, NCTAF, 1996). The National Commission on Mathematics and Science Teaching for the twenty-first century makes the case that: "better teaching is grounded in improved teacher preparation and professional development" (U.S. Department of Education, USDOE, 2000, pp. 7-8). According to the Committee on Science and Mathematics Teacher Preparation, "fundamental restructuring of teacher preparation and professional development is needed to best serve the interest of students' learning and their future success as individuals, workers, and citizens" (NRC, 2001a, pp. 1-2). In *Tomorrow's Schools of Education* (1995), the Holmes Group suggests, "the indisputable link between the quality of elementary and secondary schools and the quality of the education schools must be acknowledged—and we must respond" (p. 3).

This need for career-long professional development combined with the need to restructure K-12 schools and teacher education has "created a unique opportunity for collaborative systemic reform, where the many components of reform are addressed and their interdependencies and interrelationships are recognized" (NRC, 2001a, p. 75). PDS provide a way to initiate and sustain systemic reform. The PDS model involves educational stakeholders in inquiry and reflective practice to discover how to develop and maintain effective educational systems (Holmes Group, 1990). PDSs address the continuum of teacher development from undergraduate preparation, through early career induction, and continued professional growth. PDS participants acknowledge that improvements at any one of these levels cannot succeed without improvements at the other levels (Richardson, 1994; Teitel, 1998). Partnering institutions in a PDS model share responsibility for: (1) the clinical preparation of new teachers; (2) the continuing professional development of all educators; (3) the support of children's learning; and (4) the support of practice based inquiry directed toward the improvement of teaching and learning (NCATE, 2001).

RESEARCH FRAMEWORK

What was our response to the reform movement of the 1980s and the standards that followed? Did the science standards make a difference? To construct a more meaningful understanding of reform in KSU teacher education and its K-12 partner schools, a multifaceted, longitudinal study was initiated to examine the process and impact of change on the KSU PDS Partnership. Since 1989, a wide variety of quantitative and qualitative data have been gathered from all stakeholders using an evaluative case study design (Guba & Lincoln, 1981). Data sources include multiple surveys and interviews of PDS teachers, administrators, K-12 students, parents, KSU faculty, and KSU students; numerous institutional and project documents and records; and student assessment data. Both quantitative (descriptive and inferential statistics) and qualitative (content analysis, pattern analysis, and constant comparison) techniques were used to analyze all data. Long-term observations were made, data were triangulated and cross-checked by multiple collaborative researchers, and member checks, peer examination, and audit trails were used to increase validity and reliability (Merrian, 1998; Miles & Huberman, 1998).

The Framework for Investigating the Influence of Nationally Developed Standards in Mathematics, Science, and Technology Education (NRC, 2002) was used to reexamine and interpret data across time (1989 –2004) and organizational units (K-12 schools, the college of education and the college of arts and sciences). This framework identifies three channels of influence to describe how reform trends might move through the educational system to impact teaching and learning: curriculum, teacher development, and assessment and accountability. Our analysis of the KSU PDS reform projects focuses on teacher development, specifically initial teacher preparation and on-going professional development, as a channel of influence. Due to the complexity of the change process, the other channels of influence, curriculum and assessment and accountability, are woven into our interpretations of the data.

The framework also identifies a set of guiding questions to investigate the influence of the standards. The critical overarching questions are: 1. *How has the system responded to the standards?* and 2. *What are the consequences for student learning?* The framework questions used to guide our analysis of how the educational system has responded to the introduction of national and state standards included*: How are nationally developed standards being received and interpreted? What actions have been taken in response? What has changed as a result? and What components of the system have been affected and how?* (NRC, 2002). These questions were explored by tracking events and documenting changes in the KSU PDSs and teacher education program in relation to major teacher development initiatives conducted from 1989 to 2004. The questions used to guide our analysis of the consequences for teachers, teaching practices, and student learning included: *What actions have teachers taken in response to the standards? What about teachers' classroom practice has changed? Who has been affected and how?* and *How have student learning and achievement changed?* (NRC, 2002*)* These questions were explored by examining teachers' perceptions of impact and state student assessment data. The results of our investigation are presented as two case studies: (1) Initial Standards-Based Teacher Development Projects, and (2) Recent Standards-Based Teacher Development Projects.

Sources of Data for Initial Standards-Based Teacher Development Projects (1989—1995)

Strategies used to track events and document impacts of early teacher development reform initiatives have been reported in six dissertations and multiple papers and presentations summarized by Shroyer, Wright, and Ramey-Gassert (1996). Only the highlights of these strategies will be presented in this chapter since they have been reported previously. Data collection strategies included: (1) content analysis of project documents and records (course syllabi, district mission, goals, and scope and sequence, teacher education program requirements, planning team records, and project reports); (2) participant observations of team meetings, PDS activities, and professional development; (3) yearly surveys of teachers, administrators, faculty, and undergraduate students to assess attitudes, beliefs, and teaching practices using project constructed instruments, self-efficacy belief instruments (Enochs & Riggs, 1990; Riggs & Enochs, 1990), and attitude toward science scales (Shrigley, 1974; Shrigley & Johnson, 1974); (4) pre and post interviews with teachers and faculty; (6) focus group meetings with parents; (7) observations of teaching practices of undergraduate students, practicing teachers, and graduates using informal and formal observation protocols (Burry-Stock, 1993); and (8) student scores on the 1996 Kansas science assessments at grades 5, 8, and 10.

Sources of Data for Recent Standards-Based Teacher Development Projects (1999 –2004)

Strategies used to track events and document the impacts of recent teacher development reform initiatives will be described more thoroughly since they have not been reported previously. Data collection strategies included: (1) content analysis of district, teacher education, and project documents, records and curricular materials; (2) participant observations of all project activities, district leadership meetings, and teacher education leadership meetings; (3) surveys and follow-up interviews of science and math team members; (4) interviews with science team members; and (5) student scores on the 2001 and 2003 Kansas science assessment at grades 4, 7, and 11.

The following documents and records were analyzed to track key events and organizational impact: (1) district documents (mission statement, goals, graduation requirements, science standards, school action plans, course plans, action research reports, and teacher curriculum maps); (2) teacher education program documents and records (course syllabi, planning records, project reports, and program requirements); (3) science team documents such as team tasks, matrices aligning courses and programs with the standards, science performance-based standards, improvement recommendations, and peer consultation records. Documents were analyzed using content analysis (Guba & Lincoln, 1988). Patterns of change were identified for the Manhattan-Ogden schools and the colleges of education and arts and sciences. These patterns of change were then compared to changes over time in relation to project activities, district policies and curriculum standards, and state policies, standards, and assessments.

From 1989 to 2004, the first and second authors made participant observations during project activities, district and college meetings, and professional development sessions. The role of the third author will be described in the second case study. The first author was a codirector for three of the four teacher development initiatives described below and is one of the directors of the KSU PDS Partnership. The second author was the codirector of a fourth teacher development initiative described below, a participant in the other three initiatives, the principal of one of the first three PDSs, and the principal of the high school PDS.

The first author attended all college of education meetings, retreats, and program planning sessions as a member of the college program coordinating council, and NCATE documentation team. She attended campus-wide teaching enhancement meetings and faculty share sessions as an advisory board member for The Faculty Exchange for Teaching Excellence. The first author also attended monthly Manhattan-Ogden district leadership meetings as a member of the Goals 2000 Committee, Science Curriculum Committee, Program Planning Council, and Teacher Leadership Cadre. As the coordinator of PDSs, the first author facilitated monthly PDS teacher leadership meetings. As an elementary and secondary PDS principal, the second author coordinated PDS professional development activities, school curriculum development, and school improvement initiatives. She made frequent classroom observations and was responsible for assessing teaching practices, student learning, and all school programs. As a district administrator, the second author attended Manhattan-Ogden district leadership meetings including the Administrative Council, Teacher Leadership Cadre, curriculum planning sessions, and school board meetings. She also attended PDS on-site councils at the elementary and secondary level and district-wide PDS advisory committee meetings, project planning team sessions, and district and university partnership meetings. As the coordinator of the first Math, Science, Technology Summer Magnet School, the second author attended all planning sessions, the summer school program, year-long follow-up professional development activities, and conducted the project evaluation.

Science and mathematics team members from the colleges of education and arts and sciences were surveyed in 2000 to determine their curriculum, teaching strategies, familiarity with the standards, and the alignment between their teaching and the standards. Follow up interviews were used to clarify and add depth to survey responses and elicit specific examples of teaching practices. The surveys were analyzed in relationship to survey and interview responses from the initial teacher development projects (1989–1995) to document changes in teaching practices across time. The responses of the science team are the focus of this analysis with the math team responses serving as a point of comparison.

Interviews were conducted in 2004 with science team members including three science educators, five scientists, and eight K-12 science teachers. The third author, a science educator not connected with the science team, to maintain neutrality and encourage open responses, conducted these interviews. Participants responded to questions about science team activities such as reviewing the standards, aligning courses and programs with the standards, peer consultation, developing the KSU science performance-based standards, identifying programmatic gaps and redundancies, and professional development activities. For each of these activities, team members were asked to describe their level of participation and its impact on them professionally as a teacher, the impact on their teaching, and the impact on their students. Key responses were recorded in writing, videotaped, and transferred to a CD. All three authors independently coded the interviews and identified patterns and trends. Individual findings were compared and found to be consistent. Patterns were identified for individuals, for each team activity, and for each organizational unit (K-12 PDS and colleges of education and arts and sciences). Summary statements were then made regarding the positive and negative impact of science team activities on K-12 teachers, scientists, educators, and students.

Student achievement data used to document impact on learning include the 2001 and 2003 state of Kansas science assessments at grades four, seven, and 11. Student achievement scores were collected in science in 1996, 2001, and 2003 but the science exam changed in content, format, and grade level assessed from 1996 to 2001 making longitudinal comparisons impossible. Consequently, science achievement scores for 2001 and 2003 were compared to statewide mean achievement scores and achievement gains from 2001 and 2003 were noted. As a point of comparison, we have compared mathematics achievement in our PDSs, collected yearly from 1995 to 2004, to the science achievement gains from 2001 to 2003. Science scores are the focus of this analysis, but math scores provide a comparison to document the impact of curriculum and assessment and accountability (NRC, 2002).

CASE STUDY OF INITIAL STANDARDS-BASED TEACHER DEVELOPMENT PROJECTS (1989—1995)

How are nationally developed standards being received and interpreted? What actions have been taken in response? What has changed as a result? What compo-

nents of the system have been affected and how? What actions have teachers taken in response to the standards? Supported by the National Science Foundation (NSF) and the Kansas State Department of Education, KSU and the Manhattan-Ogden district began two interrelated reform projects in 1989 to improve elementary science and mathematics teaching while simultaneously reforming elementary science and mathematics teacher education. Both projects were designed around national standards in mathematics (National Council of Teachers of Mathematics, NCTM, 1989a), a national framework for math, science, and technology education (National Center for Improving Science Education, NCISE, 1989), and recommendations from Project 2061: Science for All Americans (AAAS, 1989).

A 2-year KSBE-funded project began as a Math-Science-Technology (MST) Summer Magnet School and Professional Development Center. The MST Summer Magnet School was designed to provide an innovative summer school experience for K-6 students to enable them to develop higher-level thinking and problem-solving skills in science, mathematics, and technology. The Professional Development Center was to be conducted simultaneously with the magnet school, to provide exemplary training and field experiences for teachers to give them the opportunity to learn and practice the philosophy and strategies of hands-on, activity-based teaching in science, mathematics and technology. The Professional Development Center also was designed to prepare teachers as peer coaches to model, evaluate, and improve the teaching strategies being attempted in the summer school. The second author was a codirector of this project; the first author was the resident science educator responsible for professional development.

An initial team of 25 teachers was selected to engage in weekly professional development and planning, from January to June of 1990, as they created and implemented the MST magnet school. These teachers were introduced to the standards and reform documents and challenged to translate this vision of reform into teaching practices. Once the month-long, half-day summer magnet school began, the teachers met each afternoon to process and evaluate the day's teaching and to continue planning and improving the summer school experience. At this point an additional 25 teachers were brought in to participate in professional development opportunities in conjunction with observations and interactions with the magnet school children and peer coaching and professional development with the magnet school teachers. The second year of the project included follow-up support as project teachers implemented the new teaching practices they had learned into their classroom teaching and disseminated these practices to additional teachers in their buildings. In this manner, professional development was on going, embedded in teaching and curriculum development, and aligned with effective K-12 practices.

Teachers' initial reactions to the standards and reform agenda were mixed. Magnet school teachers had many arguments with one another, the project directors, and the resident science educator. Some challenged the need for change in a system in which they had been successful. Others, who were less successful in the traditional system and those who already used nontraditional strategies, welcomed the reform. Tears and shouting matches hindered many professional development sessions. In the end, it was the responses of the children and their parents that made the greatest difference. Children participating in the MST Summer Magnet School were successfully learning science and mathematics concepts that some teachers were sure would be too difficult for students. Parents were pleased with their children's new learning and enthusiasm for science and mathematics. The magnet school teachers also were learning science and mathematics and having as much fun as their students!

By the end of the second project year, all of the magnet school teachers had implemented at least some new practices while others had completely revised their teaching. The MST Summer Magnet School was conducted from 1990 to 1998. It provided such an exemplary teaching site that for five summers (1994 –1998), the KSU elementary science and mathematics methods and field experiences were taught in conjunction with the district summer school.

The concept of the MST Summer Magnet School and Professional Development Center provided the basis for the MST Professional Development School model created in 1990. Three elementary PDSs were identified as sites for enhanced science and mathematics teaching and learning for future, new, and experienced teachers. With support from the NSF, a PDS teacher development model was created to prepare future and experienced elementary teachers for more effective science, mathematics, and technology teaching. Fifty-two project participants were identified to coordinate all K-16 PDS science and mathematics reform efforts. These participants included 30 elementary teachers (from the PDS), three elementary principals (from the PDS), three central office administrators (director of curriculum and instruction and coordinators of staff development and elementary education), six scientists (biologist, chemist, two geologists, physicist, and biochemist), two mathematicians, six science educators, a mathematics educator, and two technology educators. The majority of participating teachers were participants in the MST Summer Magnet School. The lead author was a coordinator of the project and the second author was one of the participating principals and team members.

All project participants were organized into three planning teams, mathematics, life science, and physical science. Each team included K-6 teachers, an administrator, scientists or mathematicians and science or mathematics educators. Each team was given the responsibility to reform content courses, methods courses and field experiences (in life science, physical science, or mathematics) based on research and the reform movement. All project participants also were challenged to help the three MST PDSs become exemplary sites of science, mathematics, and technology teaching. Monthly professional development days and annual month-long summer institutes provided opportunities for planning team members to read reform documents (The Mathematical Association of America, 1991; NRC, 1988; NSTA, 1988), examine the mathematics standards (NCTM, 1989a; NCTM, 1989b), the NCISE framework (1989), and early drafts of the *Benchmarks* (AAAS, 1993). These documents provided the foundation for changes in the KSU teacher education program and K-16 teaching practices.

Nine science and mathematics undergraduate courses and six semesters of field experiences, conducted in PDSs, were developed through this project. The majority of PDS teacher participants were magnet school teachers who had already been exposed to the reform documents, so they were very receptive to the standards. They worked diligently to implement the standards into the KSU program and their own teaching, often serving as mentors for faculty members less familiar with the reform movement. PDS teachers piloted new science and mathematics curriculum, planned and implemented yearly *innovations in action* to improve their teaching, and conducted action research to determine the effectiveness of new practices.

MST After School Clubs, Summer Magnet Schools, and family math and science nights provided opportunities for experienced teachers and future teachers in the teacher education program to jointly enhance their teaching while providing enrichment opportunities for children and parents (Shroyer, Ramey-Gassert, Hancock, Moore, & Walker, 1995). The MST After School Clubs were sponsored by KSU as field experiences for science and mathematics elementary undergraduate methods students. KSU students conducted them with support from PDS teachers and KSU faculty. MST clubs started with 100 K-6 students and 25 KSU students in three PDSs. By 1998, they were conducted at eight schools by over 100 KSU students and involved approximately 600 K-6 students a semester. In addition, PDS teachers conducted math and science family nights with support from KSU students and faculty. By 1998, three to six PDSs a year conducted these science and math family nights. Each participating school offered up to six evenings a semester to provide enrichment for children and opportunities for parents to become more familiar with standards-based teaching.

Impact on Teachers and Teaching Practices

What actions have teachers taken in response to the standards? What about teachers' classroom practice has changed? And who has been affected and how? Pre and post surveys, interviews, and observations of the 30 participating PDS teachers indicated more science was being taught and it was being taught using hands-on inquiry approaches. Parent focus groups indicated support for these new initiatives (Daisy & Shroyer, 1995. Pre and post-test comparisons also indicated significant gains in teachers' attitudes toward science and science teaching efficacy beliefs (Ramey-Gassert, Shroyer, & Staver, 1996; Willhite, 1995). In addition, these teachers received more than a dozen state and national teaching awards and gave more than 30 presentations at regional and state conferences (Shroyer, Wright, & Ramey-Gassert, 1996). From a principal's perspective, attendance at national conferences, funded by NSF and the state, provided new opportunities for growth and networking with other science teachers. Project teachers shared their enthusiasm and provided mentoring to other teachers in their schools expanding new ideas and teaching strategies within the buildings and across the district. As parents became more educated about the teaching of science, they became more assertive about requesting such teaching.

The standards-based reform projects also impacted teachers and teaching practices in the KSU teacher education program. Pre and post-surveys, interviews, observations, and project documents demonstrated an increase in inquiry teaching, group work, and a stronger focus on student understanding in content courses (Cooper, 1995; Govindarajan, 1993; Zollman, 1994). As a consequence of this project, the KSU teacher education program was completely revised into a field-based PDS program at both the elementary and secondary level and the PDS Partnership was expanded to include 27 PDSs from five districts across Kansas.

Future teachers in the KSU teacher education program also were positively impacted by these initial teacher development projects. The first cadre of 30 undergraduate students who completed the revised program demonstrated improved science and mathematics content knowledge, science teaching efficacy beliefs, attitudes toward science, and teaching (Shroyer & Wright, 1995; Stalheim-Smith & Scharmann, 1994; Wilson, 1996). Follow-up interviews and observations using the expert science teaching educational evaluation model (Burry-Stock, 1993) were conducted with a smaller segment of the first cadre of graduates teaching in Kansas. This smaller segment of graduates showed a continuous increase in teaching proficiency during the first 5 years of their teaching practice (Bolick, 1996; Shroyer & Wright, 1998).

This evidence suggests substantial impact of the standards on teaching and learning across the PDS Partnership. The partnership was recognized as one of two national "Examples of Successful Collaboration" (Robinson & Darling-Hammond, 1994) and participating project members received the *Innovation in Teaching Science Teachers* and the *Research into Practice* awards in 1996 from the Association for the Education of Teachers of Science.

Impact on Student Learning

How have student learning and achievement changed? Changes in elementary teaching practices cited above resulted in greater numbers of elementary students participating in science, mathematics, and technology activities. From 1990 to 1998, 100 to 150 students attended MST summer school programs each year and from 300 to 600 students attended MST After School Clubs every semester. Hundreds of parents and children attended family math and science nights. These activities also provided an opportunity for a more diverse group of students to engage in science and mathematics. During the first MST Summer Magnet School, it was difficult for the directors to achieve the student balance they had strived for, and the majority of students attending were white, male, and middle class. By 1994, approximately 50% of the students attending the magnet school were females and over 50% were minorities. From 1996 to 1998, the summer school program was modified and moved to Ogden to better meet the needs of a highly diverse and economically disadvantaged group of students.

It is difficult to determine if access to enriched opportunities in science impacted student learning as measured by state science assessments because only one year of data is available from 1989 to 1995. The first state assessment in science was given in 1996. It was a challenging test designed to align with state standards before national standards were available. Questions focused on experimental design and knowledge, data interpretation, and integrated process skills. The assessment included a multiweek performance measure at the elementary and middle school levels. Mean scores were reported for the district at the elementary, middle, and high school level and also for each individual school.

Mean district scores for Manhattan-Ogden students were slightly higher than the state means for students in the same grade levels. At the elementary level, district means were approximately 1% higher than the state mean while middle and high school mean scores were 2% above the state mean. An examination of individual school scores revealed that seven of the elementary schools were above the state mean (1 –10%) while three were below the mean (4 –7%). Both middle schools were above the state mean (1 –2%), and the high school was above the state mean (2%). The three elementary schools performing below the state mean had considerably greater numbers of economically disadvantaged students than the average Kansas elementary school (47%, 52%, and 71% disadvantaged compared to the state mean of 34%). Four elementary schools that performed above the state mean also had greater numbers of disadvantaged students (35%, 39%, 39%, and 47%) compared to the average elementary school (34%). Both middle schools had greater numbers of disadvantaged students (35%) compared to the average Kansas middle school (30%). The student population at the high school was approximately the same (21% disadvantaged) as the state average (20% disadvantaged). It thus appears that Manhattan-Ogden students were doing as well as expected in 1996 compared to other state students and disadvantaged students at the elementary level were performing better than expected in comparison to the state.

Long-Term Impact

KSU faculty surveys and KSU course syllabi examined in the year 2000 indicated science methods courses at both the elementary and secondary level were inquiry oriented and standards-based and the entire education program had been revised. Thus, the influence of the standards on the KSU College of Education was extensive and sustained. But, this was not the case for all science teaching at KSU or in the PDSs. The initial improvements noted above were slowed, and in some cases reversed, by organizational changes in school and university faculty, changes in educational priorities, and new state assessments and accountability measures.

Many KSU science faculty members who participated in the NSF project had moved to other institutions or retired by 1998, and the three original PDSs were expanded to include ten additional Manhattan-Ogden schools (Shroyer & Wright, 1998). Consequently, the majority of KSU science and mathematics faculty and PDS teachers and administrators had not been introduced to the standards through the teacher development projects and did not have the advantage of 5 years of intensive professional development. However, at the elementary level, PDS teachers each year were exposed to over 50 KSU field experience students in each building, who were familiar with the science standards. At the secondary level, at least half of the science teachers worked with KSU students each year who were familiar with the science standards.

The surveys and follow-up interviews conducted in 2000 indicated that change was still evident in the teaching of the remaining KSU faculty who participated in the NSF project. One faculty member started the Faculty Exchange for Teaching Excellence. Three faculty members were identified as KSU teaching scholars the first 3 years this award for excellence in teaching was given. But, only one faculty member from each science department had participated in the original NSF teacher development project and when these individuals left campus, their influence was lost. Three out of six science classes, specifically developed through the NSF project, were lost when faculty members who created them left KSU. Two additional science classes were lost due to budget considerations. As a point of comparison, one of the two NSF project mathematicians retired, but the one who remained became a leader in the department as chair of the curriculum committee, thus maintaining the impact of the NSF project. Two of the three mathematics courses designed for the NSF project are still in place, and several others have replaced the one course that is no longer taught.

A large number of the 30 original KSDE and NSF project teachers became school and district leaders. In 1999, these teachers led the majority of elementary "School Improvement Reports" presented to the Manhattan-Ogden school board. Teacher leaders were identified in each building in 1999 to enhance K-12 curriculum and instruction (The Teacher Leadership Cadre or TLC). The majority of teachers selected for the TLC were participants in the initial teacher development projects. The district science scope and sequence, written by a teacher involved in both reform projects, was based on national standards. District goals created in 1992 included: "The education of all students will be based on high standards" and "Curricular standards will be implemented consistently across the district." However, there was a large turnover in district administrators, a growing emphasis on site-based management, and the impact of the science standards at the district level was waning.

District leadership meetings and an analysis of district documents revealed that school-wide reform was still evident in the original three PDSs, but new PDS teachers were more traditional and less familiar with the standards. An analysis of district curriculum maps completed in 1999 indicated that science teaching in most buildings was not consistent within a school, let alone across the district. The majority of these curriculum maps did not align with the district scope and sequence or state or national science standards. The *Full Option Science System* (Regents, 2000), an elementary science curriculum developed with NSF funds, was adopted in 1999, but not without a struggle between teachers who had participated in the teacher development projects and other district teachers. Secondary teachers who had not participated in either the state or federal teacher development projects soundly rejected high school science curricula developed with NSF funds.

The district emphasis on science instruction was also influenced negatively by changing state mandates and assessments and accountability measures. State priorities, and thus district priorities, had shifted to reading and mathematics by 1993 with the initiation of a new state accreditation plan (Quality Performance Accreditation – QPA). According to the QPA plan, accreditation was based on student achievement in reading, mathematics, and one subject selected by each school. QPA provided the incentive for statewide testing in Kansas. Reading and mathematics were tested for the first time in 1995 and have been tested every year since 1995. Science was tested for the first time in 1996. After the initial science test, Kansas did not test students in science again until 2001. During this time, the Kansas science standards achieved national notoriety due to the state controversy over evolution and the resulting revisions in the state standards. The revised standards did not align with national standards. By 2001, a new state board of education was elected and the state standards were revised once again to align with national standards. A new state exam was created to align with state standards, but it did not include a performance measure and, as of 2004, science achievement still was not a part of the mandated adequate yearly progress under the No Child Left Behind Act.

Interpretations

Although sustainable changes were evident in the original three PDSs, the College of Education, and the teaching practices of original project participants, the influence of high stakes assessments and accountability measures on teaching practices and student learning also is evident. Although most PDSs performed above the mean on the state tests in 1996, earlier science improvement efforts in the PDSs were overshadowed by more pressing needs in reading and mathematics, subjects that were being assessed yearly. Teachers who had felt compelled to teach science as inquiry in preparation for the challenging state performance assessment lost incentive when the performance component was removed from the exam.

As a point of comparison, the impact of the initial teacher development projects on student learning was easier to document in mathematics because exams were given every year from 1995 to 1999. During these years, district means increased steadily at all three levels. An examination of individual school data demonstrated increased scores for eight of the ten elementary schools, both middle schools, and the high school. Across all schools, total mathematics scores increased from 1 to 30% while scores in mathematical problem solving, an area targeted for improvement in all schools by the state accreditation process (QPA), increased from 5 to 47%. One of the original PDSs, with 52% of its students labeled as economically disadvantaged, had the second to the lowest mathematics scores in the district in 1995 and the second to the highest scores by 1999—a 29% increase overall and a 47% increase in problem solving. Another one of the original PDSs, with 39% of its students economically disadvantaged, had the highest score in the district by 1999—a 14% increase overall and 26% increase in problem solving. This school was recognized by the USDOE as one of five national sites for "Excellence in Professional Development" (USDOE, 1998; WestEd, 2000).

CASE STUDY OF RECENT STANDARDS-BASED REFORM PROJECT (1999—2004)

In 1999, the KSU PDS Partnership began two synergistic teacher development projects. The KSU PDS Partnership was a Teacher Quality Enhancement project supported by the USDOE and the Kansas Collaborative for Excellence in Teacher Preparation (KCETP) was supported by the NSF. The lead author was a codirector of both projects. The second author was the high school principal, a leader of the new-teacher mentoring program supported by both projects, and a trainer of trainers in a PDS Partnership program to implement standards-based teaching across K-16 classrooms. The third author was and continues to be an evaluator for the PDS Partnership project.

Six of the eight leaders for these projects (an associate superintendent of curriculum and instruction, a mathematician, two scientists, two science educators, a PDS coordinator, and a teacher) were participants in the earlier NSF teacher preparation project. These leaders decided to incorporate the most successful elements of the original project while designing new strategies to initiate more sustainable organizational reform.

Through the KSU PDS Partnership, 30 education faculty, 30 arts and sciences faculty, and 80 practicing teachers and administrators set out to examine and improve teacher quality across the entire KSU teacher education program and all 27 K-12 PDSs. Nine planning teams were identified (math, science, language arts, humanities, social studies, foundations, special education, English language learners, and mentoring). The KCETP project supported the work of the math, science, and mentoring teams. Both projects focused on improvements in initial preparation, new teacher mentoring, and ongoing professional development. The science planning team included three biologists, a chemist, a geologist, two physicists, three science educators, five elementary teachers, and five secondary science teachers.

The planning teams created new K-12 teacher education performance-based standards while they simultaneously implemented standards-based curriculum and instructional strategies in their own K-16 classrooms and programs. Project directors proposed the design, adaptation, and adoption of new standards across KSU departments and all 27 PDSs, to increase K-16 understanding of the standards and promote more sustainable change through institutional level commitments as compared to individual commitments. While implementing course improvements, team members participated in a peer consultation process (Bernstein, 1996), which involved collaborative inquiry into curriculum, instruction, and assessment.

This teacher development project began by focusing on improvements in individual classrooms to enhance personal meaning and ownership. The planning teams met monthly for 5 years and during four summer institutes to engage in professional development, planning, and collaboration. Faculty members and teachers were introduced to state and national content standards for teachers and K-12 students. Teams also examined standards for beginning teachers (INTASC, 1995; NCATE, 2002) and experienced teachers (NBPTS, 1998). The teams read and discussed reform documents (Darling-Hammond, 1999; NCTAF, 1996, 1998; NRC, 2001; USDOE, 1998, 1999, 2000) and participated in book studies using *How People Learn: Brain, Mind, Experience, and School* (NRC, 2000) and *Knowing What Students Know: The Science and Design of Educational Assessment* (NRC, 2001b)

After examining the standards and reform documents, each team set out to align an individual course or grade level curriculum to the content standards. Each team member completed a matrix that identified his or her course goals then aligned each course goal with individual state and national K-12 science standards. Team members also identified how the goal/standard was taught and assessed. Team members then implemented changes in their courses to better align with the state and national science standards. Peer consultation was used to facilitate and support this classroom improvement process. Team members paired off into peer consultation teams of two to three individuals from different backgrounds. Science faculty paired with education faculty and K-12 teachers. Peer consultants examined one another's course materials (syllabi, units, lesson plans, assignments, and assessments) and the matrices that indicated how the course aligned with the standards. They reflected together on how to implement the standards, improve K-16 teaching, and enhance student learning.

From 2001 through 2004, individual course reform activities were expanded to include programmatic improvements based on science content standards (KSBE, 2002; NRC, 1996) and teaching standards for initial licensure (KSBE, 2002a, NCATE, 2002), new teachers (INTASC, 1995), and experienced teachers (NBPTS, 1998). Teams examined student achievement and program survey data from PDS teachers, faculty, administrators, K-12 students and their parents, teacher education students, and teacher education graduates. These data were used to identify programmatic strengths and weaknesses and to develop K-16 improvement plans.

The science team was responsible for developing general education standards in science (for all K-12 teachers) as well as science standards for future elementary, middle school, and high-school teachers. The science team articulated the desired standard, identified the content courses, methods courses and field experiences where the standard would be taught and assessed, planned the evidence to be collected in courses and field experiences to demonstrate the standard had been met, and designed methods and criteria (rubrics, etc.) to assess this evidence. The science team also identified areas of programmatic gaps and redundancies in terms of the standards. As a final step in the improvement process, teams made changes to the teacher education program to align existing courses and field experiences with the new standards.

Each PDS also completed a school improvement action plan. Teachers used student assessment data and state and national standards to identify school wide strengths and weaknesses. Each school then identified at least one school improvement goal to become a focus for partnership improvement efforts. This information was used each year to plan ongoing professional development opportunities. Planning team members identified ways in which the entire team could help schools realize their improvement goals.

Impact on Teachers and Teaching Practices

What actions have teachers taken in response to the standards? What about teachers' classroom practice has changed? And who has been affected and how? The science team was interviewed in the spring of 2004 after its members had examined the standards, aligned their courses with the standards, identified gaps and redundancies in their programs in relation to the standards, and developed the KSU teacher education performance-based standards. Each K-12 teacher, science faculty member, and education faculty member was asked how the standards impacted them as teachers, how the standards impacted their teaching, and how the standards impacted their students. District, KSU, and project documents and records also were examined to identify programmatic changes.

Analysis of program documents and science team interviews indicates extensive impact of the standards on the KSU teacher education program, far beyond science education. New science teacher education standards have been created and a new performance based assessment system has been implemented throughout the teacher education program. Content courses, methods courses, and field experiences have been examined and revised where necessary to improve alignment with these standards. Education faculty members felt their methods courses were in alignment with the standards before the 1999 reform efforts began, but the process of aligning the teacher education program with the standards revealed several programmatic gaps, which needed to be addressed. Content faculty members described positive impact of the standards on themselves as teachers, their teaching, or their students (future teachers). Their impact statements included: using the standards to determine concepts to be covered and how to organize those concepts; increased use of inquiry, student engagement, and questioning; greater understanding of students and learning; and new strategies to assess what students have learned including the use of rubrics and class talk. No negative comments were made.

The K-12 science team members recognized gaps in their own teacher preparation and appreciated how the new standards addressed these weaknesses. They felt the standards provided guidance for better content and pedagogical preparation. An elementary teacher stated, "I know if KSU is using these standards, then the students will be more prepared." When asked about the impact of the standards on new teachers, a high school teacher stated, "I've already seen it. I'm very impressed. I've seen a huge impact on the student teachers coming through. They are better prepared. I'm really impressed with the kids I am getting." The secondary science educator gave examples of his students' success in job interviews and the positive comments he has received from teachers who have supervised KSU field experience students and administrators who have interviewed and hired KSU secondary science education graduates.

Interview data and program records also indicate that the standards have strongly influenced K-12 teachers and their teaching. The majority of the K-12 teachers said they were already familiar with the K-12 content standards before 1999. Reviewing the standards as a team reinforced what they already knew and served as a catalyst for looking more closely at what and how they teach. The K-12 teachers said the standards focused their teaching and provided a framework for their curriculum. According to an elementary teacher, "any time you are in the classroom and you look at the standards, it brings into focus what specifically you need to make sure you are covering throughout the year." A middle school teacher stated, "Sometimes what a teacher teaches, and I am no exception, has to do with what we enjoy teaching or what we are comfortable with." She felt she now was "more sensitive to trying to match what I teach with the content standards". A high school teacher said the standards helped her "cut out the fluffy part of what I was doing." A recent graduate of the KSU teacher education program and new high school biology teacher said the standards helped her develop her course outline. She said the standards made it clear what her students needed to know, which made her teaching "more doable." She no longer feels like she has to "get through the entire textbook" but she makes sure "I hit everything in the standards first."

Elementary and secondary teachers on the science team discussed their new awareness of teaching standards for initial licensure, new teachers, and experienced teachers. None of the teachers was familiar with these teaching standards before the reform projects began. One elementary teacher felt that looking at all the standards together helped the team understand "how all the pieces of what we all do fit together." Teachers felt the new licensure standards in science impacted their teaching as well as the teaching of future teachers in the teacher education program. The process of looking at these standards "causes you to look at your own teaching and ask, am I doing the things we expect future teachers to do?" Six of the eight teachers on the science team participated in the NBPTS process as part of our teacher development initiatives. Their understanding of all standards supported the NBPTS process and participating in NBPTS deepened this understanding and helped them recognize the connections across standards. One teacher noted, "It all fit together—looking at the standards, going through NBPTS, looking at my own practice and how I do thing." Another said, "Everything I do is [now] related to the standards in some way."

Team members also stressed the value of aligning their courses with the standards. A high school teacher said, "You're looking at your curriculum and you're checking the standards and you're saying am I missing them? And if I am missing them, how am I going to include them?" Another high school teacher noted that the process "made me aware of the weaknesses and that's half the battle, knowing what you don't know." A novice teacher said she was working with a math teacher to look at gaps and redundancies across both classes. An elementary teacher said the process helped her and another elementary team member "to see some of the gaps and to find ways to bridge those gaps" in the curriculum that they use.

The standards-based teacher development projects also produced impact at the district level. Beginning in 2000, all Manhattan-Ogden teachers were exposed to the standards. New district curriculum standards were developed based on state and national standards. The teacher leadership cadre placed the standards in grade-level notebooks in each school and on the district Website, and teachers were asked to align their curriculum with these standards. New district assessments were identified or created to assess all students at the first of the year and the end of the year to measure growth in relation to the standards. District goals continue to place a focus on standards-based teaching, but new school board goals have added accountability measures and expectations that all schools meet the rigorous state standard of excellence.

As the high school science teachers aligned their courses with the standards, they determined that not all students were being exposed to all standards due to varied course options. Students could select a course of study that would be in alignment with the science standards, or they could select courses that left out key areas in life or physical science. This alignment issue was taken more seriously when state science assessment scores dropped between 1996 and 2003. As a consequence, a new biology course has been developed and will be required of all high school students starting the fall of 2004. The high school science department also is planning a required course in physics. This change in science requirements points out the influence of the standards through curriculum, assessment, and accountability. It is difficult to tell if the changes in requirements were due primarily to the curriculum alignment process. But, once the assessments revealed a problem, the alignment process helped to identify the source of the problem as the required science curriculum and changes were made.

The state assessments and No Child Left Behind Act legislation also had a negative impact on district science improvement efforts from 1999 to 2004. According to national and state mandates, districts are held accountable for adequate yearly progress in reading and mathematics at the elementary, middle, and high school levels. Districts will not use science assessments as an indicator of adequate yearly progress until 2005. Consequently, all schools in the district identified either mathematics, or reading, or both as targeted areas of improvement on their school improvement action plans. Although district school board goals set forth an expectation for each school to implement standards and reach the state standards of excellence, the professional development activities aligned with these goals focused on reading and mathematics. The mathematics and the language arts teams were heavily involved in PDS professional development but the science team received few requests for assistance.

District administrators believe science has not declined as a district area of importance, but mathematics and reading have increased in importance. Since resources, including staff time, are limited, they must be devoted to the highest priorities. The science curriculum adoption process scheduled for 2004 was delayed and a mathematics adoption process was implemented instead. District science assessments are still given at the beginning and end of each year in every grade, but accountability measures are linked to reading and mathematics rather than to science. The teacher leadership cadre has expressed great concern for the stress many elementary teachers feel "trying to fit it all in." These lead teachers have sited examples of low morale and teachers leaving the field because "teaching is not fun anymore."

Impact on Student Learning

How have student learning and achievement changed? Teachers believed students benefited from their use of the science content standards. They felt the standards helped their students learn important concepts and *big ideas*. A high school teacher said he now stresses "larger ideas" like form and function "over and over again" throughout his course to help students develop a deeper understanding of unifying concepts in biology. The novice high school teacher believes "students see the importance of big ideas because they are covered in multiple classrooms—they start to see the big idea too." Teachers also felt the stronger focus they developed in their own teaching helped students focus their own learning. An elementary teacher stated, "It made a clear picture for me and then I was able to communicate this with students." A high school teacher said, "It helps them [students] to be more focused." Another teacher spoke of how the teaching standards helped him make better use of small group discussions to check for understanding, "I can see where people have misconceptions."

Student learning at the district level also was documented by analyzing state assessment results. State science assessment data from 2001 cannot be accurately compared with data from 1996 because the exam was changed, the performance measure was eliminated, and the grade level for the assessments was changed. District results can be compared, however, to statewide results, and the 2001 data can be compared to the 2003 data. In 2001 and 2003, science assessments were given to fourth, seventh, and 10th graders. The science assessments were designed by a team of science teachers, science educators, and assessment experts based on state standards that were closely aligned with national standards. Students were asked questions about physical, life, environmental, and earth sciences. Questions included both scientific process and knowledge and scores were a combination of both fields. Mean scores were reported for the district at the elementary, middle, and high school level and for each individual school. Individual student scores were disaggregated into performance levels. In 2001 the performance levels were advanced, proficient, satisfactory, basic, and unsatisfactory. In 2003 the performance levels were exemplary, advanced, proficient, basic, and unsatisfactory.

In 2001, district mean elementary scores were higher than the state mean scores while the secondary students performed similarly to the state mean. The district elementary school score was 5% percent above the state elementary school mean; the district middle school score was the same as the state middle school mean; and the district high school score was 1% above the state high school mean. The state elementary mean scores increased by 10% between 1996 and 2001, while the district elementary mean scores increased by 14%. The 1996 to 2001 changes in district means at the secondary level were similar to changes in state means (1–2% variations).

At the individual school level, 11 of the 13 schools were above the state mean while two were below (three were below in 1996). Nine of the 10 elementary schools were above the state mean (up to 11% higher) and one was slightly below the state mean (1.5% lower). One middle school and the high school were slightly above the state mean (1% higher) and one middle school was slightly lower than the state mean (1% lower). In nine of the 10 elementary schools and the high school, a greater number of students scored at the highest performance level (advanced) and fewer students scored at the lowest performance level (unsatisfactory) compared to state means. The middle school performance levels were similar to the state means. One elementary school met the state standard of excellence.

The elementary school performing below the state mean had the greatest number of disadvantaged students (23% higher than the state mean) while the middle school performing below the state mean had a similar percentage of disadvantaged students compared to the state mean. Three of the 11 schools performing above the state mean had higher numbers of disadvantaged students (10–13% more), two had comparable numbers of disadvantaged students (within 2% of the state mean), and six had fewer disadvantaged students compared to the average for all Kansas schools.

In 2003, the district again performed better than the mean of all state districts on the science exams. Student performance at the elementary level remained strong, the middle school scores improved, but the high school scores declined. The elementary district score was 4% above the state mean, the middle school district score was 5% above the state mean, and the high school was 2% lower than the state mean.

At the individual school level, students at 10 of the 11 schools scored as well or better than the state mean, and one scored below the state mean (in 1996 three and in 2001 two schools were below the state mean). Two elementary schools that participated in the 2001 assessment were closed, leaving eight elementary schools. The two schools that closed had performed above the state mean on the 2001 science assessment. All eight remaining elementary schools performed at or above the state mean in 2003. Six of the eight elementary schools were above the state mean while two were similar to the state mean (the two buildings with the greatest numbers of disadvantaged students (15% and 25% higher than the state mean). Both middle schools scored higher than the state mean in 2003, but the high school scores were lower than the state mean (the previously mentioned decline that helped prompt the change in science requirements). In every building, except one elementary school, a greater number of students scored at the highest performance levels (exemplary and/or advanced) and fewer students scored at the lowest performance levels (basic and/or unsatisfactory) when compared to state means. The elementary school with the greatest numbers of disadvantaged students (25% higher than the state level) was the lone exception to this statement. In this school, fewer students scored at the highest proficiency levels, but fewer scored at the lowest proficiency level as well, and a greater number of students scored at the proficient level when compared to state means. Only one elementary school achieved the state standard of excellence—the same school that achieved this distinction in 2001.

It was also possible to compare the 2001 science results to the 2003 results since the tests were equivalent in content, format, and grade level assessed. At the district level, the elementary science score increased by 1% (compared to a 2% state mean increase), the middle school score increased by 6% (compared to a 1% state mean increase), and the high school score decreased by 2% (compared to a 1% state mean increase). At the individual school level, six of the eight elementary schools increased their scores, one school remained approximately the same, and the scores for one school decreased. The two schools that did not increase their scores still achieved above the state mean both years. Both middle schools increased their science scores while the high school science scores dropped. In all buildings except the high school, student proficiency levels increased between 2001 and 2003. Greater numbers of the district's students scored at the exemplary, advanced, and proficient levels while fewer students scored at the basic or unsatisfactory levels

These data demonstrate that the district's schools performed well in comparison to the state in both 2001 (11 of 13 schools above the state mean and two schools below the state mean) and in 2003 (10 of the 11 schools at or above the state mean and only one school below the state mean). In addition, eight of these eleven schools improved their scores from 2001 to 2003, while one stayed at similar level and two declined. Individual student performance levels across the district also improved.

Interpretations

While we cannot say unequivocally that student scores, or the improvement in students' scores, are related to the standards-based reform, there is strong evidence from these data to support such an interpretation. The elementary schools showed the greatest impact in 2001. Student scores in most elementary buildings were considerably higher than the state means, even for schools with greater numbers of disadvantaged students. The reform projects from 1989 to 1995 were focused on elementary science education, and the impact of these projects on elementary teachers has been previously documented. The elementary schools also adopted standards-based science curricula funded by the NSF in 1999 and many schools were piloting these materials since 1991. The 2001 performance at the secondary level was more modest. Middle school science scores were equivalent to the state mean and high school scores were only slightly above the state mean. Secondary teachers were not involved in the reform project from 1989 to 1995 and they did not adopt standards-based curricula. So, the 2001 data provide insight into the positive impact of standards-based curriculum and teacher development initiatives.

Although all eight elementary schools scored at or above the state mean in 2003 and most continued to improve from 2001 to 2003, this progress was slow (1% gain in mean scores) in comparison to gains between 1996 and 2001 (14% gain in mean scores). This may have been due to the high priority being placed on reading and mathematics since 2001. As a point of comparison, the 2004 state data indicate that all elementary schools improved their scores in mathematics from 2001 to 2004, and all but two schools also reached the state standard of excellence in 2004. Both teacher development projects from 1999 to 1994 focused on mathematics as well as science, but mathematics has been a higher priority than science at the national, state, and district level with yearly state assessments since 1998. Thus, mathematics is an example of the impact of standards through the influence of both teacher development and assessment and accountability.

Middle schools scores were at the state mean in 2001 but increased 6% by 2003 to a level 5% above the state mean. Although the middle schools did not adopt a standards-based curriculum, all eight science teachers participated in the PDS teacher education program from 2001 to 2003 as supervisors and mentors of future teachers. In addition, half of the middle school science teachers were involved in teacher development projects, participated in NBPTS, and attended graduate school at KSU from 1999 to 2004. Consequently, each middle school teacher was involved in some way in KSU's teacher education reform efforts. These middle level scores provided additional support for the positive influence of standards-based teacher development initiatives.

At the high school level, science scores were slightly higher than the state mean in 2001 but dropped slightly below the state mean in 2003. It is important to note that a group of students from an advanced placement biology class was omitted from the 2003 results because their tests were lost. Since this was an advanced class, it might explain the 4% drop in scores at the exemplary level, but it does not explain the 5% increase in scores in the unsatisfactory level. The explanation offered by the science department was that students' choice of science courses had resulted in poor preparation for many students. When high school teachers on the science team examined the high school curriculum in relation to the science content standards, they noted that students only would be able to meet all of the standards if they selected advanced courses. Since the high school required a set number of course credits but gave students the choice of which courses to take, it was highly likely that many students would not select the advanced courses needed to address all the standards. This conclusion led to the development of two new courses that will be required of all students. The high school case provides evidence of the impact of curriculum, assessment, and accountability.

LESSONS LEARNED

Through these two case studies of teacher development, we have traced the impact of the standards and the reform movements on teachers, their teaching practices, and their students. The impact has been positive but the journey has been long. KSU and the Manhattan-Ogden Unified School District have jointly invested 15 years in collaborative K-16 systemic reform based on state and national standards. These efforts have resulted in changes in teaching practices for K-16 teachers, continuous renewal in the KSU teacher education program, and expanded opportunities for K-16 students. For our final comments, we will return to the overarching questions identified in the *Framework for Investigating the Influence of Nationally Developed Standards in Mathematics, Science, and Technology Education*: *How has the system responded to the introduction of national and state standards?* and *What are the consequences for student learning?* (NRC, 2002).

How has the system responded to the introduction of national and state standards?

At the district level we have documented changes in K-12 teachers' beliefs, attitudes, and teaching practices. State and district standards are aligned with national standards, a standards-based curriculum has been adopted at the elementary level, and two new courses are being implemented at the high school level to align the enacted district curriculum with district, state, and national standards. Perhaps most importantly, all district students, particularly economically disadvantaged students, have had increased opportunities to participate in meaningful science.

At KSU, the new teacher education standards and comprehensive assessment system have been implemented. Content courses, methods courses, and field experiences have been revised to align with these teaching standards as well as state and national content and teaching standards. Initial improvements in elementary education have expanded to secondary education, the K-12 program is standards-based with extensive field experiences in PDSs, and the performance of graduates in elementary and secondary education has steadily improved.

These K-16 systemic reform experiences led to our first conclusions: *(1) significant educational change requires extensive and continuous time, resources, professional development, and implementation support across the system.* Earlier teacher development projects demonstrate that the improvement process can be slowed, or even reversed, without continuous attention. Just as we need to assess prior learning of students and plan our teaching accordingly, we need K-16 organizational mechanisms to continuously monitor teacher understanding and plan continuous professional development accordingly. Our experiences support national trends and suggest professional development involve multiple strategies and delivery formats, focus on student learning, provide opportunities to examine practice and design and implement new curriculum, and be collaborative (Loucks-Horsley, Hewson, Love, & Stiles, 1998; WestEd, 2000).

Teachers at all levels described deeper understanding of key concepts in science, student learning, effective teaching, and assessment. Both K-12 and college teachers noted their classes are more organized, focused on essential concepts and big ideas, interactive, inquiry oriented, and centered on student learning. These educators believed a variety of professional development strategies were helpful in deepening their understanding of the standards and improving their teaching practices. They specifically mentioned their exposure to several national speakers, tasks in which they were expected to align their courses with the standards and then identify gaps and redundancies, and the peer consultation process as benefiting their practice.

Reactions to professional development prompted a cross-project comparison of strategies to introduce the standards to K-16 educators. This comparison has led to our second conclusion: *(2) deeper understanding and implementation of the standards are developed through extensive and meaningful work with the standards.* Teachers in the MST Summer Magnet School used the reform documents to create their summer curriculum. Teachers and faculty involved in the USDOE supported PDS Partnership project aligned their courses to the standards, identified areas of weakness, and used the standards to develop standards for the KSU teacher education program. Both of these groups of teachers demonstrated deeper understanding of the standards and greater changes in their teaching practices as compared to the participants in both NSF projects who were introduced to the standards through readings and presentations.

What Are The Consequences For Student Learning?

All K-16 educators involved in these teacher development projects perceived their changes in practice to have positively impacted their students, although they found it difficult to identify strategies to document impact on student learning. There are limitations to using state assessments to measure student learning, particularly when only 3 years of data are available and the first exam was considerably different from the last two. But K-12 educators are mandated to use state assessments and, consequently, the KSU PDS Partnership has opted to use these assessments as well. At the district level and in the majority of schools, the mean achievement on all three state science exams was higher than statewide mean achievement. Several elementary schools and the middle schools have seen large gains in achievement and the number of schools performing below the state mean has decreased on each of the three state assessments. Most importantly, the performance levels of economically disadvantaged students in 2001 and 2003 (performance levels were not calculated in 1996) was stronger in every school in comparison to state means and schools that serve predominately low income students have made consistent progress across each assessment. However, improvement has been sporadic at the secondary level and slowed at the elementary level between 2001 and 2003 and only one school reached the rigorous state standard of excellence across all 3 testing years.

In comparison, the mean district mathematics scores are significantly higher than state means; students in all schools have consistently increased their scores; school and district gains are considerably greater than state mean gains; and all but two elementary schools achieved the state standard of excellence in 2004. This comparison led to our third conclusion: *(3) the influence of teacher development initiatives on teachers, teaching practices, and student learning can be minimized or enhanced by curriculum, assessment, and accountability measures.* None of these factors seem to be exclusively related to student learning, but they are interdependent and when positively combined, as in the mathematics example, they have a powerful impact. In the case of science, a variable high school curriculum, inconsistent state standards and assessments, and a lack of state accountability in terms of school accreditation (QPA) and adequate yearly progress have negatively impacted the influence of teacher development activities. All of these factors are related to district and state priorities and accountability measures. Science is not seen as a national, state, or district priority and improvements, although evident, are inconsistent and slow. In the case of mathematics, the high priority placed on this field, the consistent state and district focus on problem solving, yearly assessments, and the importance of mathematics in terms of state accreditation and district and state adequate yearly progress have had a positive impact on teacher development initiatives, and progress has been consistently strong.

This comparison of science and mathematics student progress also has resulted in our fourth and final conclusion: *(4) teacher development efforts need to be centered on student learning.* Teachers in our projects found it difficult to identify targets for student learning and to document student learning. This was true for K-12 teachers and college faculty. It was also true for novice teachers as well as those attempting national board certification. Too often teacher development programs have focused on identifying, implementing, and/or examining teaching behaviors without simultaneously analyzing the impact of these behaviors on student learning. Our most successful initiatives have involved action research and school improvement action plans based on the content and teaching standards, an analysis of student data, and data based decision-making. This approach has been most effective when conducted at a school-wide level (Yahnke, Shroyer, Bietau, Hancock, & Bennett, 2005). The tremendous gains in the two PDSs discussed earlier involved a combination of school-wide improvement plans and action research. As teachers, we all must be reminded that student learning is the heart of the standards, the reform movement, and teaching.

SUMMARY

The chapter presented one example of the impact of the science standards on K-16 systemic reform. This example of reform is based on a professional development school model to improve K-16 teaching and learning by enhancing teacher development across the continuum of teacher growth—from initial preparation through continuous professional development. Through two case studies of teacher development, we have traced the impact of the science standards on K-16 science teachers, science teaching practices, and K-12 student learning in science. Both cases demonstrate how science standards have influenced teacher development and how teacher development has impacted the educational system and student learning. We have documented changes in teaching practices for K-16 teachers, continuous renewal in the KSU teacher education program, and expanded opportunities and learning for K-12 and teacher education students. This evidence suggests substantial impact of the standards on teaching and learning across the PDS Partnership. The evidence also suggests the powerful influence of the interrelated and interdependent components of reform—specifically teacher development, curriculum, assessment, and accountability measures. Each of these channels of influence (NRC, 2002) must be carefully addressed to promote and maintain positive impact of the standards.

We believe that the science standards did make a difference and continue to make a difference for our K-16 PDS Partnership. The professional development school model has provided an effective mechanism for initiating and sustaining systemic reform in science based on the standards. By focusing our reform efforts on K-16 educational stakeholders and the continuum of teacher development, we have created a community of learners capable of addressing the interdependencies and interrelationships of the many components of reform (NRC, 2001a, p. 75). This community of learners has collaboratively and simultaneously enhanced K-16 science teaching and learning based on a new vision for science education provided by the science standards.

REFERENCES

American Association for the Advancement of Science. (1989). *Project 2061: Science for all Americans.* Washington, DC: Author.

American Association for the Advancement of Science. (1993). *Benchmarks for scientific literacy.* New York: Oxford University Press.

Bernstein, D. J. (1996). A departmental system for balancing the development and evaluation of college teaching. *Innovative Higher Education, 20,* 241-248.

Bolick, M. E. (1996). *Socialization influences of the elementary environment on a beginning teacher prepared as a constructivist educator: An interpretive case study.* Unpublished doctoral dissertation, College of Education, Kansas State University.

Burry-Stock, J. (1993). *The expert science teaching educational evaluation model (ESTEEM) manual.* Kalamazoo, MI: Center for Research on Educational Accountability and Teacher Evaluation at Western Michigan University.

Carnegie Forum on Education and the Economy. (1986). *A nation prepared: Teachers for the 21st century.* New York: Author.

Cooper, C. K. (1995). *Qualitative analysis of preservice elementary teachers' scientific ways of thinking, attitudes, and perceptions during collaborative earth science field-based experiences.* Unpublished doctoral dissertation, College of Education, Kansas State University.

Daisey, P., & Shroyer, M. G. (1995). Parents speak up: Examining parent and teacher roles in elementary science instruction. *Science and Children, 33*(3), 24-26.

Darling-Hammond, L. (1999). *Solving the dilemmas of teacher supply, demand, and standards: How we can ensure a competent, caring, and qualified teacher for every child.* New York: National Commission on Teaching and America's Future.

Enochs, L. G., & Riggs, I. M. (1990). Further development of an elementary science teaching efficacy belief instrument: A preservice elementary scale. *School Science and Mathematics, 90*(8), 295-705.

Fuhrman, S. H., & Massell, D. (1992). *Issues and strategies in systemic reform* (CPRE Series RR-025). New Brunswick, NJ: Consortium for Policy Research in Education.

Govindarajan, G. (1993). *Analysis of preservice elementary, school teachers' collaborative problem solving in a constructivist-based interdisciplinary science course.* Unpublished doctoral dissertation, College of Education, Kansas State University.

Guba, E. G., & Lincoln, Y. S. (1981). *Effective evaluation: Improving the usefulness of evaluation results through responsive and naturalistic approaches.* San Francisco, CA: Jossey-Bass

Holmes Group. (1986). *Tomorrow's teachers.* East Lansing, MI: Author.

Holmes Group. (1990). *Tomorrow's schools: Principles for design of professional development schools.* East Lansing, MI: Author.

Holmes Group. (1995). *Tomorrow's schools of education.* East Lansing, MI: Author.

Interstate New Teacher Assessment and Support Consortium. (1995). *INTASC core standards.* Retrieved July 16, 2004, from http://developo.ccsso.cybercentral.com/intasc.htm

Kansas State Department of Education. (2002a). *Certification and teacher education regulations and teaching standards for Kansas certification and teacher education.* Topeka, KS: Author.

Kansas State Department of Education. (2002b). *Kansas Science education standards.* Topeka, KS: Author.

Loucks-Horsley, S., Hewson, P. W., Love, N., & Stiles, K. E. (1998). *Designing professional development for teachers of science and mathematics.* Thousand Oakes, CA: Corwin Press.

The Mathematical Association of America. (1991). *A call for change: Recommendations for the mathematical preparation of teachers of mathematics.* Washington, DC: Author.

Merrian, S. B. (1998). *Qualitative research and case study applications in education: Revised and expanded from case study research in education* (2nd ed.). San Francisco: Jossey-Bass.

Miles, M. B., & Huberman, A. M. (1994). *Qualitative data analysis: An expanded sourcebook.* Thousand Oakes, CA: Sage.

National Board for Professional Teaching Standards. (1998). *What teachers should know and be able to do.* Retrieved July 16, 2004, from http://www.nbpts.org/nbpts/standards/intro.html

National Center for Improving Science Education. (1989). *Science and technology education for the elementary years: Frameworks for curriculum and instruction.* Andover, WA: The Network.

National Commission on Excellence in Education. (1983). *A nation at risk: The imperative for educational reform.* Washington, DC: U.S. Department of Education.

National Commission on Teaching and America's Future. (1996). *What matters most: Teaching for America's future.* New York: Author.

National Commission on Teaching and America's Future. (1998). *Teaching for high standards: What policymakers need to know and be able to do.* New York: Author.

National Council for Accreditation of Teacher Education. (2001). *Standards for professional development schools.* Washington, DC: Author.

National Council for Accreditation of Teacher Education. (2002). *Professional standards for the accreditation of schools, colleges, and departments of education.* Washington, DC: Author.

National Council of Teachers of Mathematics. (1989a). *Professional standards for teaching mathematics.* Reston, VA: Author.

National Council of Teachers of Mathematics. (1989b). *Curriculum and evaluation standards for school mathematics.* Reston, VA: Author.

National Research Council. (1988). *Everybody counts: A report to the nation on the future of mathematics education.* Washington, DC: National Academy Press.

National Research Council. (1996). *National science education standards.* Washington, DC: National Academy Press.

National Research Council. (2000). *How people learn: Brain, mind, experience, and school.* Washington, DC: National Academy Press.

National Research Council. (2001a). *Educating teachers of science, mathematics, and technology. New practices for the new millennium.* Washington, DC: National Academy Press.

National Research Council. (2001b). *Knowing what students know: The science and design of educational assessment.* Washington, DC: National Academy Press.

National Research Council. (2002). *Investigating the influence of standards: A framework for research in mathematics, science, and technology education.* Washington, DC: National Academy Press.

National Science Board. (1983). *Educating Americans for the 21st century: A report to the American people and the National Science Board.* Washington, DC: Author.

National Science Teacher's Association. (1988). *Science education initiatives for the 1990s.* Washington, DC: Author.

Ramey-Gassert, L., Shroyer, G., & Staver, J. (1996). A qualitative study of factors influencing science teaching self-efficacy of elementary level teachers. *Science Education, 80*(3), 283-315.

Regents of the University of California. (2000). *Full option science systems.* Naslua, NH: Delta Education.

Richardson, S. W. (1994). *The professional development school: A common sense approach to improving education*. Fort Worth, TX: Sid W. Richardson Foundation.

Riggs, I. M., & Enochs, L. G. (1990). Toward the development of an elementary teachers' science teaching belief instrument. *Science Education, 74*(6), 625-637.

Robinson, S., & Darling-Hammond, L. (1994). Change for collaboration and collaboration for change: Transforming teaching through school-university partnerships. In L. Darling-Hammond (Ed.), *Professional development schools: Schools for developing a profession* (pp 207-209). New York: Teachers College Press.

Shrigley, R. L. (1974). The attitude of preservice elementary teachers toward science. *School Science and Mathematics, 74*(3), 243-250.

Shrigley, R. L., & Johnson, T. M. (1974). The attitude of inservice teachers toward science. *School Science and Mathematics, 74*(5), 437-446.

Shroyer, M. G., Ramey-Gassert, L., Hancock, M., Moore, P., & Walker, M. (1995). Math, science, technology after school clubs and summer magnet school: Collaborative professional development opportunities for science educators. *Journal of Science Teacher Education, 6*(2), 112-119.

Shroyer, M. G., Wright, E. L. (1995, April). *Expertise in preservice elementary teaching in science, mathematics, and technology: Evaluation of an innovative model*. Paper presented at the National Association for Research in Science Teaching. San Francisco, CA.

Shroyer, M. G., & Wright, E. L. (1998, April). *A longitudinal case study of organizational change*. Paper presented at the National Association for Research in Science Teaching, San Diego, CA.

Shroyer, M. G., Wright, E. L., & Ramey-Gassert, L. (1996). An innovative model for collaborative reform in elementary school science teaching. *Journal of Science Teacher Education, 7*(3), 151-168.

Stalheim-Smith, A., & Scharmann, L. C. (1994). General biology: Creating a positive learning environment for elementary education majors. *The American Biology Teacher, 56*(4), 216-220.

Teitel, L. (1998). Professional development schools: A literature review. In M. Levine (Ed.), *Designing standards that work* (pp. 33-80). Washington, DC: National Council for Accreditation of Teacher Education.

U.S. Department of Education. (1998). *Promising practices: New ways to improve teacher quality*. Washington, DC: U.S. Government Printing Office.

U.S. Department of Education. (1999). *Teacher quality: A report on the preparation and qualifications of public school teachers*. Washington, DC: National Center for Educational Statistics.

U.S. Department of Education. (2000). *Before it's too late: A report to the nation from The National Commission on mathematics and science teaching for the 21st century*. Washington, DC: Education Publications Center.

WestEd. (2000). *Teachers who learn kids who achieve: A look at schools with model professional development*. San Francisco, CA: Author.

Willhite, K. T. (1995). *Changes in elementary science teachers during their participation in a science, mathematics and technology teacher preparation project.* Unpublished doctoral dissertation, College of Education, Kansas State University.

Wilson, J. (1996). An evaluation of the field experiences of the innovative model for the preparation of elementary school teachers for science, mathematics, and technology. *Journal of Teacher Education, 47*(1), 53-59.

Yahnke, S., Shroyer, M. G., Bietau, L., Hancock, M., & Bietau, L. (2005). *Collaborating to renew and reform K-16 education*. In J. E. Neapolitan & T. R. Berkeley (Eds.), *Staying the course with professional development schools*. New York: Peter Lang.

Zollman, D. (1994). Preparing future science teachers. *Physics Education, 29*, 271-275.

CHAPTER 12

THE IMPACT OF STATE STANDARDS ON TEACHER PROFESSIONAL DEVELOPMENT AND STUDENT PERFORMANCE IN MIDDLE SCHOOL SCIENCE

A Texas Case Study

Christy MacKinnon, Judith Fowles, Edward Gonzales, Bonnie McCormick, and William Thomann

A case study is presented that examines how national and state standards have affected middle school science education. The state of Texas changed requirements for the Texas Essential Knowledge and Skills (TEKS) beginning with the academic year 1998–1999. The new standards required concept strands in life, earth, and physical sciences to be taught at all middle school grade levels (sixth–eighth). The new state tests challenged many teachers to upgrade their science knowledge and pedagogical skills. In response to their needs, a collaborative effort between a private university

The Impact of State and National Standards on K-12 Science Teaching, 363–390

and a large urban school district resulted in the development of a masters of arts (MA) degree in multidisciplinary sciences. The goal of the degree program was to link teacher content enhancement to improved student performance in middle school science through implementation of a constructivist epistemology. This study includes discussion about the roles of both the university initiated professional development program and the school district's efforts to support the necessary changes in teacher practice. The interactions of these roles are examined in order to interpret the influence of standards-based instruction and its impact on student performance as measured by state standardized exams over a 4-year period.

INTRODUCTION

Understand the structure and function of life. This proposed standard was sent to the senior author by an agency requesting input from college science faculty. The first concern, of course, was that *understand* was not a specific or measurable action. In further reading of the document, this standard was suggested to include kindergarten students! The agency's request was quickly relegated to the wastebasket, but the extreme distortion of the usefulness of standards was much harder to discard. Was this going to be the impact of standards on K-12 science education?

A response to this question is presented as a case study analysis of a collaborative effort between a large urban school district and a private university to address the needs of middle school science teachers. These needs were manifested when the state of Texas changed requirements for student essential knowledge and skills for the academic year 1998–1999. The new standards required concept strands in life, earth, and physical sciences to be taught in all grade levels. This requirement challenged many teachers to upgrade their science knowledge and pedagogical skills. A collaborative effort between the University of the Incarnate Word (UIW) and Northside Independent School District (ISD) resulted in the development of an MA degree in multidisciplinary sciences. The goal of the degree program was to link teacher content enhancement to improved student performance in middle school science through implementation of a constructivist epistemology. The goal was to be achieved by meeting the following objectives: (1) significantly extend teacher content knowledge in biology, chemistry, earth science, and physics; (2) integrate prealgebra and algebra level mathematics skills in these content areas; (3) enhance classroom implementation of reform-based science by improving instructional strategies, individually through technology applications and action research, as well as collectively by revising a middle school curriculum used by the district to be standards-based.

LITERATURE REVIEW

In 1989 the American Association for the Advancement of Sciences (AAAS) published *Science for All Americans*, challenging the nation's schools to produce a scientifically literate society. It provided four guiding principles to achieve this goal: (1) science for all students; (2) science learning as an active process; (3) school science as contemporary science; and (4) science education improvement as part of systemic education reform. Roadmaps to accomplish this reform were documented in two subsequent publications, *Benchmarks for Science Literacy* (AAAS, 1993) and the *National Science Education Standards* (National Research Council, NRC, 1996). These two documents drew a clear picture of the science learning in elementary and secondary classrooms. They advocated the use of inquiry and constructivism as instructional strategies, the integration of math and technology into the sciences, and an increase of science content knowledge for science teachers (AAAS, 1993; NRC, 1996). The pedagogy for science teaching, then, was one that actively engaged students in reasoning about scientific phenomena (Kennedy, 1998). The responsibility for implementing these policy driven standards was mainly in the hands of the classroom teacher (Kennedy, 1998). The expected end result of this reform movement was a scientifically literate work force capable of competing in a scientifically and technologically oriented global economy.

These documents served as a guide for state and local education agencies to make widespread changes in how science was taught and assessed. They described content and assessment standards, instructional strategies, and educator professional development guidelines. Most of the 50 states have revised their state science targets and assessments to be in alignment with these documents (Atkin & Black, 2003; Burry-Stock & Casebeer, 2003; Good & Shumansky, 2001; NRC, 2001) and developed new curriculum frameworks to guide instruction, and new assessments to test students' knowledge (Darling-Hammond, 2004). Although Texas had some standards-based testing before 1998, it was in need of significant revision. The new version of TEKS was published in 1997 and was a state reform effort to reflect these standards-based recommendations. Texas also revised student achievement assessments to be aligned tightly to those standards. The new state exam (Texas Assessment of Knowledge and Skills, or TAKS) was administered from 1999 through 2002. Students in Texas were tested at the fifth, eighth, 10th and 11th grades on the science TEKS. Tests for the eighth grade were stopped in 2003, but are scheduled to begin again in 2005.

If the directive of reform-based science is to actively engage students in reasoning about scientific phenomena (Kennedy, 1998), the standards have brought mixed results in the typical classroom. Although generally there has been an improvement in the use of hands-on science, actively engaging students in reasoning about scientific phenomena has usually been found only in model-making (such as DNA, RNA) activities, structured labs from the textbook ancillary, or the district curriculum guide. These activities reinforce the concepts taught in the context of TAKS testing. It is a common complaint from teachers that they have sacrificed the themes, interconnections, interdisciplinary science and ventures into nature of science inquiry in an effort to "cover" the biology TEKS to prepare for the TAKS (Abd-El-Khalick, Bell, & Lederman, 1998). Benchmark testing in the four state-assessed content areas, and practice TAKS testing, have forced the vast majority of teachers to adhere to a required curriculum with tried and true methods in an effort to make up for lost instructional time due to test preparation and mandates. Traditional transmission of content knowledge from teacher to student therefore dominated this learning experience.

However, in addition to content knowledge Shulman (1986) offers two additional types of knowledge that are necessary for teachers to understand and transmit content. These are pedagogical content knowledge, and curricular content knowledge. Kennedy (1998) argues that "recitational knowledge" will not give science teachers the background they need to be successful at inquiry teaching; teachers themselves must have a conceptual understanding of the sciences. In addition, teachers must have accrued pedagogical content knowledge—the ability to represent important ideas in a way that makes them understandable to students (Kennedy, 1998), including metaphors, analogies, and demonstrations (Shulman, 1986). Teachers learn pedagogical content knowledge either through their own experiences as a science student, or through years teaching, accumulating ideas that work either through their own experiences, workshops, or other teachers.

PROFESSIONAL DEVELOPMENT PROGRAM

The *National Science Education Standards* (NRC, 1996) have had a profound influence on state and local school districts to review and revise science curricula. As the result of standards-based science, the classroom teacher faced a double challenge: (1) implement new science activities correlated with standards, and (2) enhance content knowledge and instructional strategies to successfully implement a standards-based curriculum. Assessment of student performance according to standards remained the responsibility of state and local school districts. The state of Texas provided leadership in measuring student achievement through its TAKS. Recently, standards included in TEKS for middle school science were revised to achieve a closer alignment of with national standards (Texas Education Agency, 1997). The new state standards, fully implemented and assessed by state exams, occurred in the academic year 1998–1999. The new state standards required concept strands in life, earth, and physical sciences to be taught in sixth through eighth grades. Previously, the TEKS and TAKS emphasized science knowledge presented in the eighth grade.

Middle school science teachers were recruited by the secondary science supervisor for the Northside ISD to participate in the program. The retention rate of teacher participants in this program was 69%; 18 of the original 26 participants completed the graduate program in 18 months, and remained with the school district to serve as change agents on their campuses for at least 4 years. Of the eight teacher participants who left the program, only four left for academic reasons (e.g., not enough time available for study and preparation). The others left for personal reasons.

STUDY PROCEDURE

The study included two phases. The first phase involved the design and implementation of a graduate level program to increase life, earth and physical sciences content knowledge of middle school science teachers. The second phase involved the analysis of student performance as measured by state standardized exams. The questions investigated were: (1) Did teachers improve content knowledge? (2) Did teachers improve skills and application of technology? (3) Did teachers apply new knowledge and skills in the revision and implementation of the school district's curriculum? (4) Were improvements in teacher professional development reflected in student performance as measured by state standardized exams? (5) What were the needs of the school district to support implementation of a standards-based curriculum? (6) How did the roles of the university and the school district interact to support change in teacher practice?

The goal of the MA in multidisciplinary sciences degree program at UIW was to respect, employ and extend the learner characteristics of adults and teacher professionals in time intensive, academically demanding courses. The MA in multidisciplinary sciences consisted of 36 credit hours: 3 hours of integrated math & science, 3 hours of science instructional technology; 6 hours each of biology, chemistry, earth science, and physics; 3 hours of action research, and 3 hours of curriculum implementation of a revised, standards-based curriculum. Specific content and process objectives were identified for each science course, and included introductory as well as advanced knowledge. The course objectives were correlated with the TEKS, the science standards determined by the state of Texas (Appendix A). All content skills were supported through technology and classroom implementation activities. The university faculty involved in the program modeled diverse instructional and assessment strategies, but all emphasized hands-on activities and/or problem-solving projects.

The background of each professor involved in the program was non-traditional for college science faculty. The diversity of faculty experiences was an important factor in the development and implementation of the MA in multidisciplinary sciences. There were two biology faculty involved in the program. One had earned secondary science education certification prior to completion of a traditional science doctorate. The other had extensive experience in informal science education as a docent for the city's zoological park. She earned her PhD in science education. The two biology faculty also had lengthy experience with the NASA undergraduate program in science and mathematics titled Opportunities for Visionary Academics (NOVA). NOVA is a model professional development program for college level science educators with two essential components: (1) a team approach (faculty and administrators) to revise and implement undergraduate courses that meet the needs of preservice science teachers under the new national standards and (2) intensive faculty change that resulted from collaboration with experienced faculty during course implementation and workshops to increase knowledge of best pedagogical practices (Sunal et al., 2001).

The earth science professor had previous experience in developing two programs for preservice teachers while he was at a different university. The programs included the bachelor of arts degree in earth science-secondary education, and a bachelor of arts degree in geography-secondary. At the UIW, he was instrumental in the development of a geology minor for preservice elementary education majors. He has given many presentations and demonstrations in earth science and physics at local elementary, middle, and high schools in the San Antonio and surrounding area, and has sponsored in-class science presentations by many UIW education majors at several local elementary and middle schools in San Antonio, Texas.

There were two chemistry faculty involved in the program. One had a PhD in chemistry, but no previous experience with a professional development program. He taught the Chemistry I course in a format that emphasized lecture, but also included demonstrations and experimental activities. The Chemistry II professor had diverse training as a pharmacist (bachelor of science and master of science), and a PhD in biochemistry. He later earned secondary education certification in chemistry, and a MA in curriculum supervision. He brought to the program a unique combination of theory and practice of bioorganic principles, in addition to his knowledge of secondary education and curriculum development. His courses incorporated chemical modeling experiences, problem solving, and laboratory activities.

The physics courses were the most problematic in meeting the needs of the teachers. Although the professor for the Physics I course had extensive experience with science education reform, he was philosophically resistant to offering graduate credit in physics courses that did not match traditional, advanced content. This viewpoint led to substantial disagreement with the validity of a graduate level professional development program in science, and also created understandable friction between the teachers and the professor. There was a change in the instructor for the Physics II course and the second course was taught in a manner consistent with reform-based learning. The physics instructor also taught the earth science courses,

The secondary science instructional specialist played an essential role in the effectiveness of the graduate courses to meet the teachers' needs. She had credibility with both the graduate faculty and teachers with her academic background, and was respected for her teaching excellence in many years experience as a middle and high school science teacher. She served as a mentor and motivator to the teachers, and provided invaluable feedback to the graduate faculty so that instructional concerns were made known and corrective actions were possible at the time difficulties were encountered. Importantly, she viewed her position within the district as an opportunity for leadership in professional development of teachers, and not just a bureaucratic supervision of budgets and evaluations.

The graduate program included course activities for "immersion in inquiry, action research, curriculum development, adaptation, and implementation" (Loucks-Horsley et al., 1998). Teacher content knowledge was assessed through prepost tests, course exams and quizzes, written analysis of current science and science education literature, and an action research project. Prepost test scores were evaluated for statistical significance by paired t-tests. Teacher changes in classroom implementation were qualitatively assessed through on-site observations. The observation rubric changed over time, but each version assessed teacher content knowledge, method of instruction (e.g., traditional, activity-based), engagement of students in activities, and integration of technology and mathematics. The teachers also completed a lengthy questionnaire about their experience with the program. Student performance was assessed by scores on the TAKS administered at the end of the eighth grade.

FINDINGS OF THE PROFESSIONAL DEVELOPMENT PROGRAM

The pre and posttest results are found in Table 12.1. Sample sizes changed due to the variation in paired prepost test scores. Some faculty reported their results in points; other reported them in percentages. Prepost tests were not administered for the Integrated Mathematics and Science for Middle School or for the Biology I courses since these courses were offered before the formal beginning of the MA program.

The pre and posttests in Biology II included multiple-choice and several discussion questions. Paired t-tests analysis indicated that a significant difference ($p < .001$) was found between the pre and posttests scores for both the multiple-choice and discussion questions. The pre and postests in Chemistry I and Chemistry II consisted of multiple-choice questions only. Paired t-test analysis indicated a significant difference ($p < .001$) between pre and posttest scores. Thus, for both biology and chemistry, there was a dramatic increase in content knowledge following the completion of the graduate courses.

The Earth Science pre and posttests were assessed in distinct content areas, each test consisting of multiple choice and discussion questions. The Earth Science I course assessed content knowledge in geology and oceanography. A nonsignificant difference was found between the pre- and posttest performance of the teachers in geology and oceanography, although noticeable increases were observed in the posttest scores. The Earth Science II course assessed content knowledge in astronomy and meteorology. A significant difference ($p < .05$) was found in the short answer portion of pre and posttests astronomy. A nonsignificant difference was observed on the multiple-choice portion. Significant differences were observed between the pre and posttests of the multiple-choice and discussion portions of the meteorology content ($p < 0.05$). The course objectives and assessments for Earth Science I and II were the most extensive of all of the courses and prepost tests had a greater number of questions and a greater range of content. Given that separate courses are typically given for each of the earth science content areas, the *density* of the information may have been too great for the time frame allotted for the courses.

A significant difference was found between the pre and posttests scores for Physics I ($p < 0.05$). The teachers' performance in the physics content areas was notable given the absence of supportive learning environment. In spite of their frustrations with the negative statements from the instructor about their lack of prior content knowledge, the teachers persisted (often with group study sessions), mastered the content, and demonstrated their ability to perform at a graduate level with material as presented. In interviews with the teachers about their experiences in other graduate courses, their worst-case scenario was the instructional program in the Physics I class; yet they had a significant gain in scores in this class. A significant difference was found between the pre- and post-

Table 12.1. MA Course Paired t-tests from Pre-Post Assessments

Content Area		*Pre-Test*	*Post-Test*	*t*	*p*
Biology II* (*N* = 20)					
Multiple-Choice	Mean	17.22	20.50	7.08	<.001
	SD	3.08	3.00		
Discussion	Mean	11.17	17.17	6.51	<.001
	SD	6.18	7.01		
Total Score	Mean	28.28	37.78	8.05	>.001
	SD	8.53	9.77		
Chemistry I# (*N* = 17)					
Total Score	Mean	58.75	74.5	–7.34	<.001
	SD	14.06	10.92		
Chemistry II# (*N* = 21)					
Total Score	Mean	30.43	77.86	14.94	>.001
	SD	13.17	12.64		
Geology* (*N* = 23)					
Written	Mean	3.15	3.89	–.60	>.05
	SD	1.07	1.04		
Multiple-Choice	Mean	23.38	30.87	–0.95	>.05
	SD	7.45	8.30		
Oceanography* (*N* = 23)					
Written	Mean	1.26	3.74	–1.64	>.05
	SD	1.23	1.76		
Multiple-Choice	Mean	19.88	35.39	–.190	>.05
	SD	7.36	8.96		
Astronomy* (*N* = 23)					
Written	Mean	9.32	21.83	-2.16	<.05
	SD	6.39	5.01		
Multiple-Choice	Mean	21.68	34.30	–1.93	>.05
	SD	6.70	6.37		
Meteorology* (*N* = 23)					
Written	Mean	6.88	17.43	–2.96*	<.05
	SD	4.20	2.77		
Multiple-Choice	Mean	26.56	43.70	–2.94	<.05
	SD	7.20	3.80		
Physics I# (*N* = 21)					
	Mean	29.5	91.1	2.14	<.05
	SD	26.0	31.1		
Physics II* (*N* = 20)					
Multiple-Choice	Mean	61.35	73.65	5.30	<.001
	SD	15.23	14.50		
Problems	Mean	11.55	79.50	19.82	<.001
	SD	16.90	13.23		
Total Score	Mean	72.90	153.15	16.66	<.001
	SD	30.11	22.42		

* = scores in points; # scores in percentages.

test scores for the Physics II assessment, which included both multiple-choice ($p < 0.001$) and problems ($p < 0.001$).

The course in science instructional technology emphasized computer-based data acquisition and analysis, technology adjuncts such as probes and censors, graphing calculators, course Web design, and internet research skills. About half of the teachers had previous courses in instructional technology and were not required to take the program's course. The prepost tests assessed teachers' self-perceived knowledge and skill of technology via short answer and multiple-choice questions. A significant difference was observed in the pre and posttests (data not shown) indicating teachers increased their confidence with the technology and therefore increased their use of it in their instruction.

All teachers conducted action research projects (Hubbard & Miller, 1993) in technology that required development of a Web page, a student activity using the information from the Web page and other sites, and a section for posting student results. The majority of teachers completed their projects during the first academic year of the degree program. A few extended projects to the next academic year. Several of the completed projects were displayed on the district's Website for at least 2 years.

All teachers participated in cooperative efforts to comprehensively redesign the district curriculum guide. Standard-based content and student activities were extensively revised through a comprehensive revision of the seventh grade science curriculum used by the district. The eighth grade curriculum had been updated just prior to the start of the program. Revision of the seventh grade curriculum engaged the sixth grade teachers to be aware of what content their students needed to perform well in the seventh grade, and the eighth grade teachers would know what students should have learned in the seventh grade. Thus, there was a vertical alignment of the curriculum for middle school science.

The standards-based curriculum developed by the teachers in the MA in the multidisciplinary sciences program was based on the learning cycle sequence (Lawson, Abraham, & Renner, 1989). Thus, the largest amount of class time was allocated to student activities. The revised middle school curriculum included instructional/learning strategies based on the learning cycle sequence many lessons were revised to incorporate Internet technology, graphing calculators, and activities utilizing scientific equipment and/or computer technology. The revised middle school science curriculum was implemented in the academic year 1998–1999.

In 1999, the Northside ISD mitigated the time pressure of a reform-based curriculum by the adoption of a block schedule in which students take four courses twice a week for a 90-minute class period. While the block scheduling was essential for implementing the standards-based curriculum, it was not sufficient to account for the increase in student performance on state standardized exams.

At the conclusion of course-work for the program, teachers completed a questionnaire about their self-perceived changes in content knowledge and classroom implementation (Table 12.2). For the majority of questions, more than 70% of the responses "agreed" or "strongly agreed" to the idea that standards based changes were made in knowledge and instructional approaches.

Classroom observations of the teacher participants occurred during the 4 years following the conclusion of the graduate program. The observa-

Table 12.2. Summary of Teacher Questionnaire (*N* =13)

	Moderately Disagree	*Strongly Disagree*	*Moderately Agree*	*Strongly Agree*
My problem-solving instructional strategies have changed.	38%	8%	0	54%
My approach to teaching specific content has changed.	0	0	62%	38%
My instructional strategies have changed in general.	69%	0	0	31%
The content presented in the earth science courses enhanced my knowledge.	8%	0	0	92%
The content presented in the physics courses enhanced my knowledge.	23%	23%	0	54%
The content presented in the chemistry courses enhanced my knowledge.	0	23%	54%	23%
The hands-on activities in the earth science courses have been used (or adapted for use) in my classroom.	15%	0	54%	31%
The hands-on activities in the physics courses have been used (or adapted for use) in my classroom.	31%	15%	0	54%
The hands-on activities in the chemistry courses have been used (or adapted for use) in my classroom.	23%	15%	15%	46%
My participation in this program has had a positive impact on my students' academic performance.	8%	0	62%	31%
I have increased technology use in my classroom.	46%	8%	0	46%
My overall professional competence has been improved.	0	0	77%	23%

tion rubric was modified several times during this time period, but all versions observed whether or not teachers demonstrated an increase in content knowledge, implemented a student-centered learning environment that emphasized hands-on activities, asked higher level questions of the students, and made connections across the sciences, mathematics, and technology. About a third of the teacher participants were observed specifically for study purposes. However, the secondary science instructional specialist observed all teachers at least once per year for district performance evaluation. Collectively, the observations indicated a qualitative improvement in the teachers' content knowledge and classroom pedagogical skills. Several teachers were aware that they needed improvement in their questioning skills to help students draw connections between subjects, and that the improvement was a function of time and practice, not attitude. Interestingly, several teachers remarked that they did not feel that they were "teaching to the state tests," but rather implementing a curriculum for which they had ownership, with a confidence in their abilities to achieve success in their reformed based instruction.

In addition to changes in teachers' knowledge and skills, several MA teachers quickly assumed leadership roles within the district, and therefore became change agents on their campuses and in the district. Two became department coordinators, with responsibilities for teacher and curriculum reviews. One assumed additional responsibilities as a middle/high school science teacher consultant, and advised teachers on instructional strategies and classroom implementation of standards based practices. Another became a teacher consultant for the city's Urban Systemic Initiative program as a mentor teacher. When the secondary science instructional specialist (second author) retired, a graduate of the MA program was selected by the district to become the secondary science instructional specialist. She has coordinated 6th–12th grade curricula, conducted annual teacher performance evaluations, managed the districts budget for science education, and continued the advocacy role for professional development of teachers.

The essential assessment of the professional development program was change in student performance on state tests (Table 12.3 and Appendix B). Seven campuses were selected for assessment of student scores on standardized state exams. Five campuses were selected based on MA teacher participants who were teaching eighth grade science, the grade level tested in the state TAKS. Four of the five had predominantly Hispanic and/or African American student populations. Two campuses were selected as controls. Both of these control schools, also, had predominantly Hispanic and/or African American student populations. All seven schools performance was assessed on total percent of students who met minimum requirements, as well as the percent of economically

Table 12.3. Pre-Post Performance Scores in 8th Grade Science Assessed by TAAS, 1999-2002*

Location	*Met Minimum Standards*	*Economically Disadvantaged*	*At Risk*	*African American*	*Hispanic*	*Anglo*
District	89-95	82-91	81-89	88-92	84-93	95-98
Campus 1	81-93	81-92	75-88	84-92	79-93	91-97
Campus 2	92-98	87-97	84-94	95-96	88-97	96-97
Campus 3	90-95	90-93	86-90	97-91	85-95	95-97
Campus 4	78-88	74-87	71-81	67-88	77-95	94-100
Campus 5	95-98	85-93	89-95	79-90	92-97	97-100
Control 1	81-89	76-87	75-78	100-85	76-88	95-97
Control 2	92-94	89-93	87-86	89-87	91-93	94-97

*Scores of special education students included in state report.

disadvantaged students, percent of at-risk for dropping out of school, African American, Hispanic, or Anglo students who met minimal standards. The 1998 exam scores reflected student performance prior to start of the MA graduate program. Scores of special education students were not included in the state reports. The revised state exams for eighth grade science were administered from 1999–2002, and included special education student scores.

During the 4 years of TAKS testing, the district overall score increased from 89% to 95% of students who met the minimum standards. The scores for economically disadvantaged students increased from 82% to 92%. The scores for at risk students increased from 81% to 88%. The scores for African American students increased from 88% to 92%. The scores for Hispanic students increased from 84% to 92%. The scores for Anglo students increased from 95% to 98%.

The five campuses with MA teacher participants also had an overall improvement in eighth grade science performance, but some had greater gains in specific student populations than the district average. For all campuses, the performance of Anglo students who met minimum standards increased from 94% to above 97% in 2002.

Campus 1 had an increase in all students who met minimum standards from 81% in 1999 to 93% in 2002, compared with the district's 89-95%. The performance of specific student populations showed a more dramatic increase. The percentage of economically disadvantaged students who met minimum standards increased from 81% in 1999 to 92% in 2002. Similarly, the percentage of at-risk students who met minimal standards increased from 75% in 1999 to 88% in 2002. The percentage of African American students who met minimum standards in 1999 was 84% and this increased to 92% in 2002. The percentage of Hispanic students who met minimum standards in 1999 was 79%, and increased to 93% in 2002. Anglo students increased performance from 91% to 97%.

Campus 2 had an increased in all student scores from 92% in 1999 to 98% in 2002, well above the district score of 93%. These higher scores in 2002 also occurred in the student population of African Americans (96%) and Hispanics (97%).

Campus 3 had an increase in the percent of students who met minimum standards from 1999 to 2002 comparable to that of the district. While the at-risk student population and Hispanic students were above district average in 1999 (86% and 85%, respectively), the strong performance was maintained through 2002 (90% and 95% vs. district 2002 scores of 89% for at-risk students and 93% for Hispanic students.

Campus 4 had a low performance of all students in 1999 who met minimum standards (78% vs. district 89%). However, there was a consistent increase in all student populations who met minimum standards in 2002. The greatest increases were seen in economically disadvantaged students (74% who met minimum standards in 1999 to 87% in 2002 vs. district 91% in 2002), and African American students (67% in 1999 to 88% in 2002 vs. district 88% in 1999 to 92% in 2002).

Campus 5 had above district scores for all student populations in 1999. Student performance continued above district scores in 2002 for all student populations, although the African American student population meeting minimum standards was lower, 90% versus the district score of 92%.

Control Campus 1 had student performance at or below district scores who met minimum standards in 1999, with the exception of the African American student population, which had a score of 100%. The scores for all student populations that met minimum standards remained below district scores in 2002. The economically disadvantaged students had low scores in 1999 (72% who met minimum standards), and showed an increase in 2002 (87%). The African American student population had 100% who met minimum performance in 1999, but declined to 85% in 2002. This decrease may be due in part to an increase in numbers of African American students over the 4 year period; in 1999 there were less than five African American students, and by 2002 the number had increased to more than 30. For the Hispanic student population, the percentage that met minimum standards in 1999 was 76%, and showed an increased to 88% in 2002. The performance of Anglo students showed an increase from 1999 of 95% to 97% in 2002.

Control Campus 2 had scores at or above district student performance for all student populations in 1999. However in 2002 there was a decrease in student performance compared to the districts for the student populations of at-risk and African American, and Anglo students.

DISCUSSION

Hargreaves and Fullum (1998, p. 83) emphasized that "ways must be found for educators to step out into wider learning networks; for school and universities to form partnerships in which teacher education and school improvement are pursued in tandem." The collaborative effort between the UIW and Northside ISD to develop the graduate program and for teacher implementation of reform based curricula illustrate necessary roles for finding these ways to step out. The university extended its view of professional development of science teachers. The development of the MA in multidisciplinary sciences, housed in the school of mathematics, science, and engineering, reflected an unusual perception of the science faculty that science education reform cannot be met at the university level by the traditional science graduate program intended for preparation for doctoral studies, or the traditional education programs, most often intended for administrative positions. Instead, a science graduate program for the professional development of teachers had to be unique to their needs for content enhancement and improved classroom implementation for reform based curricula. The NRC (2001) argued that standards based curricula that demand high student performance will not be successful until teachers are educated in the philosophical and instructional goals that are consistent with the national and state standards. This view was supported by a metaanalysis conducted by Druva and Anderson (1983), which demonstrated that teacher content knowledge correlated with higher-level thinking expected of students, as well as students' positive perceptions of science.

The university level reward system for faculty tenure and promotion had to change to include faculty activities in professional development programs. Participation in professional development efforts can be a valid scholarly endeavor that can integrate theory and practice of teaching (Brent & Hodges, 1998; House, 1990; Houston, 1989; Putnam & Borko, 2000). Faculty at UIW invested significant amounts of time not only writing and administrating grants that provided funds to support teachers' enrollment in the program, but also in developing new courses and teaching them at night, weekends and during the summer. These instructional efforts were in addition to, not in place of, the standard faculty teaching load. If these activities had not been viewed as scholarly endeavors, then it is highly unlikely that there would have been faculty commitment to engage in a professional development program even if there was strong personal interest to do so.

The role of the school district also was revisited to find successful ways to support reform based science education. The school district had a strong advocate for professional development that was long-term and sustained. Traditional in-service professional development that lasted for a few days was viewed as clearly insufficient to give teachers the comprehensive content and pedagogical knowledge and skills necessary for change. The school district was willing to allocate a significant portion of their state funds targeted for professional development to tuition support for pursuit of advanced degrees.

SUMMARY

Standards-based curricula can have a significant impact on student performance if the interactions of the university and school district roles are mutually supportive to achieve the goal of improved student performance in science. As this case study demonstrates, both roles required the presence of strong advocates for change. The university had hired faculty that had both advanced content knowledge and pedagogical content knowledge for effective science teaching. The university was also flexible in the development of programs that met the needs of teachers, a task that is often easier for private universities than state-supported ones. The university acknowledged and rewarded faculty participation in the professional development of precollege teachers by promotion, tenure, and performance-based pay increases. The school district had an administrative position designated for secondary science. The school district had a fiscal commitment to allocate funds for substantial professional development of teachers, and to provide the necessary financial and curricular support for the implementation of standards-based science education. For standards to have a significant impact, these comprehensive collaborations and commitments from both the players at a university and a school district must be present in order to provide the substrate for enhancing student performance in science.

ACKNOWLEDGMENTS

Dr. Jeff Greathouse and the late Dr. Hugh (Tom) Hudson are thanked for their participation in the development and implementation of the MA in multidisciplinary sciences. Dr. Leo Edwards, Fayettville State University,

North Carolina is thanked for his statistical analyses of prepost tests and final evaluation reports. This paper is dedicated to the memory of Dr. T. Reginald Taylor, mentor and friend, who supported and encouraged our efforts to help reform mathematics and science. This project was funded in part by grants from the THECB Eisenhower Professional Development Program. The opinions stated reflect those of the authors and do not necessarily reflect those of the funding agencies.

APPENDIX A: CONTENT COURSE OBJECTIVES AND TEKS CONCEPTS STRANDS, STUDENT KNOWLEDGE AND SKILLS

Integrated Mathematics and Sciences; Activities embedded in each content course.

TEKS Concepts	*The student is expected to:*
The student conducts field and laboratory investigations using safe, environmentally appropriate, and ethical practices.	Demonstrate safe practices during field and laboratory investigations. Make wise choices in the use and conservation of resources and the disposal or recycling of materials.
The student uses scientific inquiry methods during field and laboratory investigations.	Plan and implement investigative procedures including asking questions, formulating testable hypotheses, and selecting and using equipment and technology. Collect data by observing and measuring. Analyze and interpret information to construct reasonable explanations from direct to indirect evidence. Communicate valid conclusions. Construct graphs, tables, maps and charts using tools including computers to organized, examine and evaluate data.
The student uses critical thinking and scientific problem solving to make informed decisions.	Analyze, review, and critique scientific explanations, including hypotheses and theories, as to their strengths and weaknesses using scientific evidence and information. Draw inferences based on data related to promotional materials for products and services. Represent the natural world using models and identify their limitations. Evaluate the impact of research on scientific though, society, and the environment.
The student knows how to use a variety of tools and methods to conduct science inquiry.	Collect, analyze, and record information using tools including beakers, petri dishes, meter sticks, graduated cylinders, weather instruments, timing devices, hot plates, test tubes, safety goggles, spring scales, magnets, balances, microscopes, telescopes, thermometers, compasses, computers and computer probes. 6th grade: Identify patterns in collected information using percent, average, range and frequency. 7th grade: Collect and analyze information to recognize patterns such as rates of change. 8th grade: Extrapolate from collected information to make predictions.

Biology I Ecology and Evolution Course Objectives	*TEKS: Concepts: The student knows: The student is expected to:*
Recognize the relationship of genotype to phenotype; Apply probability rules to principles of Calculate and graph exponential growth potential; predict future population growth and determine the potential limits to growth. Compare the effect of predator type and substrate on gene frequencies in a population of individuals with different color varieties. Apply the Hardy-Weinberg principle to predict the gene frequencies in successive generations as a mechanism for evaluating evolutionary change in a population; Graph changes in the gene frequencies over time. Apply the principles of natural selection to explain real world events. Infer the evolutionary relationship of organisms based on similarities and differences of DNA sequences. Determine relationships between organisms by analyzing a classification system. Classify the major divisions of plant by structural adaptations to land and reproductive characteristics. Explain relationships between populations in a community. Evaluate the consequences of environmental change in ecosystems. Investigate the effect of humans on the world's ecosystems	6th grade: Traits of species can change through generations and that the instructions for traits are contained in the genetic material of the organism. Identify some changes in traits that can occur over several generations through natural occurrence and selective breeding. Identify cells as structures containing genetic material. Interpret the role of genes in inheritance. 7th grade: Traits of species can change through generations....Identify that sexual reproduction results in more diverse offspring and asexual reproduction in more uniform offspring. Compare traits of organisms of different species that enhance their survival and reproduction. Distinguish between dominant and recessive traits and recognize that inherited traits of an individual are contained in genetic material. There is a relationship between organisms and the environment. Identify components of the ecosystem. Observe and describe how organisms including producers, consumers, and decomposers live together in an environment and use existing resources. Describe how different environments support different varieties of organisms. Observe and describe the roles of ecological succession in ecosystems. 8th grade: Traits of species can change through generations...Identify that change in environmental conditions can affect the survival of individuals and of species. Distinguish between inherited traits and other characteristics that result from interactions with the environment. Make predictions about possible outcomes of various genetic combinations of inherited characteristics. Natural events and human activities can alter Earth systems. Analyze how natural or human events may have contributed to the extinction of species.

Biology II Topics in Cellular and Molecular Biology Course Objectives	*TEKS: Concepts: The student knows: The student is expected to:*
Classify of organisms based on cellular structures and functional requirements. Compare cellular functions to organismal functions. Analyze human systems for cellular and organ level mechanisms of homeostasis. Integrate cellular structure and function to organismal responsiveness to changes in the environment.	6th grade: The relationship between structure and function in living systems. Differentiate between structure and function; determine that all organisms are composed of cells that carry on functions to sustain life; identify how structure complements function at different levels of organizations. Responses of organisms are caused by internal or external stimuli. Identify responses in organisms to internal and external stimuli. 7th grade: The relationship between structure and function in living systems. Identify the systems of the human organism and describe their functions; describe how organisms maintain stable internal conditions while living in changing external environments. There is a relationship between force and motion. Relate forces to basic processes in living organisms including the flow of blood. 8th grade: Interdependence occurs among living systems. Describe interactions among systems in the human organism; identify feedback mechanisms that maintain equilibrium of body systems.

Chemistry I Properties and Changes of Matter Course Objectives	*TEKS: Concepts: The student knows: The student is expected to:*
Identify solid, liquid and gas as physical states of matter and measure energy changes that accompany changes of state. Express physical properties in terms of density, and apply density principles to evaluate the efficiency of recycling programs. Classify compounds, mixtures and elements and determine the percentage composition of components in a mixture. Investigate physical changes in mixed materials without chemical change. Investigate chemical changes in a reaction. Conduct calorimeter measurements, and determine the energy content of materials. Identify and investigate factors (volume, solute, percent composition) that affect phase changes in water. Demonstrate scale models of atoms and molecules. Summarize events that contribute to current theories of atomic structure. Determine the nuclear and electronic structure. Determine the percentage composition of a compound containing a metal and oxygen. Identify chemical elements in the periodic table based on their physical and chemical properties. Construct a computational database of chemical elements. Determine average atomic masses. Investigate half-life of radioactive isotopes. Compare products of fusion and fission reactions and evaluate their current applications.	6th grade: Substances have physical and chemical properties. Demonstrate that new substances can be made when two or more substances are chemically combined and compare the properties of the new substances to the original substances. Classify substances by their physical and chemical properties. Complex interactions occur between matter and energy. Define matter and energy. 7th grade: Substances have physical and chemical properties. Identify and demonstrate every day examples of chemical phenomena. Describe physical properties of elements and identify how they are used to position an element on the periodic table; recognize that compounds are composed of elements. 8th grade: The student knows that matter is composed of atoms. Describe the structure and parts of an atom. Identify the properties of an atom including mass and electrical charge. Substances have chemical and physical properties. Demonstrate that substances may react chemically to form new substances. Interpret information on the periodic table to understand that physical properties are used to group elements. Recognize the importance of formulas and equations to express what happens in a chemical reaction. Identify the physical and chemical properties influence the development an application of everyday materials.

Chemistry II: Bio-organic Chemistry Course Objectives	*TEKS: Concepts- The student knows: The student is expected to:*
Discuss chemical properties of biological compounds. Interpret acid base reactions. Solve reactions of fundamental organic compounds. Discuss and correlate biochemical reactions involved with respiration, digestions and excretions.	6th grade: Substances have physical and chemical properties. Demonstrate that new substances can be made when two or more substances are chemically combined and compare the properties of the new substances to the original substances. Classify substances by their physical and chemical properties. 7th grade: Substances have physical and chemical properties. Identify and demonstrate everyday examples of chemical phenomena. Recognize that compounds are composed of elements. 8th grade: Complex interactions occur between matter and energy. Illustrate interactions matter and energy including specific heat. Identify and demonstrate that loss or gain of heat energy occurs during exothermic and endothermic chemical reactions.

Earth Science I: Geology Course Objectives	*TEKS: Concepts: The student knows: The student is expected to:*
Identify the common minerals and common rocks of the Earth's surface Interpret and evaluate topographic and geologic maps, measure map distances, interpret mappings, covert scales to bar, verbal, and graphical notation. Identify major structural features of the earth's surface Identify, classify, evaluate major erosional and depositional features of the Earth's continents. Identify and compare basic soil types, test for common minerals, and determine pH of soils. Review the concept of plate tectonics and hypothesize how this concept can be used to show how continents and ocean basins evolved through time. Identify and classify fossils, determine how fossils are used to find geologic age of strata and interpret depositional environment of sedimentary deposit.	6th grade: The structures and functions of Earth systems. Summarize the rock cycle. There is a relationship between force and motion. Identify forces that shape features of the Earth including uplifting, movement of water, and volcanic activity. 7th grade: Natural events and human activity can alter Earth systems. Analyze effects of regional erosional deposition and weathering; describe and predict the impact of different catastrophic events on the Earth. Complex interactions occur between matter and energy. Illustrate examples of potential and kinetic energy in everyday life such as movements of geologic faults. 8th grade: Natural Events and human activities can alter Earth Systems. Predict land features resulting from gradual changes such as mountain building, beach erosion, land subsidence, and continental drift; describe how human activities have modified soil, water quality.

Earth Science I: Oceanography Course Objectives	*TEKS: Concepts: The student knows: The student is expected to:*
Identify and interpret the topographic features of the ocean floor. Discuss and evaluate the chemical properties of seawater. Calculate wave base, average wave velocity and wavelength of waves along the shoreline. Identify the world's oceans and seafloor topography. Evaluate the nature and formation of surf along the shoreline. Classify and interpret erosional and depositional features found along smooth, flat-lying, and rocky shorelines. Identify and discuss the major surface and deep ocean currents. Interpret and evaluate bathymetric maps and coastal topographic maps.	6th grade: There is a relationship between force and motion. Identify and describe the changes in position, direction of motion, and speed of an object when acted upon by force. The student uses scientific inquiry methods during field and laboratory investigations. Construct graphs, tables, maps, and charts using tools including computers to organize, examine, and evaluate data. Substances have physical and chemical properties. Classify substances by their physical and chemical properties. 7th grade: Natural events and human activity can alter Earth systems. Analyze effects of regional erosional deposition. 8th grade: There is a relationship between force and motion. Recognize that waves are generated and can travel through different media. Complex interactions occur between energy and matter. Describe interactions among solar, weather, and ocean systems.

Earth Science II Astronomy & Meteorology Course Objectives	*TEKS: Concepts: The student knows: The student is expected to:*
Identify constellations in the northern hemisphere, bright stars, planets and other celestial objects in the sky. Recognize lunar phases and identify causes of lunar and solar eclipses. Identify types of telescopes, types of spectra of stars. Identify layers of the sun and recognize features such as sunspots, faculae, granular structure, prominences. Recognize different types of galaxies Construct and interpret the Hertzsprung-Russell diagram. Evaluate the nature of stellar evolution of low mass, intermediate mass, and high mass stars. Identify and classify the nature of the terrestrial and Jovian planets. Inquire into the evolution of galaxies based on theory and observations. Summarize and evaluate the various theories of the origin of the universe. Identify major air masses and air mass weather, low and high pressure systems. Define and classify the major circulatory air patterns of the Earth's atmosphere and major layers of the Earth's atmosphere Measure temperature, relative humidity, air pressure, wind velocity and identify clouds for making a local weather forecast. Classify and compare the different kinds of and causes for such precipitation. Identify the different kinds of atmospheric phenomena such as rainbows, solar arcs, auroras. Determine how a hurricane and a tornado forms. Identify the different cloud types and Predict weather based on cloud types and physical properties of the atmosphere. Discuss and develop a model for the formation of normal and supercell thunderstorms.	6th grade: Systems may combine with other systems to form a larger system. Identify and describe a system that results from the combination of two or more systems such as the solar system. Describe how the properties of a system are different from the properties of its parts. Knows components of our solar system. Identify characteristics of objects in our solar system including the Sun, planets, meteorites, comets, asteroids and moons. Knows the structure and functions of Earth systems. Describe components of the atmosphere and identify the role of atmospheric movement in weather change. 7th grade: Knows the components of our solar system. Identify and illustrate how the tilt of the Earth on its axis as it rotates and revolves around the Sun causes changes in seasons and length of day. Relate Earth's movement and the moon's orbit to the observed cyclical phases of the moon. 8th grade: Cycles exist in Earth systems. Analyze and predict the sequence of events in the lunar cycle. Relate the role of oceans to climatic changes. Knows characteristics of the universe. Describes characteristics of the universe such as stars and galaxies. Explain the use of light years to describe distances in the universe. Research and describe historical scientific theories of the origin of the universe.

Physics I: Motion and Forces *Course Objectives*	*TEKS: Concepts: The student knows:* *The student is expected to:*
Explain relationships between motions and forces. List the fundamental physical measurement quantities and SI units to express them. Apply dimensional analysis to assess expressions of physics solutions. Apply rules of determining significant figures. Convert units within the SI systems and between the SI and British measurement system. Designate points in a Cartesian coordinate system. Apply algebraic principles and basic trigonometric formulas to solve physics problems.	6th grade: There is a relationship between force and motion. Identify and describe the changes in position, direction of motion, and speed of an object when acted upon by force. Demonstrate that changes in motion can be measured and graphically represented. 7th grade: There is a relationship between force and motion. Demonstrate basic relationships between force and motion using simple machines including pulleys and levers. Demonstrate that an object will remain at rest or move at a constant speed and in a straight line if it is not being subjected to an unbalanced force. 8th grade: There is a relationship between force and motion. Demonstrate how unbalanced forces cause changes in the speed or direction of an object's motion.

Physics II: Energy, Forces, Motion *Course Objectives*	*TEKS: Concepts: The student knows:* *The student is expected to:*
Conduct activities and solve problems that measure motion, speed, velocity, change in velocity, distance, time, and acceleration. Define displacement in mathematical terms and apply the equation to determine average velocity and average acceleration. Conduct activities and interpret graphs of distance-time, speed-time, and acceleration-time. Compare and contrast vector and scalar quantities. Conduct activities and solve problems that require vector addition, subtraction, multiplication, and division by trigonometric functions. State and apply Newton's laws of motion. Conduct activities and use Newton's second law to solve problems involving force, mass and acceleration. Conduct activities and solve problems of compound motion. Identify and provide examples of types of forces. Conduct activities and use various methods to measure forces. Identify and provide examples of types of wave motions. Conduct activities and solve problems involving velocity, frequency wavelength and period.	6th grade: There is a relationship between force and motion. Identify and describe the changes in position, direction of motion, and speed of an object when acted upon by force. Demonstrate that changes in motion can be measured and graphically represented. 7th grade: There is a relationship between force and motion. Demonstrate basic relationships between force and motion using simple machines including pulleys and levers. Demonstrate that an object will remain at rest or move at a constant speed and in a straight line if it is not being subjected to an unbalanced force. Complex interactions occur between matter and energy. Illustrate examples of potential and kinetic energy in everyday life such as objects at rest. 8th grade: There is a relationship between force and motion. Demonstrate how unbalanced forces cause changes in the speed or direction of an object's motion.

APPENDIX B: STUDENT PERFORMANCE SCORES IN 8TH GRADE SCIENCE ASSESSED BY TAKS

*1998 TAKS Data**	*Met Minimum Standards*	*Economic Disadvantaged*	*At Risk*	*African American*	*Hispanic*	*Anglo*
District	89	82	80	84	85	95
Campus 1	78	77	75	61	76	96
Campus 2	92	88	87	86	89	97
Campus 3	90	89	78	79	88	97
Campus 4	84	86	78	79	60	83
Campus 5	96	90	89	79	94	97
Control 1	78	72	70	85	75	85
Control 2	93	88	85	86	92	94
1999						
District	89	82	81	88	84	95
Campus 1	81	81	75	84	79	91
Campus 2	92	87	84	95	88	96
Campus 3	90	90	86	97	85	95
Campus 4	78	74	71	67	77	94
Campus 5	95	85	89	88	92	97
Control 1	81	76	75	100	76	95
Control 2	92	89	87	89	91	94
2000						
District	93	81	86	93	90	97
Campus 1	88	86	78	100	84	96
Campus 2	96	93	91	98	95	98
Campus 3	92	87	85	91	89	98
Campus 4	88	87	83	92	87	94
Campus 5	96	94	91	96	93	98
Control 1	85	84	79	90	83	94
Control 2	94	92	90	100	92	96
2001						
District	96	94	93	94	95	99
Campus 1	96	95	94	100	95	97
Campus 2	99	98	97	100	98	98
Campus 3	95	93	94	93	97	93
Campus 4	93	92	89	Nr	87	93
Campus 5	98	95	98	100	97	99
Control 1	87	84	81	86	85	97
Control 2	97	97	93	100	94	97
2002						
District	95	91	89	92	93	98
Campus 1	93	92	88	92	93	97
Campus 2	98	97	94	96	97	97
Campus 3	95	93	90	91	95	97
Campus 4	88	87	81	88	87	100
Campus 5	98	93	95	90	97	100
Control 1	89	87	78	85	88	97
Control 2	94	93	86	87	93	97

REFERENCES

Abd-El-Khalick, F., Bell, R., & Lederman N. (1998). The nature of science and instructional practice: Making the unnatural natural. *Science Education, 82*(4), 417-437.

American Association for the Advancement of Science. (1989). *Science for all Americans*. New York: Oxford University Press.

American Association for the Advancement of Science. (1993). *Benchmarks for science literacy.* New York: Oxford University Press.

Brent, L., & Hodges, R. (1998, February). *Preservice teacher education research on the application of the scientist-practitioner model in an undergraduate program.* Paper presented at the annual meeting of the American Association of Colleges for Teacher Education New Orleans, LA.

Burry-Stock, J., & Casebeer, C. (2003, March). *A study of the alignment of national standards, state standards, and science assessment.* Paper presented at the annual meeting of the NARST, Philadelphia. ED 475159.

Darling-Hammond, L. (2004). Standards, accountability, and school reform. *Teachers College Record, 106*, 1047-1085.

Druva, C., & Anderson, R. (1983). Science teacher characteristics by teacher behavior and by student outcome: A meta-analysis of research. *Journal of Research in Science Teaching, 20*, 467-479.

Good, R., & Shymansky, J. (2001). Nature-of science literacy in Benchmarks and Standards: Post-modern/relativist or modern/relativist. *Science & Education, 10*, 173-185.

Kennedy, M. (1998). Education reform and subject matter knowledge. *Journal of Research in Science Teaching, 35*(3), 249-263.

Hargreaves, A., & Fullan, M. (1998). *What's worth fighting for beyond your school.* New York: Teachers College Press.

House, P. (1990). Mathematical connections: A long overdue standard. *School Science and Mathematics, 90*, 517.

Houston, R. (1989). Teacher education as a field of scholarly inquiry. Action in Teacher Education, *11*, 19-24.

Hubbard, R. S., & Miller, B. M. (1993). *The art of classroom inquiry.* Portsmouth, NH: Heinemann.

Lawson, A. E., Abraham, M. R., & Renner, J. W. (1989). *A theory of instruction: Using the learning cycle to teach science concepts and thinking skills* (NARST Monograph No. 1). Kansas State University: National Association for Research in Science Teaching.

Loucks-Horsley, S., Hewson, P. W., Love, N., & Stiles, K. E. (1998). *Designing professional development for teachers of science and mathematics*. Thousand Oaks, CA. Corwin Press.

National Research Council. (1996). *National science education standards*. Washington DC: National Academy Press.

National Research Council. (2001). *Educating teachers of science, mathematics and technology: New Practices for the new millenium.* Washington, DC: National Academy Press.

Putnam, R., & Borko, H. (2000). What do new views of knowledge and thinking have to say about research on teacher learning? *Educational Researcher, 21*, 4-15.

Shulman, L. (1986). Those who understand: Knowledge growth in teaching. *Educational Researcher, 15*, 4-14.

Sunal, D., Bland, J., Sunal, C., Whitaker, K., Freeman, M., Edwards, L., Johnson, R., & Odell, M. (2001). Teaching science in higher education: Faculty professional development and barriers to change. *School Science and Mathematics, 101*(5), 246-257.

Texas Education Agency. (1997). *Texas essential knowledge and skills. Middle school science* (19 TAC Chapter 112). Austin, Texas: Author.

CHAPTER 13

IMPACT OF SCIENCE STANDARDS ON CURRICULUM AND INSTRUCTION IN THE EARTH SCIENCES

Fred Finley and Larry Enochs

The impact of the *National Science Education Standards (NSES)* has been limited in K-12 earth science teaching in the United States. There were numerous reasons that K-12 curriculum and teaching moved away from the application and implementation of the national standards in the past decade. Because of this limited impact, the authors began to investigate the reasons for the lack of impact of *NSES* on earth science education. The authors found several features of the science standards that may be limiting their impact. These limiting features were found to be useful for those who work on revisions of the science standards, revisions that will be useful for successful implementation in K-12 classrooms.

INTRODUCTION

Most states do not require an earth science course for graduation (American Geological Institute, 2002). Given that state assessment of science

The Impact of State and National Standards on K-12 Science Teaching, 391–410

outcomes is generally standards-based, it follows that students are tested on the science education standards without necessarily having a stand-alone earth science course at the secondary level. This policy, at best, leads to the integration of earth science content into a general science courses. The results of state science assessments are not disaggregated to get a picture of student earth content knowledge alone. Rather, the state level assessments are of general science content outcomes.

The original purpose of this chapter was to consider the extent of the impacts of the *NSES* (National Research Council, NRC, 1996) on earth science education. However, as we considered that question, we learned quickly that the reports of impacts have been limited. Based on a search professional and governmental Internet Websites, including a sample of 25 state department of education Websites, this was unexpected given that the recently developed *NSES* were intended to guide science education in the United States for many years. The limited impact was especially surprising since there has not been such a rigorous attempt to set standards since the *National Education Association Report of the Committee of Ten of Secondary School Studies* with the reports of the conferences (National Education Association, 1894) and the U.S. Bureau of Education (1920), *Reports of the Commission of the Reorganization of Secondary Education*.

Given the limited impact of the standards, the purpose of the chapter became to examine the reasons for the lack of impact of *NSES* on earth science education. There were numerous reasons that K-12 curriculum and teaching moved away from the application and implementation of the standards in the past decade. One was that our attention shifted to matters of international security. A second reason was an economic down turn that limited the availability of resources that would be allocated to education. A third was the advent of conservative taxation policies that further limited educational resources. A fourth and more direct reason was the acceptance of the No Child Left Behind Act policy and the attendant emphasis on traditional testing (U.S. Department of Education, 2001). When testing took center stage, thoughtful deliberations regarding the curriculum—what to teach, instruction—how the teaching should be done, and how school systems should act to support the use of the standards became secondary matters at best.

These events almost certainly had dramatic impacts on the application and implementation of the *NSES*. However, as we investigated the standards in their own right, we found other reasons that are part of the explanation of the science standard's limited use that are attributable directly to the *NSES* document. The identification of problems inherent in the standards does not lead to the claim that the standards are poorly done, quite the opposite. In fact, we have assumed that the incredibly complex task of writing national science standards was exceptionally well done. Our effort was to identify the kinds of problems that remain in the hope that we can alert users to challenges they may face, thus making them easier to use. It is also our hope that what we found may be useful in future revisions of the *NSES*.

THE IMPACT OF THE *NATIONAL SCIENCE EDUCATION STANDARDS* ON EARTH SCIENCE EDUCATION

Given the dearth of actual studies on the ways in which earth science is typically taught in U. S. public schools, there is little empirical evidence on which to base claims about the impacts of the *NSES* on earth science education. However, given that the most commonly used textbooks usually stand as the teachers' curriculum guide, the texts can serve as a reasonable surrogate for empirical research on classroom teaching practices.

Substantive and extensive impacts of the *NSES* on secondary school earth science teaching have been almost nonexistent in mainstream textbooks. The most widely selling earth science textbooks and programs are still organized as if there are four separate and unrelated disciplines—geology, meteorology, astronomy and oceanography—to be taught. The texts remain encyclopedic in nature, give a little attention to many topics, and employ a topic-by-topic organization that lacks any semblance of conceptual integration (American Association for the Advancement of Science, 2002).

The only changes in the mainstream textbooks and programs that may by related to the advent of *NSES* have been modest at best and include such things as the:

1. Addition of textbook and program overviews stating that the earth can be considered as a system in which there are many interactions.
2. Occasional use of the word system in chapter and section headings.
3. Occasional reference to the *parts of a system* or processes of the system in the text itself.
4. Referencing various chapters and sections of the textbooks to the standards.

Perhaps the only substantive change has been an increased emphasis on including earth hazards as an identifiable topic and the limited but noteworthy addition of considering human impacts on the environment. Another important change has been an updating of the actual content to account for changes in our knowledge of the earth. However, this is probably not attributable to the science standards as much as it is attributable to the usual sorts of textbook revisions that occur as new editions are developed.

By and large the textbooks and programs or assessments that are most commonly used today have not been substantially influenced by *NSES*. There are, however, notable exceptions such as *EarthComm*, *NASA's Earth Observatory*, the Globe Program, *Exploring Earth: Infusing Innovation into Mainstream Classrooms*, and *Visualizing Earth* that are described in *Revolution in Earth and Space Science Education* (Barstow, 2002), and the new 2005 edition of the *Pearson/Prentice Hall Earth Science* textbook.

EARTH SCIENCE EDUCATION IN THE *NATIONAL SCIENCE EDUCATION STANDARDS*

Substantial portions of the *NSES* (1996) are related to earth science education, most notably the section on content standards. That section includes standards for unifying concepts and processes in science, science as inquiry, earth and space science, physical science, life science, science and technology, science in personal and social perspectives, and the history and nature of science.

There are many standards within each of these categories that are relevant to earth sciences. In fact, nearly all the content standards could be considered as related to earth science education in one way or another. Examples of important interrelationships of general science standards in *NSES* (NRC, 1996) that should guide earth science education in our nation's classrooms follow. For additional details see Appendix.

Properties of earth materials (Grades K-4) (p. 134)
Structure of the earth system (Grades 5-8) (pp. 159-161)
Energy in the earth System (Grades 9-12) (p. 189)
Geochemical cycles (Grades 9-12) (p. 189)
The origin and evolution of the earth system (Grades 9-12) (pp. 198-190)

One additional standard from another category was included. That standard was from the unifying concepts and processes in science category (pp. 115-119) that describes the integrative schemes that bring together students' experiences in science education across grades K-12. The essential standard form this category is using the concept of systems. According to the standards,

> The natural and designed world is complex; it is too large and complicated to investigate and comprehend all at once. Scientists and students learn to define small portions for the convenience of investigation. The units of investigation can be referred to as 'systems.' A system is an organized group of related objects or components that form a whole. (p. 116)

The system concept is essential because the community of earth scientists is using the concept to redefine their research. To determine interrelationships of the other science standards with the earth sciences, we used the systems concept to develop a tool for analysis.

THE FRAMEWORK FOR CRITICISM AND SUGGESTION

"Why have the standards had so minimal an impact?" Addressing that question required a statement of the principles of developing curriculum standards that we used and an analytic framework against which the specific content standards could be compared.

The principles of developing curriculum standards we used that standards should be:

1. A current reflection of the structure and organization of the disciplines. The reflection involves using the themes that are presently used by the scientific community to organize its knowledge and research activities.
2. Grounded in the disciplines from which they are extracted.
3. Conceptually coherent when each is considered separately.
4. Conceptually coherent when considered as a set within each grade level grouping.
5. Conceptually coherent across grade level groupings.
6. Parsed among grade level groupings according to the "best we know" about students' abilities to learn them.

The analytical framework that is used in the analysis and in formulating our tool is based on our attempt to represent what has become established as conceptual framework for research in the earth sciences. The central idea is that the earth is a set of interacting systems. This idea is not new within the scientific community, as evidenced by the 1988 National Aeronautics and Space Administration (NASA) document: *Earth System Science: A Closer View* (1988). In that document, NASA provides an endorsement through its International Biosphere-Geosphere Program which was designed to

> describe and understand the interactive physical, chemical, and biological processes that regulate the total Earth system, the unique environment it provides for life, the changes that are occurring in that system, and the manner by which these changes are influenced by human actions.

Earth science groups in education have also recommended the use of this theme Mayer (2002), and the American Geological Institute (Finley & Heller, 1991). Barstow (2002) stated that innovative educational "experiences help students to understand Earth as dynamic system—rather than simply a collection of topics to read about" (p. 6).

EARTH SYSTEMS SCIENCE CONCEPTUAL FRAMEWORK

The theme used in developing the Earth Systems Science Conceptual Framework (ESSCF) is that the Earth can be understood as a set of interacting natural systems (the lithosphere, atmosphere, hydrosphere, and biosphere) and social systems (e.g., agricultural, economic, legal, communications, transportation, moral, political, and cultural). The subthemes are:

- The Earth has evolved over a period of 4.6 billion years and will continue to do so.
- Changes in the natural systems can be described in terms of the transfers and transformations of matter and energy within and across systems.
- The rates of change in the natural systems have ranged from virtually instantaneous to nearly imperceptible.
- The dimensions of the changes in natural systems range from submicroscopic to global.
- Humans are a part of the Earth system, in particular the biosphere, and have created the social systems.
- Changes in social systems can be described largely in terms of humans' transformations and transfers of natural resources.

Given the above theme, what is needed is a conceptual framework that can be used to describe and explain earth science phenomena.

The approach that was taken was to create a set of analytic concepts that are the types of concepts involved in descriptions, explanations and predictions of natural phenomena that can be characterized in terms of systems (Finley, 1981; Finley & Stewart, 1982). The natural systems can be described in terms of the following types of concepts:

- *Materials*—the physical materials that are the basic components of the systems, for example, rocks, minerals and water.
- *Structures*—the identifiable physical units into which the materials are arranged, for example, rock layers, atmospheric layers, and lakes,
- *Intrasystem processes* that change or move materials and structures within the system, for example, water, flow in streams, weathering, convection, and wind.
- *Intersystem processes* that change and move the materials from one system to another, for example, evaporation, volcanism, and $CaCo_3$ precipitation from seawater.
- The *forms of energy* that drive or result from the intra and intersystem processes, for example, thermal, potential-kinetic, and nuclear.
- *Variables* that describe the materials, structures, processes and energy transfers and transformations, for example, mineral hardness, aquifer depletion and recharge rates, temperature and pressure.
- *Relational rules for variables*—quantitative and qualitative propositions (often called scientific laws) that provide specific relationships among variables, for example, slower cooling rates in otherwise comparable magma result in larger mineral grain sizes; $T = (N+1)/M$ where T is the recurrence interval of floods of a certain magnitude, n is the number of years over which the records are available and M is the rank of the severity or magnitude of the flood in terms of discharge rates.
- *Models*—physical models, maps, digitally constructed images, and pictures that show relationships among materials, structures and variables.

This set of analytic concepts is considered to represent the set of actual scientific concepts that are involved in understanding natural phenomena in terms of systems. Complete and coherent descriptions, explanations and justified predictions of earth system phenomena can be expected to include most, if not a complete set of these analytic concepts. For example, atmospheric systems cannot be adequately understood with out reference to materials, structures, variables, processes, *laws*, and forms of energy, and models of that system. If one were to develop curriculum related to the atmosphere, then the analytic concepts would indicate the kinds of concepts needed. Conversely, if one analyzed a curriculum, or in the present case curriculum standards, the absence of such concepts would indicate that the curriculum was incomplete. This is not to say that every curriculum or set of standards must have every type of concept to be good. Some types of concepts may be intentionally omitted because they are too complex for the students or because only some particular portion of an earth system is being taught with other parts to be taught at other times. However, if their absence is discovered the question of why they were not included must be considered.

While the above provides a list of what kinds of concepts should or at least could be included in curriculum standards, it does not describe relationships among the concepts clearly. An alternative form of presenting the analytic concepts is as an analytic concept map. The use of this map is as a basis for evaluating a specific scheme of earth science concepts. If the scheme of actual earth science concepts were congruent with the analytical concept map, one could claim that the earth science concepts were complete and conceptually coherent. Conversely, if they differed, then one would be able to identify what was missing—specific concepts or relationships among concepts. Figure 13.1 is the analytical concept map that was used in the analysis of the Earth science portion of the national education standards. The next section presents the comparisons that were judged to be informative with respect to the question of why the impact of the standards has been so marginal.

ANALYSIS OF THE STANDARDS

The analysis of the standards was conducted by comparing a concept of selected *NSES* statements to the ESSCF to determine what claims could be made about the quality of the science standard under consideration. The first standard we considered was the K-4 grade set, properties of earth materials (*NSES*, 1996, p. 134). The concept map is shown in Figure 13.2. Several features of this standard are notable.

1. The earth description of earth materials is limited to solids, liquids and gases with soils and rocks as the specific solids. The variables used to describe the materials are only vaguely referenced as physical and chemical properties. Given the age of the children in this grade level group, this is acceptable. However, what is not evident is that the nature of the earth's materials and the variables that describe them are never elaborated in the more advanced science standards. If the standards were integrated across grade levels, then more sophisticated concepts about the earth's materials should be presented later.
2. There is an off-topic segment on the idea that soils support plants and animals the older of which inform us about past environments. Allocating instructional time to this idea when the students at this

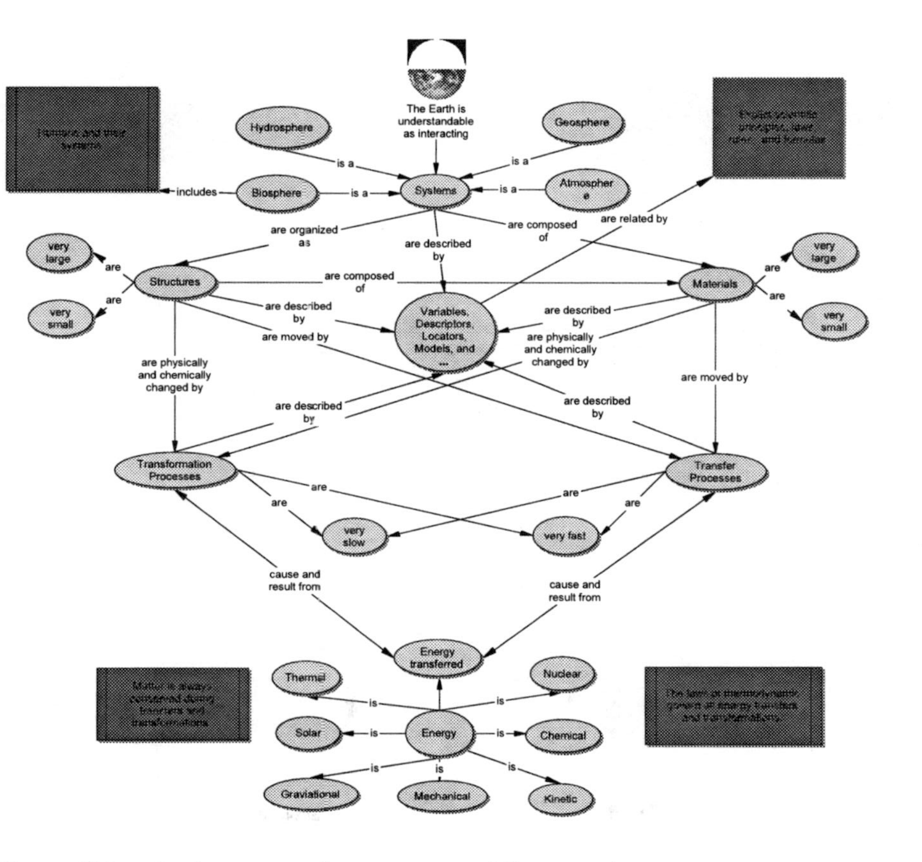

Figure 13.1. Earth systems science conceptual framework map.

age cannot understand geologic time seems inappropriate. A better understanding of properties of the Earth's materials would have been more consistent with the overall intent of the standard.

3. The section on human interactions with the environment, the dependence of humans on the Earth's materials seems valuable in that it is directly related to standards from the science in personal and social perspectives (*NSES*, 1996, pp. 138-141) sections of the content standards.
4. There is no mention of the unifying concept of a system or apparent organization of what is included that seems to be drawn from that concept.

The second standard we include here is the structure of the earth system (Grades 5-8) (pp. 159-161). Of all the standards, this one is perhaps the most fundamental and important in that the ideas included are those to which most others would be related. It also may be the most important in that grades 5-8 are the grades in which most earth science is taught. The full concept map for this standard includes four parts, the ideas to be taught about the geosphere, hydrosphere, atmosphere and biosphere. Only two of these are presented here because the comments about the others would repeat the same comments and criticisms that become evident by examining the geosphere (Figure 13.3) and atmosphere maps (Figure 13.4). Comments about this standard also indicate ways in which the standards are valid and yet may be somewhat problematic.

1. This map shows a comprehensive presentation of the geospheric portion of the Earth's structure in that the major subsurface structures (lithosphere, mantle, core, and plates) are included. Plate movement is shown as the mechanism for mountain formation, earth quakes, and volcanoes. Convection within the mantle is presented as causing the plate motion. However, the lack of any mention of what occurs at plate boundaries renders the presentation incomplete in a critical way as shown in the next comment.
2. The rock cycle, a traditional and long standing idea from geology is also represented. However, there are at least three significant problems. First, the rock cycle does not show that various rock types, in addition to sediments, are involved. Second, the standard indicates that only sediments are buried. Igneous, metamorphic and sedimentary rocks can and frequently are buried and returned to the Earth's surface. Third, not all buried materials are recrystallized before being returned to the surface. Entire mountain ranges are primarily sedimentary rocks that have not been recrystallized.

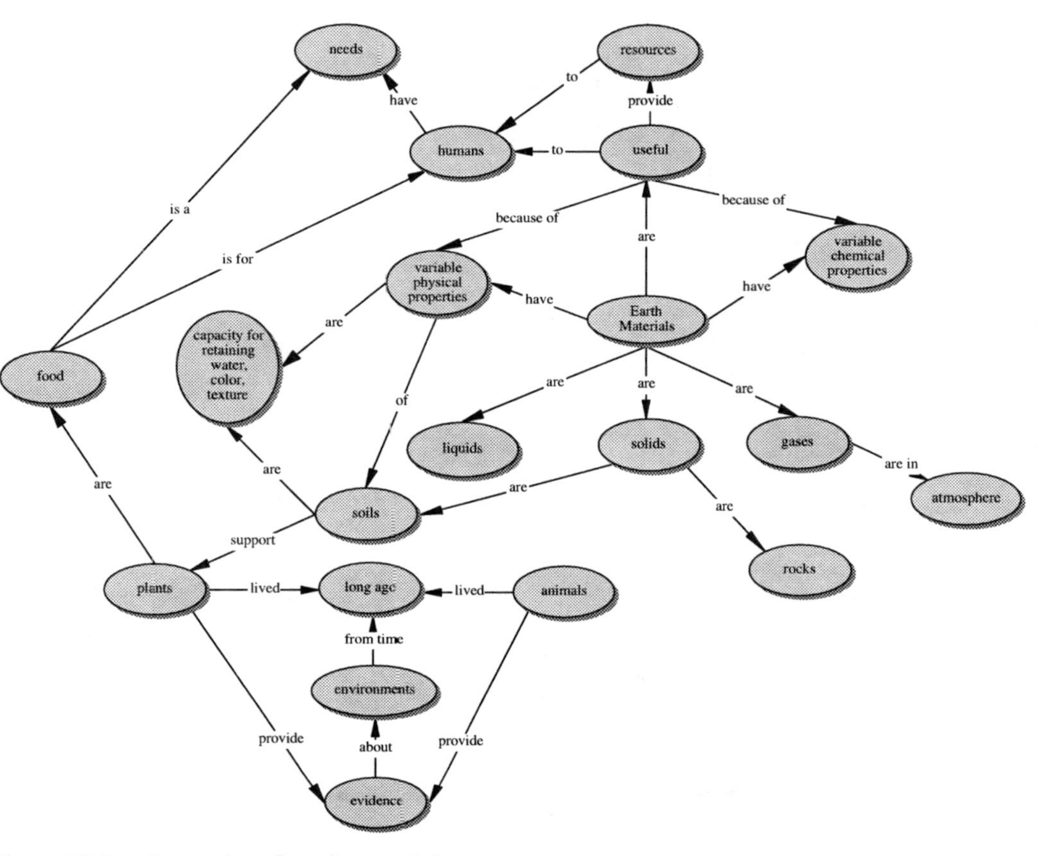

Figure 13.2. Properties of earth materials.

Fourth, and most important, the rock cycle is not clearly related to the idea of plate movements. Plate movements result in collisions, subduction, and rifting that cycle the Earth's materials. This problem indicates that the geosphere standards omit critical information and do not relate one set of ideas, plate tectonics, to the another, the rock cycle as well as is required for students to understand the structure of the earth.

3. The structure of the Earth system cannot be considered properly without more information being included about the Earth's materials. At the very least igneous, metamorphic and sedimentary rocks, their properties, and the variables that make them identifiable must be included. This is essential if, as is indicated in other standards, students are to be able to inquire about the Earth systems. The primary information needed to understand our planet is in the concepts we use to describe and interpret the rocks we see.
4. The concept map does not provide much guidance about what *landforms* are important in addition to mountains and volcanoes. Rivers, desserts, valleys, rivers, lakes, coast lines, and various glacial features are important in understanding the earth system. These surface features and others are especially in the area of understanding human interactions with the Earth. Where we live and what we do there is largely a matter of the surface features and landforms that are present in the area. It is difficult to see how this critical human-Earth interactions aspect of the national standards can be implemented with out key concepts related to landforms and surface features.

The concept map of the atmosphere is instructive in a way that is different from the others. The atmosphere (as are the other spheres) is a complex topic. One problem revealed here is that the standards are sometimes vague or at too high a level of generality. For example, the idea that the atmosphere has different properties at different elevations does not provide a curriculum developer much guidance as to what properties are essential. Composition, temperatures, density, the altitude of the boundaries between layers, layer thicknesses, differences within layers such as those at the poles verses the equator, and many other variables could be considered. Similarly, a statement like clouds affect weather, or heat held in the oceans effects weather, are too general to be useful. This is the case not only for the curriculum developer, but also for students. These statements do not provide students the conceptual tools they need to describe, explain, predict, or inquire about atmospheric phenomena.

The intention of the standards to provide guidelines that reduce the she quantity of information students are to be taught is important. There

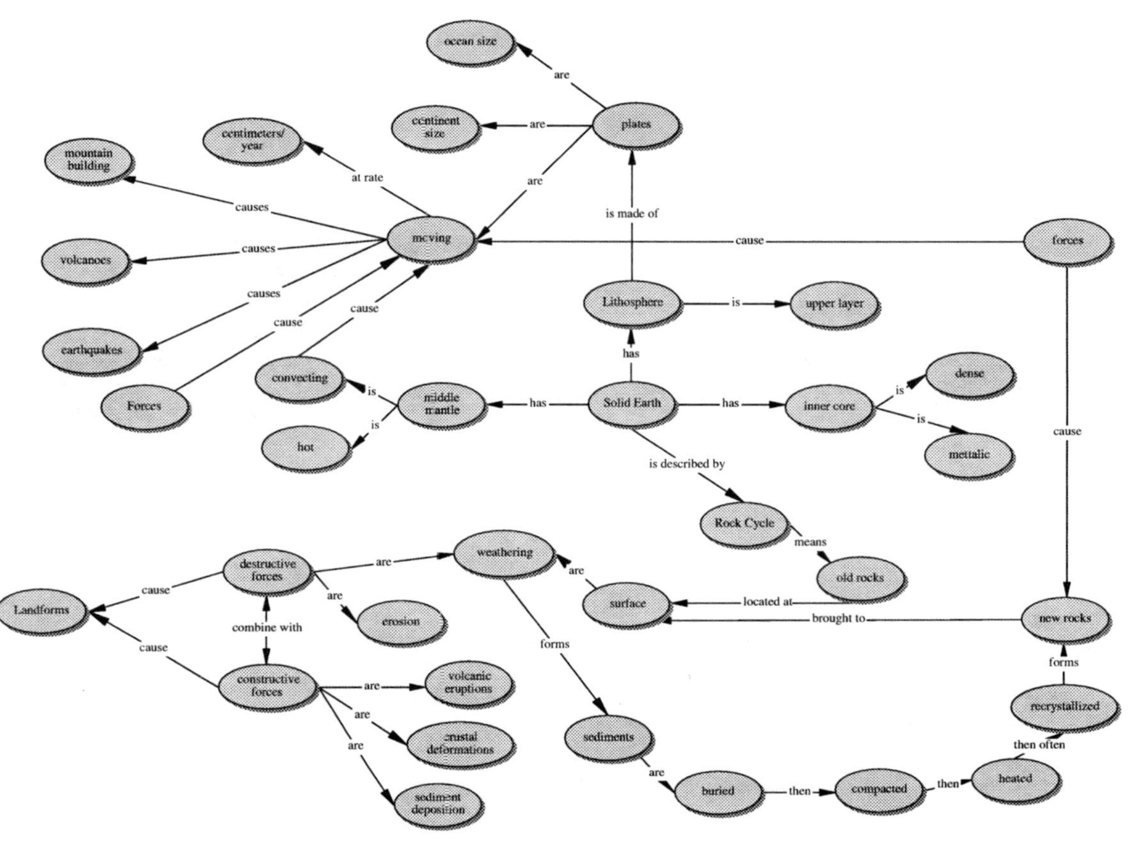

Figure 13.3. Structure of the earth system: Geosphere.

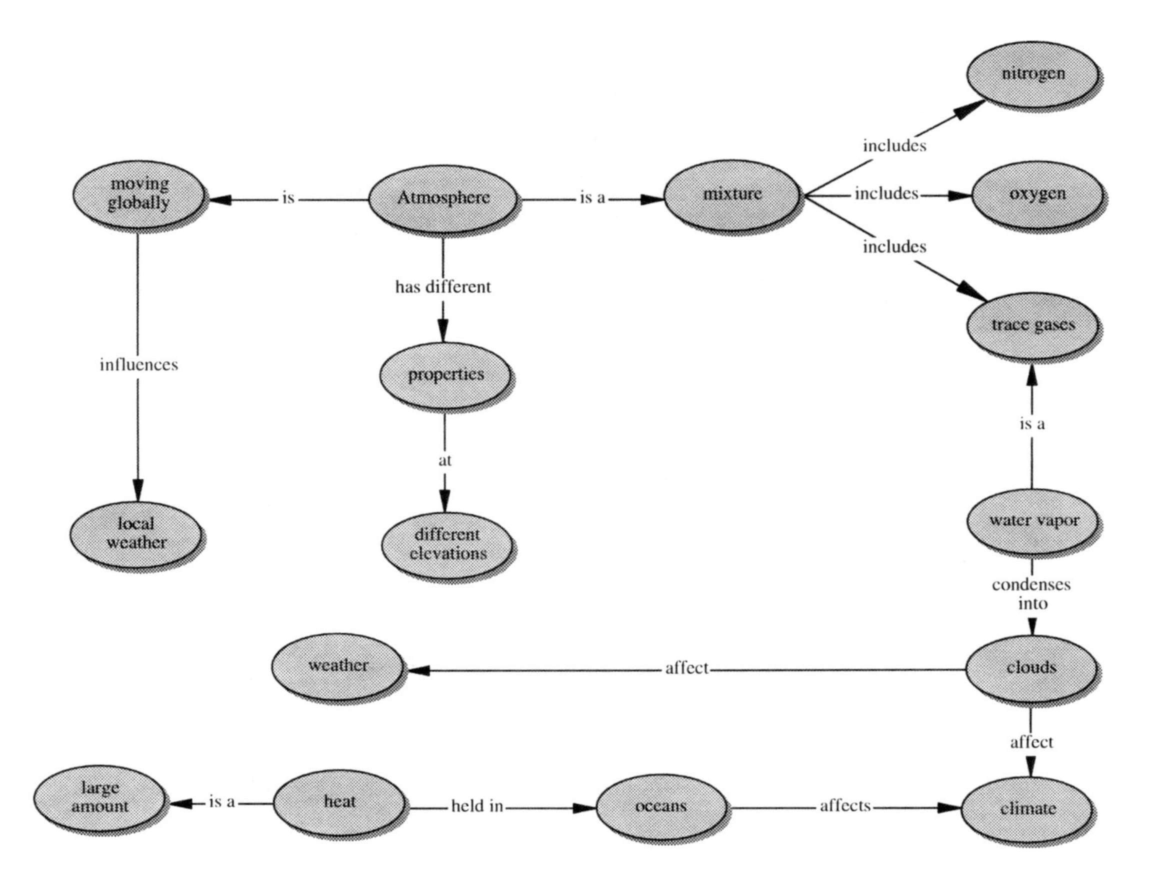

Figure 13.4. Structure of the earth system: Atmosphere.

is no argument with the idea that we try to teach far too much too fast. However, if the response to that problem is to create standards that are as general as these, then we run the risk of not providing students with ideas that are powerful enough to allow them to be scientifically literate.

DISCUSSION AND IMPLICATIONS

The original intent of this chapter was to describe the impact of *NSES* on earth science education. In the absence of research on the impacts, we instead developed an argument that the standards have had little impact to-date based on examining the textbooks that are currently in use. We then attempted to try and understand if there were features of the standards themselves that may contribute to their having a limited impact. Selected standards were examined under guidelines for what would constitute good standards and an analytic conceptual framework for earth science education.

Our findings are the following:

1. A current reflection of the structure and organization of the earth sciences as a set of interacting systems was not used as explicitly as would be desirable. First, the terms associated with thinking about natural systems were seldom used. Second, when the concept maps of the *NSES* were compared to the ESSCF, conceptual components of systems, for example the variables needed to describe the materials and structure and process of systems were limited. This is somewhat surprising in that one of the unifying concepts was systems.
2. The standards we examined were by and large accurate when compared to the disciplines in which they are grounded. Yet, the presentation of the rock cycle was wrong in that it indicated that sediment and no other kinds of earth materials were cycled. Errors such as this can tax the credibility and utility of the standards. A more subtle problem was the struggle to be accurate and yet understandable. An example of this is the way in which the term heat is used in the grades 9-12 *NSES* (p. 189), which was examined but is not shown earlier in the chapter. The concept of heat that is used in the description seems to be more of a layman's meaning than a scientifically accepted one. This idea is mentioned only because, if we are not careful, our own use of less than scientific terms can contribute to establishing or reinforcing students' misconceptions.

3. There were problems with the conceptual coherence of each standard taken on its own. Occasionally, there was as segment of a concept map that was at best weakly related to the rest of the ideas. Critical connections such as the connection between the rock cycle and plate tectonics were also not included. This problem is critical if it is wide spread throughout the standards. Anyone using the science standards to guide their educational practices needs access to standards that are conceptually coherent.
4. We found places where the standards were missing critical ideas. The best example of this is the absence of including igneous, metamorphic and sedimentary rock types in any graded level grouping. This kind of omission cannot be set aside by stating that *it would be too much to include*. When critical concepts are missing, the ultimate goal of developing scientifically literate students cannot be attained. Standards must be complete and specific enough to provide curriculum developers, teachers and assessment designers with concrete guidance regarding what students should learn so that they can describe, explain, predict and inquire about natural phenomena.
5. Finally, we saw instances where key words used in the standards were too general to be useful. This was the case when the terms *physical properties* and *chemical properties* were used and in the very general nature of the standards about the atmosphere. If students are to come to understand the earth system, then learning specific critical concepts and not their placeholders is required.

We found several features of the science standards that may be limiting their impact. However, this is not to say finding problematic features with some of the standards implies that the standards are fundamentally or extensively flawed. We did not complete a systematic analysis and accounting of how often, when, and where the potentially limiting features occurred. Instead all we can claim is that there are matters to be considered by those who use *NSES* to guide professional decisions. Perhaps being aware of the kinds of problems we identified will go some way toward making their use easier. Those who use them can at least know some of the challenges they will need to meet. We also hope that what we discovered will be useful to those who work on revisions of the science standards, a task that the original authors envisioned and endorsed. If this is undertaken, then perhaps using our guidelines for what science standards must do and using an analytic conceptual framework such as the systems framework we employed, will make the revision easier and even better than the excellent work that has been already done.

APPENDIX

Details of interrelationships between the general standards in *NSES* and the earth science.

Properties of Earth Materials (Grades K-4)

- Earth materials are solid rocks and soils, water, and the gases of the atmosphere. The varied materials have different physical and chemical properties, which make them useful in different ways, for example, as building materials, as sources of fuel, or for growing the plants we use as food. Earth materials provide many of the resources that humans use.
- Soils have properties of color and texture, capacity to retain water, and ability to support the growth of many kinds of plants, including those in our food supply.
- Fossils provide evidence about the plants and animals that lived long ago and the nature of the environment at that time.

Structure of the Earth System (Grades 5-8)

- The solid earth is layered with a lithosphere; hot, convecting mantle; and dense, metallic core.
- Lithospheric plates on the scales of continents and oceans constantly move at rates of centimeters per year in response to movements in the mantle. Major geological events, such as earthquakes, volcanic eruptions, and mountain building, result from these plate motions.
- Landforms are the result of a combination of constructive and destructive forces. Constructive forces include crustal deformation, volcanic eruption, and deposition of sediment, while destructive forces include weathering and erosion.
- Some changes in the solid earth can be described as the *rock cycle*. Old rocks at the earth's surface weather, forming sediments that are buried, then compacted, heated, and often recrystallized into new rock. Eventually, those new rocks may be brought to the surface by the forces that drive plate motions, and the rock cycle continues.
- Soil consists of weathered rocks and decomposed organic material from dead plants, animals, and bacteria. Soils are often found in

layers, with each having a different chemical composition and texture.

- Water, which covers the majority of the earth's surface, circulates through the crust, oceans, and atmosphere in what is known as the *water cycle*. Water evaporates from the earth's surface, rises and cools as it moves to higher elevations, condenses as rain or snow, and falls to the surface where it collects in lakes, oceans, soil, and in rocks underground.
- Water is a solvent. As it passes through the water cycle it dissolves minerals and gases and carries them to the oceans.
- The atmosphere is a mixture of nitrogen, oxygen, and trace gases that include water vapor. The atmosphere has different properties at different elevations.
- Clouds, formed by the condensation of water vapor, affect weather and climate.
- Global patterns of atmospheric movement influence local weather. Oceans have a major effect on climate, because water in the oceans holds a large amount of heat.
- Living organisms have played many roles in the earth system, including affecting the composition of the atmosphere, producing some types of rocks, and contributing to the weathering of rocks.

Energy In the Earth System (Grades 9-12)

- Earth systems have internal and external sources of energy, both of which create heat. The sun is the major external source of energy. Two primary sources of internal energy are the decay of radioactive isotopes and the gravitational energy from the earth's original formation.
- The outward transfer of earth's internal heat drives convection circulation in the mantle that propels the plates comprising earth's surface across the face of the globe.
- Heating of earth's surface and atmosphere by the sun drives convection within the atmosphere and oceans, producing winds and ocean currents.
- Global climate is determined by energy transfer from the sun at and near the earth's surface. This energy transfer is influenced by dynamic processes such as cloud cover and the earth's rotation, and static conditions such as the position of mountain ranges and oceans.

Geochemical Cycles

- The earth is a system containing essentially a fixed amount of each stable chemical atom or element. Each element can exist in several different chemical reservoirs. Each element on earth moves among reservoirs in the solid earth, oceans, atmosphere, and organisms as part of geochemical cycles.
- Movement of matter between reservoirs is driven by the earth's internal and external sources of energy. These movements are often accompanied by a change in the physical and chemical properties of the matter. Carbon, for example, occurs in carbonate rocks such as limestone, in the atmosphere as carbon dioxide gas, in water as dissolved carbon dioxide, and in all organisms as complex molecules that control the chemistry of life.

The Origin and Evolution of the Earth System (Grades 9-12)

- The sun, the earth, and the rest of the solar system formed from a nebular cloud of dust and gas 4.6 billion years ago. The early earth was very different from the planet we live on today.
- Geologic time can be estimated by observing rock sequences and using fossils to correlate the sequences at various locations. Current methods include using the known decay rates of radioactive isotopes present in rocks to measure the time since the rock was formed.
- Interactions among the solid earth, the oceans, the atmosphere, and organisms have resulted in the ongoing evolution of the earth system. We can observe some changes such as earthquakes and volcanic eruptions on a human time scale, but many processes such as mountain building and plate movements take place over hundreds of millions of years.
- Evidence for one-celled forms of life—the bacteria—extends back more than 3.5 billion years. The evolution of life caused dramatic changes in the composition of the earth's atmosphere, which did not originally contain oxygen.

REFERENCES

American Association for the Advancement of Science. (2002). *How well do middle school science programs measure up? Findings from Project 2061's curriculum review*

Retrieved February 15, 2005, from http://www.project2061.org/tools/textbook/mgsci/readings.htm

American Geological Institute. (2002). *2002 National status report on K-12 earth science education, AGI report.* Arlington, VA: Author.

Barstow, D. (Ed.). (2002). *Revolution in earth science and space science education*. Cambridge, MA: TERC.

Finley, F. N. (1981). A philosophical approach to describing science content: An example from geologic classification. *Science Education, 65*(5), 513-519.

Finley, F. N., & Heller, R. (1991). *Earth science education for the 21st Century,* Washington, DC: National Center for Earth Science Education, American Geological Institute.

Finley, F. N., & Stewart, J. (1982). Representing substantive structures. *Science Education*, *66*(4), 593-611.

Mayer, V. (Ed.). (2002). *Global science literacy.* Dordrecht, Netherlands: Kluwer Academic.

National Education Association. (1894). *Report of the committee of ten of secondary school studies with the reports of the conferences arranged by the committee*. New York: American Book.

National Aeronautics and Space Administration Advisory Council. (1988). *Earth system science: A closer view.* Washington, DC: Author.

National Research Council. (1996). *National science education standards*. Washington, DC: National Academy Press.

U.S. Bureau of Education. (1920). *Reports of the commission of the reorganization of secondary education, Bulletin 26*. Washington, DC: Author.

U.S. Department of Education. (2001, August 21). *No child left behind.* Retrieved August 8, 2003, from http://www.ed.gov/offices/OESE/esea/nclb/titlepage.html

CHAPTER 14

THE INFLUENCE OF SCIENCE STANDARDS AND REGULATION ON SCIENCE TEACHER QUALITY AND CURRICULUM RENEWAL

An Australian Perspective

Warren Beasley

According to Darling-Hammond (2001), the United States is characterized by a "morass of teaching standards" each set developed by different groups often for different purposes and often in isolation of each other. Australia is emerging every bit as complex. There are many calls for national standards frameworks. Many are arguing for national frameworks for establishing standards and using them for licensing/registration and certification, which draw on other professions such as medicine, engineering, and architecture (Ingvarson, 2002). In Australia, the professional standards discussions and debates sometimes run parallel to each other, often with no resolutions. Probably the largest debate relates to how standards should be used. These debates during the late 1990s for the establishment of professional teaching

The Impact of State and National Standards on K-12 Science Teaching, 411–428

standards, like in the United States and the United Kingdom, were premised on the political belief that linking student learning outcomes and teacher quality will enhance the image and status of teachers in the eyes of the community at large. The first part of this chapter concentrates on this issue of professional teaching standards, their development and ownership, and the competing models from professional associations and major employer groups. The second half is focused on curriculum and assessment standards that have a much longer public history than teaching standards.

INTRODUCTION

Debates in Australia during the late 1990s for the establishment of professional teaching standards, like in the United States and the United Kingdon, were premised on the political belief that linking student learning outcomes and teacher quality will enhance the image and status of teachers in the eyes of the community at large. This followed a period in the 1980s and early 1990s when educators in Australia, the U.S. and the United Kingdon were considering competencies in teachers work (Kennedy, 1993). Even though it was acknowledged that teaching is a very complex activity, professional bodies, registration authorities and academics set about developing competencies for teacher's work (National Project for the Quality of Teaching and Learning, 1996).

Competencies were criticized for their potential to technicize and decontextualize the work of teachers (Hattam & Smyth, 1995). Many also questioned whether competencies could be educative and at the same time useful for teacher appraisal (Kennedy, 1993). The teacher competencies agenda never really came fully to fruition (Louden, 2000); however, the international discourse about competencies was soon replaced by standards (Reynolds, 1999). The first part of this chapter concentrates on the issue of professional teaching standards, their development and ownership, and the competing models from professional associations and major employer groups. The second half is focused on curriculum and assessment standards that have a much longer public history than teaching standards.

TEACHER PROFESSIONALISM, STANDARDS AND REGULATION

Australia is a commonwealth of states and as a consequence the educational process is governed by state legislation. Even though the political responsibility for schooling resides constitutionally with state governments, the federal government has a strong hold on the financial strings. It is this financial influence that allows the Australian government to pursue a national agenda in education by using financial grants to states as leverage for state governments to implement a national vision of education.

Kennedy (2002) claims that the development of standards has been compromised by "the complexity of the political contexts around teachers and teacher education in this country" (p. 2). For example, the Australian senate inquiry into the status of the teaching profession (Senate Employment, Education and References Committee Inquiry, 1998) focused particularly on professional standards and described the characteristics of professionalism as including control of standards, admission, career paths, and disciplinary issues. The federal government has attempted to exercise the kind of influence on the teaching profession that in other professions would be the prerogative of the profession itself (Goodrum, Hackling, & Rennie, 2001).

According to Darling-Hammond (2001), the U.S. is characterized by a "morass of teaching standards" each set developed by different groups often for different purposes and often in isolation of each other. Australia is emerging every bit as complex. There are many calls for national standards frameworks. Many are arguing for national frameworks for establishing science standards and using them for licensing/registration and certification, which draw on other professions such as medicine, engineering, and architecture (Ingvarson, 2002). The involvement of the large professional associations in these projects leads Ingvarson and Wright (1999) to suggest that:

> Perhaps this represents a critical mass that will lead eventually to a truly national professional body for teachers, embracing, of course, employer and union stakeholders as well. Its functions would stand outside, but complement, the role of industrial relations in making determinations about pay systems, career structures and working conditions. (p. 29)

In Australia, the professional standards discussions and debates sometimes run parallel to each other, often with no resolutions. Probably the largest debate relates to how standards should be used. There are many suggestions:

- as quality assurance and accountability mechanisms in initial teacher education;
- as frameworks to guide induction programs;
- as reflective tools;
- as frameworks for determining and prioritizing areas of professional growth;

- for assisting selection and participation in professional development activities;
- for performance management frameworks (teachers know exactly what they are expected to do for performance appraisal (to judge the standard of performance)
- for selection of teachers for higher levels of accreditation/appointment—linked to career structures and incentives. (Mayer, Mitchell, Macdonald, Land, & Luke, 2002, p. 6)

As yet, no consensus has been reached.

PROFESSIONAL TEACHING STANDARDS: COMPETING MODELS

There now exists general agreement across many stakeholders such as the unions, professional associations, bureaucrats and politicians (state and federal) about the value of professional standards, but the structure and content of these standards remains problematic (Department of Education, Science and Training, 2003). There is little agreement about who should develop them, how they should be authenticated, and whether successful teachers should be rewarded.

The Australian Science Teachers Association (ASTA) (2002) has developed a three dimensional model which is labeled *Professional Standards for Highly Accomplished Teachers of Science*. A summary of the ASTA standard is reproduced in Table 14.1.

In comparison the National Science Teachers Association (NSTA) (2003) *Standards for Science Teacher Preparation* are claimed to be consistent with the vision of the *National Science Education Standards* (National Research Council, 1996). The NSTA sees teachers of science at all grade levels demonstrating competencies consistent with the achievement of this vision. These NSTA competencies are intended as the foundation for a performance assessment system. The competencies address the knowledge, skills and dispositions that are deemed important by the NSTA for teachers in the field of science and are claimed by NSTA to be also consistent with the standards of the National Board for Professional Teaching Standards (NBPTS) and the Interstate New Teachers Assessment and Support Consortium. The science teaching competencies are detailed over 42 pages of text and are classified under the headings:

1. Content
2. Nature of Science
3. Inquiry

Table 14.1. A Summary of the Professional Teaching Standards Developed by the Australian Science Teachers Association

A. Professional Knowledge

Highly accomplished teachers of science have an extensive knowledge of science, science education and. students.

1. They have a broad and current knowledge of science and science curricula, related to the nature of their teaching responsibilities.
2. They have a broad and current knowledge of teaching, learning and assessment in science.
3. They know their students well and they understand the influence of cultural, developmental, gender and other contextual factors on their students' learning in science.

B. Professional Practice

Highly accomplished teachers of science work with their students to achieve high quality learning outcomes in science.

4. They design coherent learning programs appropriate for their students' needs and interests.
5. They create and maintain intellectually challenging, emotionally supportive and physically safe learning environments.
6. They engage students in generating, constructing and testing scientific knowledge by collecting, analyzing and evaluating evidence.
7. They continually look for and implement ways of the major ideas of science.
8. They develop in students the confidence and ability to use scientific knowledge and processes to make informed decisions.
9. They use a wide variety of strategies, coherent with learning goals, to monitor and assess students' learning and provide effective feedback.

C. Professional Attributes

Highly accomplished teachers of science are reflective, committed to improvement and active members of their professional community.

10. They analyze, evaluate and refine their teaching practice to improve student learning.
11. They work collegially, within their school community and wider professional communities to improve the quality and effectiveness of science education.

4. Issues
5. General skills of Teaching
6. Curriculum
7. Science in the Community
8. Assessment
9. Safety and Welfare

These comprehensive lists of desired teacher behaviors accompanied by rationales together represent a teacher competency model to science teacher development. These competencies are designed to inform the teacher assessment processes for the accreditation or certification of science teachers nationally. Beyer (2002) cautions an approach to teaching in this system is based on a technical-rational model of teacher behavior and largely ignores social, political and philosophical understandings.

In contrast the major employer groups in Australia, that is the state departments of education, are creating models which are much more generic than those produced by the professional teacher associations. The Queensland state department of education is at the forefront of employer-generated standards and has subsequently documented its version of professional standards for teachers (The State of Queensland Department of Education, 2002). The actual standards are articulated under the twelve headings:

1. Structure flexible and innovative learning experiences for individuals and groups.
2. Contribute to language, literacy and numeracy development.
3. Construct intellectually challenging learning experiences.
4. Construct relevant learning experiences that connect with the world beyond school.
5. Construct inclusive and participatory learning experiences.
6. Integrate information and communication technologies to enhance student learning.
7. Assess and report on student learning.
8. Support the social development and participation of young people.
9. Build relationships with the wider community.
10. Contribute to professional teams.
11. Commit to professional practice.

The standards found under the headings are intended to be used as a framework to review teaching practice, to formulate goals to strengthen teaching practice, to establish personal and team professional learning and development plans, and to monitor the achievement of professional development goals (Smith, 2002). A joint Education Queensland–Queensland Teachers' Union task force now steers the standards program following a pilot study (Mayer, et al., 2002).

In the pilot study, designed to investigate the capacity of the standards to "encourage, engage, and support teachers in immersion, reflective practice, charting professional pathways, and responsive practice" (Mayer et al., 2002, p. 1), the respondents reported learning most about the following standards:

3. Construct intellectually challenging learning experiences
11. Contribute to professional teams
12. Commit to professional practice

When asked to nominate one professional standard they found *most* useful and learned from, more respondents identified standards directly related to their classroom practice and student learning. They nominated the following standards (in order from most identified):

3. Construct intellectually challenging learning experiences
6. Integrate information and communication technologies to enhance student learning
12. Commit to professional practice
1. Structure flexible and innovative learning experiences for individuals and groups
2. Contribute to language, literacy and numeracy development

Those standards nominated *least* as the ones they found most useful and learned from include ones related to aspects of teachers' work outside the classroom:

4. Construct relevant learning experiences, which connect with the world beyond school
5. Construct inclusive and participatory learning experiences
7. Assess and report on student learning
8. Support the social development and participation of young people
10. Build relationships with the wider community their professional identity, responsibility and efficacy.

Although the pilot study confirmed that for the majority of participants the standards framework captured the complex nature of teachers work and would also probably provide a valid framework for their work in the future, there was a range of opinions about how the standards should be used in relation to appraisal, promotion and remuneration.

> The Pilot in 2002 provided strong validation of the standards as a definition of teachers' work and a framework for professional learning and further recommended a shift in focus to learning in community. (Smith, 2002, p. iv)

It is suggested by the Queensland state director general of education (Smith, 2002) that the application of these standards will enable teachers to structure flexible and innovative learning experiences that:

- foster language, literacy and numeracy development are intellectually challenging;
- connect with the world beyond the school;
- are inclusive, participatory and socially critical;
- incorporate the use of information and communication technologies;
- are underpinned by valid and reliable assessment processes.

It is envisaged that these learning experiences will be designed within curriculum frameworks and be provided by teachers who have the skills, knowledge and commitment to:

- support the social development and participation of young people;
- create safe and secure learning environments;
- build relationships with families, community and business;
- contribute to professional work teams;
- engage in professional practice.

The Queensland standards framework was seen as being instrumental in helping raise the status of the teaching profession but there remains much less consensus about what that recognition should be (Mayer et al., 2002). These findings can be compared to the United States situation where Darling-Hammond (2001) cautions that "within states enormous variability often exists in the requirements associated with the many types of licenses, endorsements, and certification that are issued" (p. 754). In attempting to address these issues the NBPTS certification process is providing national certification for highly accomplished teachers. However, Serafini (2002) is fearful that such a process could result in a hierarchy within the teaching profession where one style of teaching is valued over another. Though Yinger and Hendricks-Lee (2000) offer a counter argument that standards provide a shared public language of practice and thus become a means to development and empowerment. To be of any long term benefit to professional education standards frameworks must become more than just slogan systems that offer little but are difficult to refute (King, 1994).

OWNERSHIP OF PROFESSIONAL STANDARDS

The ownership of professional teaching standards will ultimately influence the effectiveness of the standards movement in enhancing the practice of teachers. Professional associations like ASTA are staking their claim for this ownership. Teachers live busy lives and the issue of professional teaching standards does not currently figure too brightly on their day-to-day radar screen. Unless teachers themselves develop a personal sense of the importance of standards and become involved in their development then they will be regarded as being representative of employers' campaigns to discipline the profession.

The recommendations of the Australian College of Education (2001) formed from a national summit on teacher standards, quality and professionalism seems particularly appropriate at this juncture. They proposed that future work on standards must:

- be owned and driven by the teaching profession in partnership with key stakeholders
- be in the interests of the profession as well as the public interest
- be firmly grounded in an accurate and comprehensive understanding of teachers work

The intent of such activity must:

- affirm the status of and integrity of teacher qualifications
- be transparent and accessible to the profession and the wider community
- be implemented with the view to strengthening the public perception of and respect for teachers and their work
- promote teaching as a desirable career (Lovat, 2003)

CURRICULUM AND ASSESSMENT STANDARDS: A CATALYST FOR TEACHING INNOVATION

Until the 1980s Australian science education practices at the junior high school and senior high school levels were dominated by state-wide centrally controlled public examination systems that certified student achievement at age 15 and 17 years respectively. These subject-based examinations were administered with test items requiring written responses over a 2–3 hour time frame. The items sampled topics from the mandated syllabuses promulgated by each state examination authority. In the senior high school particularly these examinations had a history of university control and were used for the purpose of selection of students into university faculties, which had their own lists of required prerequisite subjects.

Access to a university placement was and remains highly competitive and therefore the assessment of high school students' academic performance attracts close political and community attention. The unmet demand for a university placement was in the order of 30–40%. In the past students entering a bachelor of science program were required to pass senior high school chemistry, physics and a mathematics subject to satisfy the prerequisite subject demands of university faculties of science. High school students were given grades based on a norm-referenced model and it was not unusual to hear of instances particularly in senior physics examinations that the pass grade had to be set at 35% in order to satisfy the statistical distribution of the normal curve. This approach to curriculum and assessment was very much framed in the English public examination tradition and teachers unashamedly taught to the examination. Teacher effectiveness had as much to do with the rehearsal of examination questions as it did with meaningful learning.

THE MOVE TO CRITERION-STANDARDS MODELS OF ASSESSMENT

The power of the format of student assessment in curriculum reform cannot be underestimated. For the compulsory years of schooling (to age 15 or 16), all Australian states and territories have adopted in the last 10 years a science curriculum framework or syllabus that describes essential learning for all students in terms of outcomes usually expressed as standards related to a concise number of criteria. The outcomes are stated with varying degrees of specificity and there are differences in the scope of the content knowledge and skills required. A description of the curriculum in terms of outcomes expressed as standards means that teachers are required to plan learning experiences that focus on supporting students to achieve these outcomes.

The public's confidence with educational assessment is often closely related to the reliability of those assessments. Criterion referencing is an attempt to interpret a student's performance by referring, not to the performance of other students, but to specified domains of knowledge. The appeal of criterion-referenced assessment is the hope it holds out for a system of noncompetitive assessment, in which students pit themselves against defined levels of achievement (which incorporate *standards*) rather than against one another. Criterion-referenced assessment will inevitably discriminate among students, given an appropriate set of standards, and students of mixed abilities.

The success of the system rests, to a large extent, on how well the criteria and the performance levels on each of the criteria are defined. This involves the subject experts in three major undertakings:

- the identification of criteria
- the nomination of a set of standards on each criterion along the continuum from lowest to highest proficiency; and
- the specification of the combinations of standards to be met before a student can be judged to have reached a particular exit Level of Achievement. This involves listing the trade-offs that are acceptable at each achievement level.

> The three tasks outlined above are based on an understanding of how subject disciplines are taught and learnt at secondary school level. However, the conceptualisation of knowledge and learning implied by the three tasks above is distinctly different.... The performance criteria are, generally speaking, continuously relevant throughout the whole period that expertise is being acquired, even though some capabilities may be more developed than others at any point in time, and some may remain latent or dormant for long periods of time. This seems to be in accord with the way in which many teachers perceive educational growth to occur. (McMeniman, 1986, p. 7).

However, Goodrum et al. (2001) have cautioned that for reporting the achievement of individuals, standards referenced forms of assessment are adequate for the compulsory years of schooling (junior high school). In the post compulsory years (senior high School) where assessment needs to serve a selection function, the small number of levels along a standards continuum may not provide sufficient discrimination for selecting students for courses where there is strong competition for entry. Given the huge competition for student places at universities throughout Australia, the discrimination required to select students becomes very fine grained and subject to community scrutiny. The use of qualitative standards descriptors for such fine-grained analysis becomes problematic when universities are required to select the most elite students from a large student cohort.

THE IMPACT OF CURRICULUM AND ASSESSMENT STANDARDS ON CLASSROOM PRACTICE

Teaching practice in science classrooms has been characterized for a very long time by the dominance of a transmission style of teaching (Goodrum et al., 2001). This practice is represented in Figure 14.1 below. The most common determining factor in this practice has been the valuing by teachers, university professors and the professional scientific societies of knowledge and conceptual understanding over all other dimensions of science learning. Although *learning through inquiry* has been an international catch cry in school science framework documents or syllabuses for the past decade, the impact on classroom practice has been limited. Changing science teaching behavior to be more consistent with ideal teaching standards remains a pipe dream for many education systems. Some insights in to what may be necessary can be gleaned from a trial project in 50 Queensland senior high school chemistry and physics classrooms. An innovative curriculum design, incorporating a "learning in context" theme with associated assessment standards emphasizing scientific investigations and scientific skills over knowledge and conceptual understanding (Queensland Studies Authority, 2001) is being trialed over a 3 year time frame.

Prior to the adoption of a new science teaching framework or syllabus in 2001 these classrooms were very much like that described in Figure 14.1.

However given the requirements of a new subject framework or syllabus which requires science to be taught and learned in contexts which have social or technological relevance to students, quite radical changes are appearing in these trial classrooms (Lucas, 2003). Although the context remains the central tenet in the learning environment, it is the different teaching and assessment standards that are forging very different practices to that which has characterized these classrooms forever.

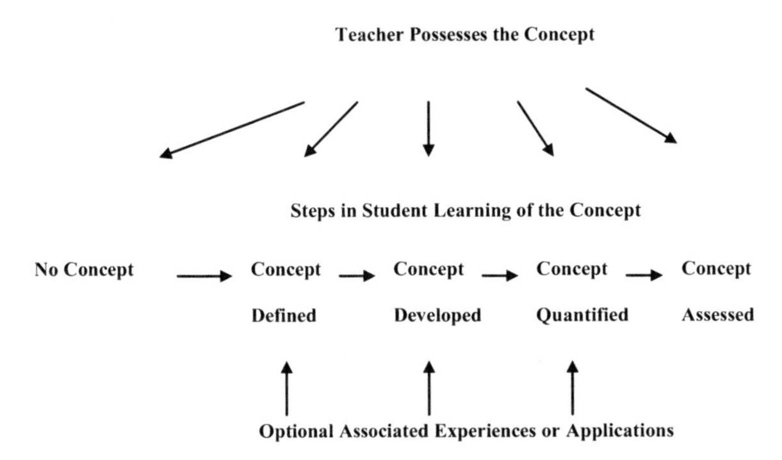

Figure 14.1. The transmission model of teaching in senior science classrooms.

Authentic student centered scientific investigations have become the major learning activity and the major source of reward for course assessment in these fifty senior high school physics and chemistry classrooms statewide (Lucas, 2003). Standards framed in terms of knowledge and conceptual understanding (KCU) and scientific investigation (SI) scientific together present new horizons for classroom activities. Knowledge and conceptual understanding standards are attained "on a need to know" basis during teachable moments in a completely different learning environment. When student exit achievement for the course of study is

Table 14.2. Criterion-Standards Descriptors for the Assessment of High School Chemistry

A. Professional Knowledge

Highly accomplished teachers of science have an extensive knowledge of science, science education and. students.

1. They have a broad and current knowledge of science and science curricula, related to the nature of their teaching responsibilities.
2. They have a broad and current knowledge of teaching, learning and assessment in science.
3. They know their students well and they understand the influence of cultural, developmental, gender and other contextual factors on their students' learning in science.

B. Professional Practice

Highly accomplished teachers of science work with their students to achieve high quality learning outcomes in science.

4. They design coherent learning programs appropriate for their students' needs and interests.
5. They create and maintain intellectually challenging, emotionally supportive and physically safe learning environments.
6. They engage students in generating, constructing and testing scientific knowledge by collecting, analyzing and evaluating evidence.
7. They continually look for and implement ways of the major ideas of science.
8. They develop in students the confidence and ability to use scientific knowledge and processes to make informed decisions.
9. They use a wide variety of strategies, coherent with learning goals, to monitor and assess students' learning and provide effective feedback.

C. Professional Attributes

Highly accomplished teachers of science are reflective, committed to improvement and active members of their professional community.

10. They analyze, evaluate and refine their teaching practice to improve student learning.
11. They work collegially, within their school community and wider professional communities to improve the quality and effectiveness of science education.

based on these new standards then classroom practice must be radically different. Students cannot achieve these standards in a transmission style classroom. A sample of these assessment standards for chemistry is provided in Table 14.2 (Queensland Studies Authority, 2001).

The generalized learning sequence for these trial classrooms is represented in Figure 14.2. This sequence of classroom events requires a very different type of classroom environment and the driving force is the attainment of different curriculum standards, which are forged in the classroom by new assessment standards. These assessment standards prioritize scientific investigation over knowledge and conceptual understanding when course exit levels of achievement are awarded.

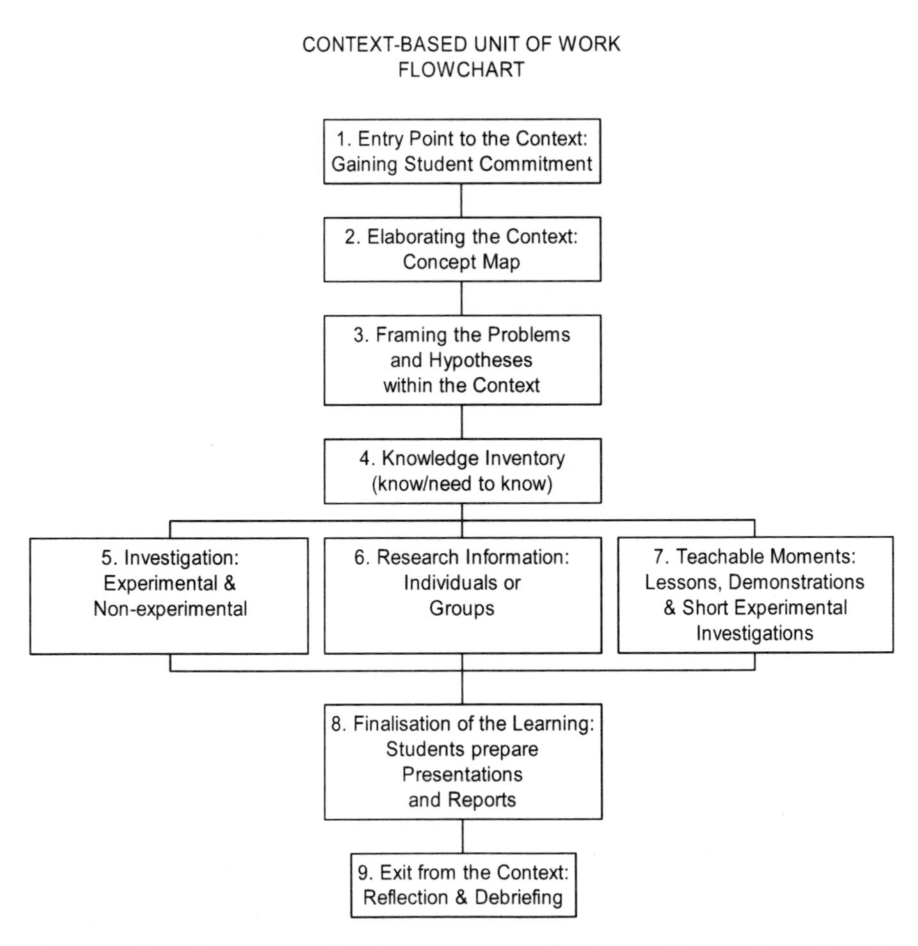

Figure 14.2. The sequence for classroom events for the teaching and learning of science in context.

Such a radical change to curriculum and assessment standards is not without very serious challenges. The 50 trial teachers have reported that serious hurdles remain in adapting current practice to suit the required curriculum and assessment standards (Lucas, 2003). This is to be expected. To change ones teaching practice in such a radical way will be a substantial journey for most science teachers and will require time, professional development support and different classroom resources. However classroom practice in these classrooms is decidedly different and that difference has been driven by two complementary changes to do with curriculum and assessment standards respectively.

IMPLICATIONS OF INFLUENCE OF STANDARDS AND REGULATION ON TEACHER QUALITY AND CURRICULUM RENEWAL

This chapter has touched upon the impact of standards on the teaching of science in Australian schools. The question of teacher professional standards and classroom practice remains very much work in progress. Competing claims of ownership have the potential to derail the process. Classroom teachers have not embraced the concept that teaching standards are a catalyst for change even though their employers and federal politicians have faith in this approach. Professional associations of teachers have attempted to bridge the gap with their own sets of professional teaching standards but again these standards are not claimed by science teachers nation wide. If there is to be an emergence of a new professionalism in science education, then all the major stakeholders will need to forge a very different relationship where teachers are convinced that standards are the cornerstone to ongoing professional renewal and support.

In contrast the impact of curriculum and assessment standards has been much more emphatic. However to be fair these standards have been part of the educational scene for over 15 years when their generation and use by teachers arose as the power of the external examinations was diminished. In some states such as Queensland, Western Australia, Tasmania, and the Australian Capital Territory, the articulation through subject framework documents of curriculum and assessment standards by the relevant boards of studies provided the initial levels of support which teachers needed to implement totally school based curriculum development and assessment protocols. In these three systems the growth of teacher expertise in developing school-based programs consistent with the framework documents has been impressive. The acceptance by the profession, the politicians and society at large that teachers can be trusted with program development and the assessment of students for university entrance contrasts significantly with the practices of the 1980s.

This huge difference could be attributed to two major factors—ownership of the standards by the profession and the relationship of the standards to the day-to-day work of teachers. The lessons to be learned are essentially two fold. In the case of professional teaching standards teachers must see these standards becoming the cornerstone of their career and as such becoming the goals for ongoing professional development initiatives. This can occur when the standards reflect the day to day work of teachers and are more than generic motherhood statements generated by employers and/or government.

The other major lesson that can be gleaned is the potential of curriculum and assessment standards to catalyze the practice of science teachers to change in ways more consistent with learning of science through inquiry. This direction has been promoted for decades yet teachers have chosen to adopt teaching approaches more consistent with transmission. This approach is consistent with assessment practices which value only conceptual knowledge and its simple application—the typical template for most pencil and paper examinations. Here lies the contradiction. High stake pencil and paper tests that have proliferated throughout the United States school systems encourage transmission style teaching yet it is well known that in science classrooms this practice produces lower level thinking outcomes for students as well as an increasing dislike for science. If science education is to become more inquiry driven as promoted by the learned professional scientific societies internationally then curriculum and assessment standards must mirror these standards. The opportunity for students to learn these skills will take substantial time and the curriculum needs to reflect this. Less is more!

The assessment of such standards cannot be achieved through standardized tests. Unless the ownership of the assessment standards resides with the profession and enacted by teachers in individual classrooms then the nature of science education will remain forever under challenge. Currently assessment is ideologically driven by politicians and vested interest groups whose motives seem contrary to those goals cherished by the wider scientific community.

The word has always been more powerful than the sword.

SUMMARY

Professional development is a journey not a destination. Thereby a substantial change in attitude by employers and government is required. There is no quick fix to the renewal and enhancement of classroom practice. Teachers on the other hand need to accept that continuous change and professional development are normal expectations of teachers work. Resources are the appetites of change and consequently budgets need to also reflect this.

REFERENCES

Australian College of Education. (2001). *Teacher standards, quality and professionalism: Towards a nationally agreed framework.* A statement by the Australian College of Education. Canberra, Australia: Author.

Australian Science Teachers Association. (2002). *National professional standards for highly accomplished teachers of science*. Deakin West ACT, Australia: Author.

Beyer, L. (2002). The politics of standards and the education of teachers. *Teaching Education, 13*(3), 305-316.

Darling-Hammond, L. (2001). Standard setting in teaching: Changes in licensing, certification and assessment. In V. Richardson (Ed.), *Handbook of research on teaching* (4th ed., pp. 751-776). Washington DC: American Educational Research Association.

Department of Education, Science and Training. (2003). *Report of the national forum on standards, quality and teacher professionalism.* Canberra, Australia: Commonwealth of Australia.

Goodrum, D., Hackling, M., & Rennie, L. (2001). *Research report: The status and quality of teaching and learning of science in Australian schools.* Canberra, Australia: Department of Education, Training and Youth Affairs.

Hattam, R., & Smyth, J. (1995). Ascertaining the nature of competent teaching: A discursive practice. *Critical Pedagogy Networker, 8*(3 & 4), 1–12.

Ingvarson, L. (2002). *Development of a national standards framework for the teaching profession* (Issue 1). Melbourne, Australia: Australian Council for Educational Research.

Ingvarson, L., & Wright, J. (1999). Science teachers are developing their own standards. *Australian Science Teachers Journal, 45*(4), 27-34.

Kennedy, K. (1993). National standards in teacher education—Why don't we have any? *South Pacific Journal of Teacher Education, 21*, 101–110.

Kennedy, K. (2002, April). *Making professional practice explicit and public: A commonsense approach.* Paper presented at the Teacher Standards, Quality and Professionalism: Towards a Common Approach, National Meeting of Professional Educators, Canberra, Australia.

King, M. B. (1994). Locking ourselves in: National standards for the teaching profession. *Teaching and Teacher Education, 10*(1), 95–108.

Louden, W. (2000). Standards for standards: The development of Australian professional standards for teachers. *Australian Journal of Education, 44*(2), 118–134.

Lovat, T. (2003). *The role of the 'teacher' coming of age*. Melbourne, Australia: Australian Council of Deans of Education.

Lucas, K. B. (2003). *Evaluation of the chemistry trial-pilot senior syllabus 2002–2003. Final report to the science subject advisory committee.* Brisbane, Australia: Queensland Studies Authority.

Mayer, D., Mitchell, J., Macdonald, D., Land, R., & Luke, A. (2002). *From personal reflection to professional community. Education Queensland professional standards for teachers evaluation of the 2002 pilot.* Brisbane, Australia: Brisbane Queensland Government.

McMeniman, M. (1886). *A standards schema*. Brisbane, Australia: Queensland Studies Authority

National Project on the Quality of Teaching and Learning. (1996). *National competency framework for beginning teaching*. Sydney, Australia: Australian Teaching Council.

National Research Council. (1996). *National science education standards*. Washington, DC: National Academy Press.

National Science Teachers Association. (2003). *Standards for science teacher preparation.* Washington DC: Author.

Queensland Studies Authority. (2001, May). *Chemistry, trial-pilot syllabus*. Queensland, Australia: Queensland Sudies Authority.

Reynolds, M. (1999). Standards and professional practice: The TTA and initial teaching training. *British Journal of Educational Studies, 47*(3), 247–260.

Senate Employment, Education and References Committee Inquiry. (1998). *A class act: Inquiry into the status of the teaching profession*: Canberra, Australia: Commonwealth of Australia.

Smith, K. (2002). *Professional standards for teachers: Guidelines for professional practice.* Statement by the director general of education. In the state of Queensland (Department of Education). Brisbane, Australia: Queensland Government.

The State of Queensland, Department of Education. (2002). *Professional standards for teachers: Guidelines for professional practice*. Brisbane: Queensland Government.

Yinger, R. J., & Hendricks-Lee, M. S. (2000). The language of standards and teacher education reform. *Educational Policy, 14*(1), 94–106.

ABOUT THE AUTHORS

Eric R. Banilower is a senior research associate at Horizon Research, Inc. (HRI). He received a bachelor's degree in physics from Haverford College. He has also completed his doctoral coursework in curriculum and instruction at the University of North Carolina at Chapel Hill. In addition to teaching science at the high school level, Mr. Banilower worked with the California Scope, Sequence, and Coordination Project developing curriculum and assessment materials for the California Science Reform Project. Mr. Banilower joined HRI in 1997 and has worked on a number of evaluation projects. He also assisted with data analysis and reporting for the 2000 National Survey of Science and Mathematics Education. Currently, Mr. Banilower directs the evaluation of American Association of Physics Teachers rural physics teaching resource agents, a national-level physics-teacher enhancement project aimed at improving physics and physical science instruction in rural areas. He is heavily involved in the core evaluation of the Local Systemic Change (LSC) Project—leading the data analysis team and supervising a study of the impact of the LSC on student achievement. He has also assisted on the evaluation design, instrument development, and data analysis for a number of research and evaluation projects at HRI. He can be contacted at erb@horizon-research.com

Warren Beasley has been a professor of science education at the University of Queensland (School of Education, Australia, 4072) for 25 years. He holds a doctorate from the University of Maryland and subsequently has led many professional development projects throughout South East Asia. His research and teaching interests are in the areas of teacher professional development and includes both preservice education and work place learning environments. In 1999–2001 he was a program director at the National Science Foundation in Washington DC and on return to Australia became a foundational member of the ministerial task force on the future of science education in the state of Queensland. He was granted life membership of the Science Teachers Association of Queensland in 1988 for his services to science education. He can be contacted at w.beasley@mailbox.uq.edu.au

George E. DeBoer is deputy director for Project 2061 of the American Association for the Advancement of Science. He joined Project 2061 from the division of elementary, secondary, and informal science of the National Science Foundation (NSF), where he served as program director. He is also professor of Educational Studies Emeritus at Colgate University. He holds a PhD in science education from Northwestern University, a master of arts in teaching in biochemistry and science education from the University of Iowa, and a bachelor of arts in biology and chemistry from Hope College. In his long career at Colgate he held positions as director of the master of arts in teaching program, chair of the education department, and acting director of the division of social sciences. Prior to becoming a university professor, he taught chemistry, biology, and earth science at Glenbrook South High School, Glenview, Illinois, and chemistry, biochemistry, and microbiology at the Evanston Hospital School of Nursing, Northwestern University Medical School. DeBoer is the author of *A History of Ideas in Science Education: Implications for Practice* (Columbia University Teachers College Press, 1991) as well as numerous articles, book chapters, and reviews. His primary scholarly interests include clarifying the goals of the science curriculum, analyzing the various meanings of science literacy, researching the history of science education, and investigating ways to effectively assess student understanding in science and mathematics. He is principal investigator or coprincipal investigator on a number of NSF-funded grants, including two projects on student assessment in science and an Interagency Education Research Initiative project on professional development in middle school mathematics, and is associate director for the Center for Curriculum Materials in Science, an NSF-funded center for learning and teaching. He is currently a member of the American Educational Research Association, the National Association for Research in Science Teaching, the National Science Teachers Association, the American Association for the Advancement of Science, and the American Association of University Professors. He can be contacted at gdeboer@aaas.org

Larry G. Enochs is currently professor in the Department of Science and Mathematics Education at Oregon State University. His doctorate in education is from Indiana University. He also holds a masters of science degree in earth science education from the University of Rochester. During the past 20 years, he has been involved in several teacher enhancement efforts, as principal investigator, project evaluator and advisory board member. In the area of science teacher development he has published several articles and technical reports focused on teacher enhancement in earth science education and science teaching and learning in general. Prior to joining the faculty at Oregon State University, Enochs served as director of the Center for Mathematics and Science Education Research at the University of Wisconsin–Milwaukee. Enochs also served 2 years as a program officer for the Teacher Enhancement and Research On Teaching and Learning Programs at the NSF. While at the foundation, Enochs handled earth science education proposals, and assisted in the development of new teacher enhancement guidelines. He can be contacted at enochsl@onid.orst.edu

Fred Finley holds a PhD in science education from Michigan State University and is an associate professor at the University of Minnesota in the Department Of Curriculum And Instruction. He is the past chair of that department. His research interests are in the foundations of science and environmental education and the influences of prior knowledge on students' learning. He is currently conducting research on what changes occur in the geological knowledge of college students in an introductory geology course using their essay responses to a Web-based assessment tool. He is also conducting research with colleagues in Thailand on the development of environmental social capital under a U.S. State Department educational partnership program. He has international experience in Hong Kong, Japan and China as well. He has published research in the *Journal of Research in Science Teaching*, *Science Education*, *Science and Education*, the *Chulalongkorn University Review*, several book chapters and other reports. He can be contacted at finle001@umn.edu

Judy Fowles is currently pursuing an EdD in education leadership at University of Texas at San Antonio, and serves as an adjunct faculty in the Uteach Program, a preparatory program for secondary preservice teachers in mathematics and science. She earned a bachelor of science in secondary education from Loyola University of the South, New Orleans, a master of arts in administration from University of Texas-San Antonio, and a master of arts in biology from the University of the Incarnate Word. She served as the elementary and secondary science instructional specialist for 14 years at Northside Independent School District. She taught middle and high school science in Texas, Connecticut, North Dakota, as well as in Germany and Turkey.

Connie Gabel is both a scientist and educator. During her career, she has held positions in both teaching and educational administration as well as chemical research. She has taught at all educational levels—preschool, elementary, middle, high school, and college. Her educational pursuits have included schools in both the public and private sectors as well as the government's department of defense schools overseas. Her research has been conducted in both science and education. She graduated with a bachelor of science in chemistry from James Madison University, Virginia. She obtained an master of arts in educational administration, supervision, and curriculum development from the University of Colorado, Boulder. Gabel received her PhD in educational leadership and innovation with an emphasis in curriculum, learning, and technology from the University of Colorado, Denver. She is currently teaching chemistry and teacher education at Metropolitan State College of Denver. She can be contacted at cgabel@mscd.edu

M. Jenice "Dee" Goldston, PhD is an associate professor of Science Education at the University of Alabama. Dr. Goldston's research interests focus on teacher beliefs and self efficacy related to inquiry models of pedagogy and the nature of science, conceptual change related to teacher content acquisition and pedagogy adoption, preparation of high quality science teachers and their on-going professional development, reform in teacher education and in undergraduate science teaching in higher education, and sociopolitical influences affecting the status of science teaching in K-12 schools. She has authored articles in journals including *Science Teacher, Science and Children, Physics Teacher, Journal of Science Teacher Education, Teacher Education and Practice*, and the *Journal of Research in Science Teaching.* She conducts conference presentations and workshops that range from exploring science inquiry to action research for K-12 teachers. Dr. Goldston is an associate editor for the *Journal of Science Teacher Education*. She is an active member of the National Association for Research in Science Teaching, American Educational Research Association, National Science Teachers' Association, Association for Educators of Teachers in Science, and Alabama Science Teachers Association. She is the President of the Council of Elementary Science International and is a council member for the National Science Teachers Association. Dr. Goldston has been involved in numerous National Science Foundation grants for professional and curriculum development. She is also involved in research within NASA NOVA to reform undergraduate science content courses. She can be contacted at dgoldsto@bamaed.ua.edu

Edward Gonzales, PhD is an associate professor at the University of the Incarnate Word in San Antonio, Texas. He earned his PhD in chemistry from the University of Texas-Austin, a bachelor of science and master of science in pharmacology from the University of Texas-Austin, and a master of arts in education leadership from the University of Texas-San Antonio. He has taught elementary and secondary science in San Antonio, and now teaches introductory and organic chemistry for undergraduates. His current research interests are the professional development of in-service teachers in sciences. He has authored several grants for the Texas Higher Education Coordinating Board, Eisenhower Professional Development Program, and is an active faculty member of the Multidisciplinary Sciences Graduate Program. He can be contacted at gonzales@universe.uiwtx.edu

Cecilia Hernandez is a graduate student in science education at Kansas State University (KSU) and a project coordinator on "Seeing Gender: Tools for Change" supported by the National Science Foundation. She has a bachelor of science and master of science in biology. Hernandez has taught eighth grade earth science, undergraduate biology lab courses, was a research associate at Texas Tech University, and has been an English as a second language instructor at KSU. Her research interests are the under representation of women and minorities in science. She can be contacted atcecilia@ksu.edu

Christy MacKinnon, PhD is a professor of biology at the University of the Incarnate Word, in San Antonio, Texas. She received her bachelor of science in biology from the University of Michigan-Flint, a master of science in botany from Michigan State University, and a PhD in plant sciences from Colorado State University. She currently teaches introductory cell and molecular biology and genetics courses. She had conducted research in molecular mechanisms of carcinogenesis and protein folding. Her current research interests are the professional development of preservice and in-service teachers in sciences, and best practices for teaching undergraduate students in biology. She developed and directs the master of arts program in multidisciplinary sciences. She has authored numerous grants for the Texas Higher Education Coordinating Board, Eisenhower Professional Development Program and has participated as a coprincipal investigator in a NASA Opportunities for Visionary Academics (NOVA) Project grant. She has served as a project NOVA fellow since 2001. She has presented inquiry-based laboratory activities for the Association for Biological Laboratory Education, and several papers at the annual meetings of American Education Research Association, and the National Association for Research in Science Teaching. She was also a coauthor of two chapters in (Eds.) D. W. Sunal, E. L. Wright, and J. Bland Day, *Reform in Undergraduate Science Education for the 21st Century*. She can be contacted at mackinno@universe.uiwtx.edu

Stephen Marlette holds a PhD in curriculum and instruction with an emphasis in science education from Kansas State University, Manhattan, Kansas. He is currently an assistant professor at Southern Illinois University Edwardsville (SIUE). His research interests include preservice and inservice teacher development, and change theory as it relates to science education and the standards. His professional activities include advising the SIUE National Science Teachers Association student chapter and serving the Illinois Science Teachers Association as a key leader. He can be contacted at smarlet@siue.edu

Bonnie McCormick is an associate professor of biology and Chair of the Department of Biology at the University of the Incarnate Word, San Antonio. She received a bachelor of arts in accounting from the University of Texas-Austin, a master of science in biology from the University of the Incarnate Word, and her PhD in science education from the University of Texas-Austin. She teaches both majors and nonmajors in introductory and advanced biology courses. Her research interests are in the area of undergraduate teaching and learning in the biological sciences. She has made numerous presentations to professional organizations on integrating technology to engage undergraduate students in inquiry learning and on student learning in inquiry-based classrooms. Her grant awards from the Department Of Defense and the Department Of Education have supported curriculum revisions in the introductory biology course, training of preservice science teachers, and the acquisition of technological tools for improving undergraduate science teaching through authentic research experiences. She can be contacted at mccormic@universe.uiwtx.edu

Teresa Miller, EdD, is associate professor in the Department Of Educational Leadership at Kansas State University. Prior to coming to the university last year, Miller spent 15 years in partnership with KSU's professional development schools (PDS) as both an elementary and secondary school principal at three of the PDS sites. In her current assignment, she continues the public school collaboration by working with superintendents to prepare new principals through 2-year, field-based, academy programs. Her research is focused on the cycles of leadership, leadership reproduction and the identification of leadership characteristics associated with meeting the new federal accountability requirements for schools. She can be contacted at tmiller@ksu.edu

Joan D. Pasley is a senior research associate at Horizon Research, Inc. (HRI). She received a bachelor's degree in biology from the University of Cincinnati, a master's degree in educational administration from Xavier University, and a PhD in curriculum and instruction, from the University of North Carolina at Chapel Hill. In addition to teaching science at the high school level, Dr. Pasley served as a high school administrator in an urban school district where she gained experience in curriculum development, assessment and instructional evaluation. Dr. Pasley has been working with HRI since 1994 on a number of research and evaluation projects, including the evaluation of the Ohio, South Carolina, and New Jersey statewide systemic initiatives. Dr. Pasley currently coordinates the standardized evaluation system for NSF's local systemic change through teacher enhancement project and directs the evaluation of e-mentoring for student success, an on-line mentoring program for beginning science and mathematics teachers. In addition, Dr. Pasley manages the Increasing the Availability of Materials for the Professional Development of Science and Mathematics Teachers project which aims to increase the availability of materials for the professional development of science and mathematics teachers. She can be contacted at jpasley@horizon-research.com

Jo Ellen Roseman received her PhD in biochemistry from Johns Hopkins University, her master of science degree in biology from Michigan State University, and her bachelor of science degree with high honors in zoology from the University of Michigan. Dr. Roseman is director of the American Association for the Advancement of Science's Project 2061 in Washington, DC. Dr. Roseman joined Project 2061 for its release of *Science for All Americans* in 1989 and has since been involved in the design, testing, and dissemination of all subsequent reform tools. She heads the project's 4-year effort to develop a valid and reliable procedure for evaluating both the content and instructional design of science, mathematics, and technology curriculum materials in light of national standards. Prior to joining Project 2061 Dr. Roseman was a faculty member in arts and science and education at the Johns Hopkins University, where she designed and directed graduate programs to prepare scientists and engineers to teach in K-12 classrooms and to enhance K-12 teachers' understanding of science, mathematics, and technology. She has extensive experience teaching biology and chemistry at high school, college, and graduate levels and has served on the Biological Sciences Curriculum Study Board of Directors and numerous NSF review panels in teacher enhancement and instructional materials development. She can be contacted at jroseman@aaas.org

Gail Shroyer is professor of Education (Science) and Director of the Professional Development Schools (PDS) Partnership at Kansas State University. She earned her master of science and PhD degrees from KSU in science education/curriculum and instruction. Dr. Shroyer's bachelor of arts was awarded from the University of California-Santa Cruz with a major in biology. Before coming to Kansas State University, she served as a teacher, director, and department head for Soquel High School, Santa Cruz, California. She is responsible for field experiences and professional development opportunities for inservice and preservice teachers working within 27 PDS in five diverse Kansas school districts. She has used action planning and action research to facilitate school improvement projects in all 27 PDS. Her research interests are in teacher education, school improvement, teacher leadership, professional development, action research, and equity issues. Shroyer has extensive experience with national teacher education reform as a former board member for the Association for Educators of Teachers in Science, former editor of the *Journal of Science Teacher Education*, and member of the National Research Council's Committee on Science and Mathematics Teacher Preparation. She has directed 11 national and state teacher preparation and enhancement projects, and has served as writer, field test coordinator and advisory board member for seven national K-12 curriculum and teacher development projects. She can be contacted at gshroyer@ksu.edu

P. Sean Smith is a senior research associate at Horizon Research, Inc. (HRI). He received a bachelor's degree in chemistry, a master's degree in science teaching, and a PhD in curriculum and instruction from the University of North Carolina at Chapel Hill. Prior to joining HRI in 1991, Dr. Smith taught high school chemistry and physics. In addition, he was a member of the education studies department at Berea College, where he taught courses in elementary science methods and the philosophical foundations of education. Dr. Smith worked on the evaluation of the National Science Foundation (NSF) funded statewide systemic initiative for North and South Carolina. He was involved in the design of the core evaluation for the local systemic change through teacher enhancement project. He was the project manager for the 2000 National Survey of Science and Mathematics Education and a member of the study team for *Inside the Classroom*, a nationally representative observation study of science and mathematics instruction. He currently serves as principal investigator on an NSF-funded project that is developing instruments to test models of professional development. He also directs the evaluations of various teacher enhancement and materials development projects. He can be contacted at ssmith@horizon-research.com

John R. Staver, EdD, is professor of education (science) and director of the Center for Science Education at Kansas State University. Staver has conducted extensive research on the development and construct validation of a group-administered test of Piaget's formal schema, the effects of methods and formats of Piagetian task presentation on responses and the influence of reasoning on learning in science, on students' understanding of the mole concept and its use in stoichiometric problem solving, and on activity-based K-6 science instruction. His research presently focuses on constructivist epistemology and its implications for improving science teaching and learning. He is also examining the interface between science and religion within a constructivist perspective, with a focus on the nature of each discipline and current conflicts between them. This work stems from his involvement in the continuing controversy over school science standards in Kansas. In 1994, he was elected a fellow in the American Association for the Advancement of Science for his work in behalf of a national reform agenda in science education. He can be contacted at staver@ksu.edu

Luli Stern received her PhD in molecular biology and master of science in microbiology from Tel Aviv University. She is currently a faculty at the Department of Education in Technology and Science at the Technion – IIT, Haifa 32000, Israel. Her main areas of interest include the implementation of standards-based assessment and the cognitive implications of curricular interventions in the biology classroom. Dr. Stern heads the development of two standards-based units that focus on biological evolution. Prior to joining the Technion, Dr. Stern was a research associate at the American Association for the Advancement of Science's Project 2061 and was deeply involved in a leading project designed to develop a valid and reliable tool for evaluating how well K-12 science curriculum materials address specific learning goals in the national standards documents (http://www.project2061.org/tools/textbook/mgsci/INDEX.HTM). In addition, she was in charge of evaluating students' assessment included in curriculum materials. This included developing criteria and indicators to evaluate this assessment and applying these criteria to middle and high school curricula. She can be contacted at lstern@tx.technion.ac.il

Cynthia Szymanski Sunal holds a PhD in early childhood social studies education from the University of Maryland. She currently professor of Elementary Education and previously served as department head of elementary education programs at the University of Alabama. Her research interests are in cognition, curriculum development, the effects of the online environment on conceptual change, and longitudinal investigation of the development of primary school education in developing nations. Dr.

Sunal has authored eight books, 19 chapters, and nearly 200 refereed articles in a wide range journals including *Theory and Research in Social Education, The International Social Studies Forum, Journal of Research in Childhood Education, Journal of Research in Science Teaching, Journal of College Science Teaching, African Studies Review, and School Science and Mathematics.* She has recently presented papers and organized symposia at the annual meetings of several professional organizations including the American Educational Research Association, the American Association of Colleges of Teacher Education, National Council for the Social Studies, National Association for Research in Science Teaching, and Association of Educators of Teachers of Science. Among her other professional activities is her role as executive editor of the *Journal of Interactive Online Learning.* She can be contacted at cvsunal@bama.ua.edu

Dennis W. Sunal received a PhD in science education, master of arts in interdisciplinary science, and a bachelor of science in physics all from the University of Michigan. He currently is a professor of Science Education at the University of Alabama. His teaching experiences include undergraduate and graduate courses in physics, astronomy, engineering, research in curriculum and instruction, teaching in higher education, and science teaching and learning. He holds both secondary, 6 –12, and elementary K-6, teacher certification and has taught extensively on both levels. His research interests are in professional development, alternative conceptions and conceptual change in teachers and faculty, student conceptions of the nature of science, and Web course design, pedagogy, and contextual factors in interactive online learning. He has been project director and codirector in numerous grants (e.g. NSF, NASA, Department of Education, United States Information Agency, Eisenhower Professional Development Program, U.S. Department of Energy) related to research in online interactive learning, professional development of science teachers, and in creating faculty change in teaching science, engineering, and mathematics courses in higher education. He has published in the *Journal of Research in Science Teaching*, *Science Education, School Science and Mathematics, Journal of College Science Teaching, Journal of Interactive Online Learning, Journal of Technology and Teacher Education, African Studies Review, Journal of Computers in Mathematics and Science Teaching, Science and Children, Science Scope, Science Activities*, and *Journal of Research in Childhood Education* among others. Recent research presentations have been at the annual meetings of the National Association for Research in Science Teaching, Association of Educators of Teachers of Science, American Association of Colleges of Teacher Education and the American Educational Research Association. His books have included *Teaching Elementary and Middle School Science* and *Integrating Academic Units in the Elementary*

School Curriculum, and Reform in Undergraduate Science Teaching for the 21st Century. He can be contacted at dwsunal@bama.ua.edu

Iris R. Weiss is president of Horizon Research, Inc. in Chapel Hill, NC specializing in science and mathematics education research and evaluation. Dr. Weiss received a bachelor's degree in biology from Cornell University, a master's degree in science education from Harvard University, and a PhD in curriculum and instruction from the University of North Carolina at Chapel Hill. She has directed a large number of education research, development, and evaluation projects, including several national surveys of science and mathematics teachers, and an observation study of a large, nationally-representative sample of science and mathematics teachers. Dr. Weiss participated in the evaluation of NSF's model middle school mathematics and science teacher preparation and triad curriculum programs, served on the assessment working group for the *National Standards of Science Education*, and chaired an NRC committee to develop a framework for investigating the influence of national standards. She has been involved in the evaluation of a wide variety of mathematics/science professional development and systemic reform initiatives, and is currently principal investigator for the core evaluation of NSF's local systemic change through teacher enhancement mathematics/science initiative, and Co-PI of a project to develop measures of teacher and student learning in science. She can be contacted at iweiss@horizon-research.com

Emmett L. Wright, PhD is a professor of science/environmental education and curriculum in the College of Education at Kansas State University. During the 2003–2004 and 2004–2005 academic years, he served as a program director in the division of elementary, secondary education and informal education at the National Science Foundation, Arlington, Virginia. Dr. Wright works with both the Teacher Professioanl Continuum and Information Technology Experiences for Students and Teachers programs at NSF and serves as the TPC section head. He holds a bachelor of science from the University of Kansas (biology/chemistry), master of arts from Wichita State University (biology/educational psychology) and the PhD from Pennsylvania State University (science education/environmental biology). Dr. Wright's research interests include decision-making attitudes, problem solving, misconceptions and scientific discrepant events, and international education. He has over 130 publications including research articles, yearbooks, science curriculum guides and secondary and college-level textbooks, and other reports. Dr. Wright has served as president of National Association for Research in Science Teaching and on the board of directors of NSTA and SCST. He has received major curriculum-development, teacher-education, and research grants, from EPA, NSF, U.S. Department of Education, U.S. Department of State and U.S. Department of Energy. Recently, he served as the director of research for the National Commission on Mathematics and Science Teaching for the 21st Century. Dr. Wright currently teaches doctoral-level science education seminars, and staff development and curriculum theory courses. At the undergraduate level he has taught science methods for elementary and secondary teachers, general biology, general ecology, ethnobotany, conservation of natural resources, earth systems science, and environmental decision-making. He can be contacted at birdhunt@ksu.edu

Printed in the United States
48337LVS00001B/54

9 781593 113650